Amanullah

Integrated Agriculture

Integrated Global STEM

Series Editors
Robert Krueger, Wole Soboyejo and Anita Mattson

Volume 3

Amanullah

Integrated Agriculture

An Approach for Sustainable Agriculture

DE GRUYTER The WPI Press

Author
Prof. Amanullah
Department of Agronomy
Faculty of Crop Production Sciences
The University of Agriculture Peshawar
Peshawar-25130
Pakistan
amanullah@aup.edu.pk

ISBN 978-3-11-144801-5
e-ISBN (PDF) 978-3-11-144807-7
e-ISBN (EPUB) 978-3-11-144883-1

Library of Congress Control Number: 2024935937

Bibliographic information published by the Deutsche Nationalbibliothek
The Deutsche Nationalbibliothek lists this publication in the Deutsche Nationalbibliografie; detailed bibliographic data are available on the internet at http://dnb.dnb.de.

Cover image: witsarut sakorn/iStock/Getty Images Plus
Typesetting: Integra Software Services Pvt. Ltd.
Printing and binding: CPI books GmbH, Leck

www.degruyter.com

Contents

Chapter 1
Agriculture: the backbone of nations

Introduction

Agriculture stands as the bedrock of human civilization, serving as the primary source of sustenance, economic stability, and cultural identity for nations across the globe. From ancient civilizations to modern societies, agriculture has played an indispensable role in shaping the trajectory of human development and shaping landscapes. In this chapter, we acknowledge the pivotal role of agriculture in national economies and global food security, setting the stage for a comprehensive exploration of integrated agricultural approaches on a global scale.

Agriculture, often referred to as the backbone of human civilization, has been fundamental to the survival and advancement of societies since the dawn of humanity. It is not merely a means of producing food; rather, it embodies the essence of human existence, intertwining with every aspect of life – from economics to culture, and from history to environment. From the ancient river valleys of Mesopotamia and the Indus to the sprawling fields of the American Midwest and the rice paddies of Southeast Asia, agriculture has been the driving force behind the rise and sustenance of civilizations.

Since its inception, agriculture has served as the primary source of sustenance for human populations. It has provided the essential nourishment needed for survival, enabling communities to thrive and expand. The cultivation of crops and the domestication of animals marked a significant shift in human societies, allowing for the development of settled communities and the specialization of labor. Agriculture laid the foundation for the development of trade networks, the establishment of urban centers, and the emergence of complex social structures.

Moreover, agriculture has been a key driver of economic stability and prosperity throughout history. In many countries, agriculture remains a major contributor to gross domestic product (GDP) and employment, particularly in rural areas where farming is the primary livelihood for millions of people. The agricultural sector encompasses a wide range of activities, from crop cultivation and livestock rearing to agribusiness and food processing, generating income and livelihood opportunities for individuals and communities worldwide.

Beyond its economic significance, agriculture is deeply intertwined with cultural identity and heritage. Traditional farming practices, passed down through generations, reflect the unique customs, beliefs, and values of different societies. Agricultural festivals, rituals, and celebrations serve as expressions of cultural identity and community cohesion, reinforcing the bond between people and the land. Agriculture has shaped culinary traditions, folk arts, music, and literature, enriching the cultural tapestry of nations and fostering a sense of belonging and pride.

https://doi.org/10.1515/9783111448077-001

In the context of contemporary global challenges, such as population growth, climate change, and food insecurity, the importance of agriculture has never been more pronounced. Agriculture holds the key to addressing these challenges by increasing food production, improving access to nutritious foods, and promoting sustainable farming practices.

As we confront the complexities of the twenty-first century, it is vital to acknowledge the pivotal role agriculture plays in national economies, global food security, and environmental sustainability (Figure 1.1).

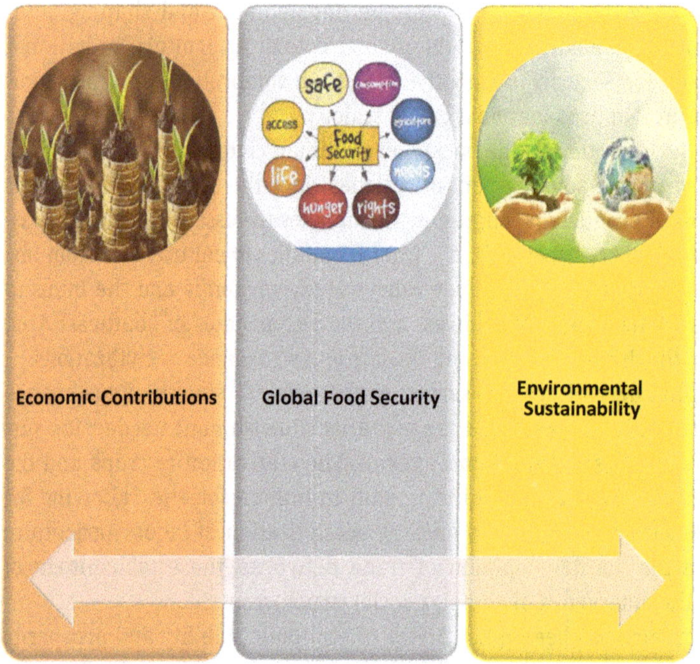

Figure 1.1: The role of agriculture in national economies, food security, and environmental sustainability.

1.1 The historical significance of agriculture

Throughout history, agriculture has been the foundation upon which civilizations have flourished. From the fertile banks of the Nile River to the vast plains of the American Midwest, agricultural innovations have fueled population growth, urbanization, and the rise of complex societies. The advent of agriculture marked a pivotal moment in human history, transitioning societies from nomadic hunter-gatherer lifestyles to settled agricultural communities. Ancient agricultural practices laid the groundwork for modern farming techniques and continue to influence agricultural systems today.

The historical significance of agriculture cannot be overstated, as it has been the driving force behind the rise and sustenance of civilizations throughout human history. Agriculture emerged independently in multiple regions of the world, marking a pivotal shift in human societies from nomadic hunter-gatherer lifestyles to settled agricultural communities.

One of the earliest examples of agriculture can be traced back to Fertile Crescent in the Middle East, where ancient people began cultivating crops such as wheat, barley, and legumes around 10,000 years ago. The domestication of plants and animals allowed for a more reliable and abundant food supply, leading to population growth and the establishment of permanent settlements. This agricultural revolution laid the groundwork for the development of complex societies, as surplus food production enabled specialization of labor, the growth of trade networks, and the emergence of social hierarchies.

Similarly, in other parts of the world, such as the Nile River Valley in Egypt, the Indus River Valley in South Asia, and the Yellow River Valley in China, agricultural innovations played a crucial role in the rise of ancient civilizations. The cultivation of crops like rice, millet, and maize supported large populations and facilitated the development of sophisticated urban centers, monumental architecture, and complex political systems.

In the Americas, indigenous civilizations such as the Maya, Aztec, and Inca developed advanced agricultural techniques suited to their respective environments. They engineered terraced fields, irrigation systems, and crop rotation methods to maximize food production in diverse landscapes ranging from tropical rainforests to high-altitude mountain regions.

The transition from hunting and gathering to agriculture represented a significant milestone in human history, marking the beginning of a more sedentary way of life. With the advent of agriculture, humans no longer needed to constantly roam in search of food, allowing them to establish permanent settlements and develop more complex social structures. This shift laid the foundation for the development of civilization as we know it.

Moreover, the agricultural practices developed by ancient civilizations continue to influence modern farming techniques. Many of the crops cultivated thousands of years ago are still staples of our diets today, and agricultural innovations such as crop rotation, irrigation, and domestication of livestock remain integral to contemporary farming practices.

The historical significance of agriculture lies in its role as the foundation upon which civilizations have flourished. Agricultural innovations have fueled population growth, urbanization, and the rise of complex societies throughout human history. The transition from nomadic hunter-gatherer lifestyles to settled agricultural communities marked a pivotal moment in human development, shaping the course of human history and laying the groundwork for modern civilization.

1.2 Economic contributions of agriculture

In addition to its historical significance, agriculture remains a cornerstone of national economies worldwide. Agriculture contributes significantly to GDP, employment, and trade in many countries, particularly in developing regions where the agricultural sector serves as the primary livelihood for rural communities. Beyond economic metrics, agriculture also plays a vital role in poverty alleviation, food security, and rural development, making it a critical component of sustainable development agendas.

The economic contributions of agriculture extend far beyond mere monetary value, encompassing a wide range of socioeconomic benefits that are integral to the well-being of nations and communities worldwide.

Advanced countries with agriculture as their backbone demonstrate how this sector contributes significantly to their economies and societies. For instance, the United States has a highly developed agricultural sector, with sophisticated technological innovations, efficient production methods, and extensive research and development efforts. Agriculture plays a crucial role in the US economy, contributing billions of dollars annually and supporting millions of jobs across various related industries.

Similarly, countries like the Netherlands and Japan have leveraged advanced agricultural practices to overcome land constraints and achieve high levels of productivity. The Netherlands is renowned for its precision agriculture techniques, innovative greenhouse technology, and sustainable farming practices, which have made it a global leader in agricultural exports despite its small size. Japan, on the other hand, utilizes advanced robotics, automation, and vertical farming methods to maximize yields and ensure food security in densely populated urban areas.

In Australia, agriculture is a major economic driver, contributing significantly to export earnings and rural livelihoods. The country's vast agricultural lands, coupled with advanced irrigation systems and sustainable farming practices, allow for the production of a diverse range of high-quality agricultural products, including grains, livestock, and horticultural crops.

These examples demonstrate how advanced countries with a strong agricultural backbone leverage technology, innovation, and sustainable practices to drive economic growth, ensure food security, and enhance the overall well-being of their populations.

First and foremost, agriculture remains a significant contributor to GDP in many countries, particularly in developing regions where the agricultural sector constitutes a substantial portion of the economy. Agriculture's contribution to GDP is substantial in many countries, especially those with primarily agrarian economies.

Here are a few examples:

India: Agriculture is a cornerstone of India's economy, contributing around 15% to its GDP. The country is one of the world's largest producers of rice, wheat, and cotton. Agriculture also provides employment to a vast rural population, supporting millions of livelihoods across the country.

Brazil: Agriculture plays a crucial role in Brazil's economy, contributing about 5% to its GDP. The country is a global powerhouse in agricultural production, particularly in soybeans, sugarcane, coffee, and beef. Brazil's agricultural sector has experienced significant growth due to technological advancements and large-scale commercial farming operations.

Vietnam: Agriculture is a vital sector in Vietnam, accounting for over 15% of its GDP. The country is a major exporter of rice, coffee, and seafood. Agriculture also provides employment to a significant portion of Vietnam's population, particularly in rural areas where small-scale farming is prevalent.

Kenya: Agriculture is the backbone of Kenya's economy, contributing approximately 25% to its GDP. The country is known for its diverse agricultural products, including tea, coffee, flowers, and fresh produce. Agriculture is the main source of income for many rural households in Kenya, supporting livelihoods and food security.

These examples highlight the significant contribution of agriculture to GDP in various countries, underscoring its importance as a driver of economic growth, employment, and food security.

The production of crops, livestock, and other agricultural commodities generates revenue streams that contribute significantly to national wealth and economic growth. For example, in the **United States**, agriculture plays a crucial role in driving economic growth and prosperity. The country is a leading producer of a wide range of agricultural commodities, including corn, soybeans, wheat, beef, and dairy products. The revenue generated from agricultural production contributes substantially to the nation's GDP and overall economic activity.

For instance, the export of agricultural products contributes billions of dollars to the US economy annually. In 2020, the total value of US agricultural exports reached over $135 billion, supporting jobs and businesses across the agricultural supply chain, including farmers, processors, exporters, and transportation companies. These exports also contribute to reducing the country's trade deficit and strengthening its position in the global market.

Moreover, the agricultural sector in the United States serves as a key driver of rural development and employment. It provides livelihoods for millions of Americans working in farming, ranching, agribusiness, and related industries. Additionally, agricultural activities stimulate economic activity in rural communities by creating demand for goods and services, such as equipment, machinery, and infrastructure.

Overall, the revenue generated from agricultural production in the United States exemplifies how the agricultural sector contributes to national wealth and economic growth, highlighting its importance in sustaining livelihoods, supporting rural communities, and driving prosperity at the national level.

Moreover, agriculture serves as a source of employment for millions of people, especially in rural areas where farming is the primary livelihood for communities.

The agricultural sector provides jobs not only in primary production but also in related industries such as food processing, transportation, and agribusiness, further stimulating economic activity and creating opportunities for entrepreneurship and innovation.

For example, in India, agriculture is the primary source of employment for a significant portion of the population, particularly in rural areas. Millions of farmers and agricultural workers are engaged in crop cultivation, livestock rearing, and other agricultural activities, providing livelihoods for themselves and their families.

In Brazil, the agricultural sector employs millions of people across various segments of the value chain. From small-scale farmers cultivating crops like soybeans and sugarcane to workers in food processing plants and transportation companies, agriculture plays a crucial role in providing jobs and income opportunities for people in both rural and urban areas.

In Kenya, agriculture is a key driver of employment and economic development, particularly in rural communities. Smallholder farmers grow a diverse range of crops such as maize, tea, coffee, and horticultural products, while agribusinesses and food processing industries create additional job opportunities along the value chain.

Furthermore, agriculture plays a crucial role in global trade, as countries exchange agricultural products to meet domestic demand and access international markets. Agricultural exports contribute to foreign exchange earnings, bolstering national economies and facilitating economic development. Moreover, trade in agricultural commodities helps to diversify sources of income and mitigate the risks associated with fluctuations in domestic markets.

For example, the United States exports a significant amount of soybeans, corn, and wheat to countries around the world, contributing to its foreign exchange earnings and strengthening its agricultural trade relationships. Thailand is a major exporter of rice, seafood, and fruits, which are in high demand in international markets. These exports play a crucial role in generating foreign exchange and supporting the country's economy. Australia's agricultural exports, including beef, wheat, and wool, are essential for diversifying its sources of income and mitigating the impact of domestic market fluctuations. These exports contribute significantly to the country's trade balance and economic stability.

Beyond its direct economic contributions, agriculture plays a vital role in poverty alleviation, food security, and rural development, making it a linchpin of sustainable development agendas. In many developing countries, where a significant portion of the population relies on agriculture for their livelihoods, investments in agricultural development have been shown to have a multiplier effect on poverty reduction and food security. By increasing agricultural productivity, improving access to markets and financial services, and promoting sustainable farming practices, agriculture can empower smallholder farmers and rural communities to lift themselves out of poverty and build more resilient livelihoods.

Moreover, agriculture is intricately linked to food security, ensuring that populations have access to an adequate and nutritious diet. For example, Canada has a strong agricultural sector that produces a diverse range of crops, including wheat, barley, canola, and pulses. Its efficient farming practices and modern infrastructure contribute to food security for its population. The Netherlands is renowned for its advanced agricultural techniques and high productivity levels. Despite its small size, the country is a leading exporter of agricultural products such as dairy, vegetables, and flowers, ensuring food security for its citizens. Brazil's vast agricultural land and favorable climate make it a major producer of soybeans, sugarcane, coffee, and beef. Its agricultural exports contribute significantly to the country's food security and economic prosperity.

Sustainable agricultural practices, such as conservation agriculture, agroforestry, and crop diversification, can enhance resilience to climate change, reduce vulnerability to food shortages, and improve nutritional outcomes. Additionally, investments in agricultural research and extension services can lead to technological innovations that boost productivity and improve crop yields, further contributing to food security and nutrition.

The economic contributions of agriculture are manifold, encompassing not only its direct contributions to GDP, employment, and trade but also its role in poverty alleviation, food security, and rural development. By recognizing the multifaceted benefits of agriculture and investing in sustainable agricultural development, policymakers can harness the potential of agriculture to drive economic growth, reduce poverty, and build more resilient and inclusive societies.

1.3 Global food security

One of the most pressing challenges facing humanity today is ensuring global food security in the face of population growth, climate change, and environmental degradation. Agriculture holds the key to addressing this challenge by increasing food production, improving access to nutritious foods, and promoting sustainable farming practices. However, achieving food security requires more than just increasing yields; it necessitates addressing inequalities in access to land, water, and resources, as well as investing in agricultural research, infrastructure, and market systems.

Global food security is a critical issue that encompasses ensuring that all people, at all times, have access to sufficient, safe, and nutritious food to meet their dietary needs and preferences for an active and healthy life. Despite significant progress in reducing hunger and malnutrition in recent decades, millions of people around the world still suffer from food insecurity, with challenges exacerbated by factors such as population growth, climate change, and environmental degradation..

Population growth is a key driver of increased demand for food, placing pressure on agricultural systems to produce more food to feed a growing global population.

With the world's population expected to reach nearly **10 billion** by **2050**, the need to enhance food production and distribution systems has never been more urgent. Agriculture holds the key to addressing this challenge by increasing food production through sustainable intensification, improving agricultural productivity, and enhancing resilience to climate change.

Climate change poses a significant threat to global food security, as rising temperatures, changing precipitation patterns, and extreme weather events disrupt agricultural production and food supply chains. Shifts in growing seasons, increased frequency of droughts, floods, and storms, and the spread of pests and diseases threaten crop yields and livestock productivity, exacerbating food insecurity in vulnerable regions. Sustainable farming practices, such as agroecology, conservation agriculture, and climate-smart agriculture, offer promising solutions for adapting to climate change while minimizing its adverse impacts on food production.

Environmental degradation, including soil erosion, deforestation, water scarcity, and biodiversity loss, further undermines global food security by degrading the natural resources upon which agriculture depends. Addressing these environmental challenges requires holistic approaches that promote sustainable land management, ecosystem restoration, and biodiversity conservation. By protecting and restoring ecosystems, enhancing soil health, and conserving water resources, agriculture can become more resilient and productive, contributing to long-term food security and environmental sustainability.

Achieving food security also requires addressing inequalities in access to land, water, and resources, which disproportionately affect smallholder farmers, women, indigenous people, and marginalized communities. Ensuring equitable access to productive resources, markets, and technologies can empower smallholder farmers to improve their livelihoods, increase their resilience to shocks, and contribute to food security at the local and national levels. Moreover, investing in social protection programs, nutrition-sensitive interventions, and food assistance programs can help vulnerable population's access to nutritious foods and reduce their vulnerability to food insecurity.

In addition to increasing food production, improving access to nutritious foods is essential for achieving food security and addressing malnutrition. Sustainable food systems that prioritize diversity, equity, and inclusion can ensure that all people have access to a nutritious and culturally appropriate diet. This requires promoting the production, distribution, and consumption of diverse foods, including fruits, vegetables, pulses, nuts, and animal-source foods, while reducing reliance on highly processed and unhealthy foods. Investing in nutrition education, food labeling, and consumer awareness campaigns can also empower individuals and communities to make healthy food choices and improve their nutritional status.

Promoting sustainable farming practices is crucial for achieving food security while safeguarding the environment and promoting inclusive and equitable development. This requires investing in agricultural research, extension services, and farmer education programs to disseminate knowledge and technologies that enhance productivity, resilience, and sustainability. Strengthening agricultural value chains, market systems, and rural infrastructure can also improve access to markets, inputs, and financial services for smallholder farmers and agribusinesses, fostering economic growth, job creation, and poverty reduction in rural areas.

Achieving global food security requires a holistic and multidimensional approach that addresses the complex challenges of population growth, climate change, and environmental degradation. By promoting sustainable farming practices, improving access to nutritious foods, and addressing inequalities in access to land, water, and resources, agriculture can play a transformative role in ensuring food security, reducing poverty, and promoting sustainable development for present and future generations.

1.4 The role of agriculture in environmental sustainability

While agriculture is essential for meeting human needs, it also has significant environmental impacts. Unsustainable agricultural practices contribute to deforestation, soil erosion, water pollution, and biodiversity loss, exacerbating environmental degradation and climate change. Recognizing the interdependence of agriculture and the environment, there is a growing imperative to adopt integrated approaches that promote both productivity and sustainability. Integrated agriculture emphasizes holistic solutions that balance economic, social, and environmental objectives, ensuring that agriculture remains viable for future generations.

The role of agriculture in environmental sustainability is paramount, as the health and productivity of agricultural systems are intricately linked to the health and resilience of the natural environment. While agriculture is essential for meeting human needs, it also has significant environmental impacts that can undermine the long-term sustainability of food production systems. Unsustainable agricultural practices, such as intensive monoculture farming, excessive use of chemical inputs, and land conversion, can contribute to deforestation, soil erosion, water pollution, and biodiversity loss, exacerbating environmental degradation and climate change.

Deforestation is a major environmental consequence of unsustainable agriculture, as forests are cleared to make way for agricultural expansion, particularly in tropical regions. Deforestation not only destroys critical habitats for biodiversity but also releases large amounts of carbon dioxide into the atmosphere, contributing to climate change. Moreover, deforestation can lead to soil degradation, loss of ecosystem services, and disruptions to local water cycles, further exacerbating environmental degradation and undermining the resilience of agricultural systems.

Soil erosion is another significant environmental challenge associated with unsustainable agricultural practices, such as intensive tillage, monoculture cropping, and inadequate soil conservation measures. Soil erosion not only depletes soil fertility and reduces agricultural productivity but also leads to sedimentation of water bodies, degradation of aquatic habitats, and increased risk of flooding and landslides. Addressing soil erosion requires implementing soil conservation practices, such as contour plowing, terracing, cover cropping, and agroforestry, that help to protect soil structure, reduce erosion, and improve soil health.

Water pollution is a pressing environmental issue linked to agricultural activities, particularly the use of chemical fertilizers, pesticides, and herbicides. Runoff from agricultural fields can carry pollutants such as nitrogen, phosphorus, and pesticides into rivers, lakes, and coastal waters, leading to eutrophication, algal blooms, and contamination of drinking water sources. To mitigate water pollution, farmers can adopt integrated pest management (IPM) practices, nutrient management strategies, and conservation tillage methods that minimize chemical inputs, reduce runoff, and protect water quality.

Biodiversity loss is another consequence of unsustainable agricultural practices, as monoculture farming, habitat destruction, and pesticide use can degrade or destroy wildlife habitats and disrupt ecological processes. Loss of biodiversity can reduce ecosystem resilience, increase vulnerability to pests and diseases, and undermine the long-term sustainability of agricultural systems. Adopting agroecological approaches, such as crop diversification, habitat restoration, and integrated pest management, can help to conserve biodiversity, enhance ecosystem services, and promote ecological resilience in agricultural landscapes.

Recognizing the interdependence of agriculture and the environment, there is a growing imperative to adopt integrated approaches that promote both productivity and sustainability. Integrated agriculture emphasizes holistic solutions that balance economic, social, and environmental objectives, ensuring that agriculture remains viable for future generations. By promoting sustainable farming practices, such as agroecology, conservation agriculture, and organic farming, farmers can reduce environmental impacts, enhance ecosystem resilience, and improve the sustainability of food production systems.

Moreover, investing in agricultural research, extension services, and farmer education programs can facilitate the adoption of sustainable farming practices and promote the transition to more environmentally friendly agricultural systems. By adopting integrated approaches that promote both productivity and sustainability, agriculture can play a transformative role in addressing environmental challenges, mitigating climate change, and promoting the long-term resilience and sustainability of food production systems.

The profound significance of agriculture is more than just an economic endeavor. Instead, it emphasizes agriculture's pivotal role in shaping human societies across time. By delving into the exploration of integrated agricultural approaches, we recognize that agriculture is deeply intertwined with the past, present, and future of humanity. It serves as a foundational element that connects us to the land, the environment, and each other in profound ways. Agriculture transcends mere economic activity; it embodies a fundamental aspect of human existence. It symbolizes the symbiotic relationship between humanity and the natural world, reflecting our dependency on the land for sustenance, livelihoods, and cultural identity. Moreover, agriculture fosters connections between people, communities, and ecosystems, highlighting the interdependence and collective responsibility inherent in our relationship with the environment.

In essence, this perspective acknowledges agriculture as a cornerstone of human civilization, with its impacts extending far beyond the realms of food production and economic growth. It underscores the intrinsic value of agriculture in shaping our societies, cultures, and identities, while also emphasizing the need for sustainable and integrated approaches to ensure the resilience and well-being of both humanity and the planet.

Throughout history, agriculture has been the catalyst for the development and evolution of human civilizations. It has provided the sustenance needed for survival, facilitated the growth of communities and cultures, and enabled the establishment of complex societies. From the earliest agricultural practices of ancient civilizations to the modern industrialized farming methods of today, agriculture has continually adapted and transformed, shaping landscapes, economies, and social structures.

Moreover, agriculture serves as a bridge between humanity and the natural world, highlighting the intricate interdependence between human societies and the environment. The health and productivity of agricultural systems are intimately linked to the health and resilience of ecosystems, emphasizing the importance of adopting sustainable and regenerative practices that promote both productivity and environmental stewardship.

By acknowledging agriculture as the backbone of nations, we underscore its critical role in ensuring the continued prosperity and sustainability of agricultural systems worldwide. Integrated agricultural approaches offer a holistic framework that recognizes the interconnectedness of social, economic, and environmental factors, seeking to optimize agricultural productivity while safeguarding natural resources and promoting resilience to climate change.

As we move forward, it is imperative to prioritize the adoption of integrated agricultural practices that enhance productivity, improve livelihoods, and protect the environment. This requires collaboration and cooperation among farmers, policymakers, researchers, and stakeholders across sectors to develop and implement innovative solutions that address the complex challenges facing agriculture and food systems.

In embracing agriculture as the cornerstone of human societies, we pave the way for a more sustainable and resilient future, where agriculture not only meets the needs of present generations but also preserves the resources and ecosystems essential for the well-being of future generations. By recognizing and honoring the foundational importance of agriculture, we embark on a journey toward a more harmonious and sustainable relationship between humanity and the natural world.

1.5 Agriculture is the backbone of Pakistan's economy

Agriculture has long been recognized as the backbone of Pakistan's economy due to its significant contributions to various facets of the nation's development.

Some of highlights of agriculture in Pakistan are listed further.

1.5.1 Employment and livelihoods

Agriculture is the largest sector in terms of employment, providing jobs for a substantial portion of the population, especially in rural areas. It supports the livelihoods of millions of farmers, farmworkers, and their families.

1.5.2 GDP contribution

Agriculture is a major contributor to Pakistan's GDP. The sector's output directly contributes to the overall economic performance of the country.

1.5.3 Food security

Agriculture is the primary source of food for Pakistan's population. The sector's productivity directly affects the availability and affordability of food items, which is essential for ensuring food security and reducing poverty.

1.5.4 Raw materials

Agriculture provides raw materials for various industries, such as textiles, food processing, and leather. This integration between agriculture and industries helps drive economic growth and diversification.

1.5.5 Export earnings

Pakistan exports a significant portion of its agricultural products, including textiles, rice, cotton, fruits, and vegetables. These exports contribute to foreign exchange earnings and balance of trade.

1.5.6 Rural development

Since a substantial part of Pakistan's population resides in rural areas, agriculture plays a pivotal role in rural development by improving living standards, infrastructure, and access to basic services.

1.5.7 Support for other sectors

The success of sectors like textiles, which is a crucial export industry, relies heavily on raw materials from agriculture, particularly cotton.

1.5.8 Foreign exchange

The revenue generated from agricultural exports provides much-needed foreign exchange, which is essential for meeting import demands and servicing external debts.

1.5.9 Sustainable resource management

Agriculture is closely linked to the management of natural resources, including land, water, and biodiversity. Sustainable agricultural practices promote resource conservation, soil health, and water management, leading to environmental sustainability and mitigating the impacts of climate change.

1.5.10 Market linkages

Agriculture fosters linkages between rural and urban economies. It provides a market for rural products and, in turn, enables urban areas to access fresh produce and other agricultural goods.

1.5.11 Economic stability

A robust agricultural sector helps stabilize the economy by providing a buffer against external shocks and price fluctuations in international markets.

1.5.12 Poverty alleviation

Agriculture has the potential to reduce poverty and income disparities by providing livelihood opportunities to smallholder farmers. Supporting small-scale farmers through improved access to credit, technology, and market linkages can enhance their income-generating capacity and lift them out of poverty.

1.5.13 Social stability

Agriculture contributes to social stability by providing a reliable source of food, income, and employment. A thriving agricultural sector helps reduce social inequalities, alleviate rural poverty, and foster social cohesion and stability within communities.

It is important to note that while agriculture is a significant contributor to Pakistan's economy, the sector faces challenges such as water scarcity, outdated farming practices, land degradation, and climate change. Addressing these challenges through sustainable agricultural practices, technological advancements, and policy support is crucial to ensuring the continued growth and resilience of the agricultural sector and, consequently, the overall economy of Pakistan.

1.6 Components of agriculture

The components of the agriculture sector play a crucial role not only at the national level but also on a global scale, influencing food production, economic growth, and livelihoods worldwide. Agriculture serves as the backbone of many economies, particularly in developing countries, where it is a primary source of employment and income for millions of people. Moreover, the agricultural sector contributes significantly to global food security by providing the essential crops, livestock, and other agricultural products needed to feed a growing population.

Furthermore, agriculture is closely linked to global economic growth and trade. Countries heavily rely on agricultural exports to generate revenue and support economic development. For example, agricultural exports from countries like the United States, Brazil, and China contribute substantially to global trade and influence market dynamics worldwide. Additionally, agriculture fosters innovation and technological advancement, driving progress in areas such as crop genetics, mechanization, and precision farming techniques, which have implications beyond national borders.

Moreover, the agriculture sector plays a vital role in addressing pressing global challenges, such as climate change, environmental degradation, and poverty alleviation. Sustainable agricultural practices, including conservation agriculture, agroforestry, and organic farming, have the potential to mitigate greenhouse gas emissions, enhance ecosystem resilience, and promote rural development. By adopting these practices on a global scale, countries can work toward achieving the Sustainable Development Goals (SDGs) set forth by the United Nations, particularly those related to zero hunger, climate action, and sustainable communities.

The components of the agriculture sector hold significant global importance, contributing to food security, economic prosperity, and environmental sustainability across nations. Recognizing the interconnectedness of agricultural systems and their impacts on a global scale is essential for fostering cooperation, innovation, and policy action to address the challenges facing agriculture and ensure a more sustainable and equitable future for all.

The components of agriculture varies among different countries. The key components of agriculture in Pakistan, for example, include the following components (Figure 1.2).

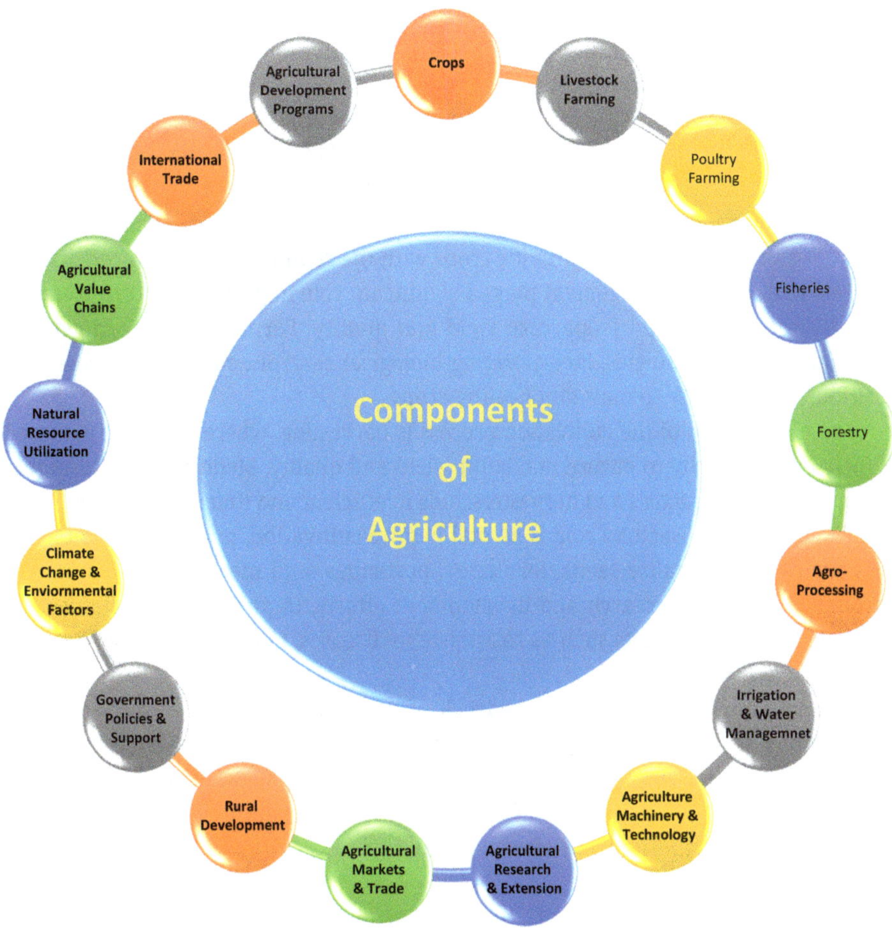

Figure 1.2: The components of agriculture.

1.6.1 Crops

Crops are the cornerstone of Pakistan's agricultural sector, playing a pivotal role in ensuring food security, generating income, and contributing to the nation's economy. The country's diverse agroclimatic zones provide an optimal environment for cultivating a wide array of crops, each suited to specific regions and conditions.

Among the major crops cultivated in Pakistan, wheat stands as a staple food and a vital source of nutrition for the population. Rice, another essential staple, contributes significantly to both domestic consumption and export earnings. Maize is utilized for various purposes, including food, fodder, and industrial applications. Cotton, a crucial cash crop, serves as the backbone of Pakistan's textile industry, a significant contributor to export revenues.

Sugarcane is pivotal for producing sugar and other by-products, while pulses and oilseeds contribute to protein-rich diets and edible oil production. The cultivation of fruits and vegetables not only enhances dietary diversity but also offers opportunities for local consumption and export.

The process of crop cultivation encompasses several stages, starting with land preparation to ensure optimal soil conditions for planting. Sowing follows, where seeds are strategically placed in the soil to facilitate germination and growth. Adequate irrigation, often supported by canal systems, tube wells, and modern drip or sprinkler systems, is crucial to provide crops with the required water for growth.

Pest and disease management plays a significant role in protecting crops from potential threats that could jeopardize yield and quality. Farmers employ integrated pest management strategies, incorporating biological controls, cultural practices, and judicious pesticide use to maintain healthy crops.

The culmination of the cultivation process is harvesting, where crops are gathered at their peak maturity to ensure maximum yield and quality. Modern agricultural machinery, including tractors and harvesters, aids in efficient and timely harvesting.

The agricultural sector's role is pivotal in crop cultivation, providing essential inputs such as high-quality seeds, fertilizers, pesticides, and mechanized equipment. Through continuous research and development efforts, improved crop varieties are developed, exhibiting traits such as resistance to diseases, tolerance to adverse conditions, and enhanced yields.

The cultivation of crops is the bedrock of Pakistan's agriculture, anchoring the nation's food security, economic growth, and livelihoods. The intricate cycle of preparing, sowing, nurturing, and harvesting crops, supported by the agricultural ecosystem, underscores the significance of this component in shaping Pakistan's agrarian landscape and contributing to its overall prosperity.

1.6.2 Livestock and poultry farming

Livestock, encompassing a variety of animals, such as cattle, buffaloes, sheep, goats, and poultry, holds a pivotal position in Pakistan's agricultural sector. This multifaceted component not only contributes to the nation's food production but also plays a critical role in supporting livelihoods, economic growth, and rural development.

Cattle and buffaloes are central to Pakistan's livestock landscape, serving as sources of milk, meat, and leather. Milk production is a vital aspect, with cows and buffaloes yielding a significant portion of the country's dairy output. The dairy sector provides nutritious milk for consumption, contributing to the population's dietary requirements while also catering to the robust dairy processing industry.

Sheep and goats contribute to both meat and wool production. These animals are particularly suited to Pakistan's diverse landscapes, with pastoral communities relying on them for subsistence and income. The wool industry, though relatively smaller compared to other livestock products, still provides employment and raw materials for various cottage industries.

Poultry farming, a rapidly growing segment of the livestock sector, produces eggs and meat, contributing to protein-rich diets. Poultry rearing is not only efficient in terms of resource utilization but also offers opportunities for smallholder farmers to engage in income-generating activities. The sector's adaptability and relatively low entry barriers make it accessible to a wide range of farmers, including those with limited resources.

Livestock rearing serves as a lifeline for smallholder farmers, offering them diverse income sources and a buffer against crop failures. This is particularly important in a country like Pakistan, where a significant portion of the population depends on agriculture for their livelihoods. The sector contributes to rural economies by generating employment opportunities, supporting local markets, and fostering entrepreneurship in related industries such as dairy processing and leather production.

Beyond economic considerations, livestock also holds cultural and social significance, often being integrated into traditional practices and festivals. Furthermore, they play a role in nutrient cycling by contributing manure that enhances soil fertility and supports sustainable farming practices.

Livestock forms an integral pillar of Pakistan's agriculture, contributing to food security, income generation, and rural development. The diverse products derived from livestock, coupled with the sector's employment opportunities, exemplify its multifunctional role in supporting both individual livelihoods and the nation's agrarian economy.

1.6.3 Fisheries

The fisheries sector holds a distinct and valuable position within Pakistan's economy, food security, and sociocultural fabric. Exploiting the nation's extensive coastline along the Arabian Sea and abundant freshwater resources, the fisheries industry makes substantial contributions to livelihoods, export revenues, and the provision of essential protein-rich sustenance.

Marine fishing forms a crucial component of Pakistan's fisheries sector, with coastal communities engaged in capturing a diverse array of marine species. Fishermen venture into the Arabian Sea to catch species like sardines, mackerels, tuna, and shrimp. These catches not only supply the domestic market but also constitute a significant portion of the country's seafood exports. The industry not only offers direct employment opportunities for fishing communities but also fuels ancillary sectors such as processing, transportation, and marketing.

Inland fishing, conducted in rivers, lakes, and reservoirs, is equally vital. The Indus River system, with its tributaries and associated water bodies, provides an environment conducive to the breeding and cultivation of freshwater fish species. Carp fish, in particular, contributes substantially to inland fish production. Inland fisheries offer a valuable source of income for communities residing along riverbanks and contribute to the nutritional diversity of local diets.

Aquaculture, a rapidly expanding subsector, involves the controlled cultivation of aquatic organisms such as fish, shrimp, and prawns. This practice complements traditional fishing methods by supplementing the supply of fish and seafood. By harnessing technology and scientific know-how, aquaculture presents opportunities for sustainable fish production and increased seafood availability.

Sustainable management and development of fisheries are paramount to preserving the well-being of fishing communities and aquatic ecosystems. Overfishing, habitat degradation, and illegal practices can threaten the sector's viability and the biodiversity of marine and freshwater environments. Initiatives such as regulated fishing seasons, mesh size regulations, and protected areas aim to ensure the longevity of fish stocks and aquatic biodiversity.

The fisheries sector's contributions extend beyond economic realms, encompassing cultural and nutritional dimensions. Fish and seafood constitute an important source of protein and essential nutrients for many Pakistanis. By enhancing fish production and access to affordable protein, the sector plays a crucial role in addressing malnutrition and supporting public health.

The fisheries sector's significance is multidimensional, intertwined with economic growth, food security, livelihoods, and environmental sustainability. Embracing responsible fishing practices, investing in aquaculture, and safeguarding aquatic ecosystems are pivotal steps to ensure the sector's continued contributions to the nation's well-being and prosperity.

1.6.4 Forestry

Pakistan's forestry sector occupies a critical niche within the nation's environmental, economic, and social fabric. The country's varied landscapes, spanning from the Himalayan forests in the north to the coastal mangroves in the south, collectively host a rich diversity of plant and animal species, making them invaluable for biodiversity conservation.

Forests in Pakistan hold a multifunctional role, contributing to a range of ecological services. Watershed management is a primary function, as forests play a pivotal role in regulating water flow, reducing soil erosion, and maintaining the quality of water resources. The northern mountain ranges serve as natural water reservoirs, supplying water to rivers that sustain ecosystems downstream and provide freshwater resources for agriculture, industry, and domestic use.

Forests also play a vital role in carbon sequestration and climate regulation. By absorbing carbon dioxide from the atmosphere, trees help mitigate the impacts of climate change, making afforestation and reforestation key strategies in global efforts to combat climate change.

Despite these ecological benefits, Pakistan's forests have faced numerous challenges. Deforestation, driven by factors such as illegal logging, land conversion, and population pressure, has led to habitat loss and environmental degradation. However, the country is actively pursuing sustainable forestry practices through various initiatives.

Afforestation and reforestation projects are central to Pakistan's efforts to restore forest cover. These endeavors involve planting native tree species in degraded areas, which not only aids in conserving biodiversity but also helps restore ecosystem services. Community-based forestry initiatives empower local communities to participate in forest management, creating a sense of ownership and incentivizing sustainable resource use.

These efforts contribute not only to environmental conservation but also to the livelihoods of local communities. Forests provide valuable non-timber forest products like fruits, medicinal plants, and honey, offering alternative income sources for rural populations. Furthermore, ecotourism in regions like the Northern Areas generates economic opportunities, encouraging the protection of forests and biodiversity.

Strengthening Pakistan's forestry sector is paramount for fostering ecological resilience and sustainability. Sustainable forest management practices, coupled with robust law enforcement against illegal logging, are crucial steps in conserving this precious resource. Through these measures, the forestry sector can continue to serve as a bulwark against climate change, a haven for biodiversity, and a source of sustenance for communities while ensuring a healthier environment for current and future generations.

1.6.5 Agroprocessing

Agroprocessing is a pivotal step in the agricultural value chain, where the focus shifts from raw produce to enhanced products. It encompasses various activities such as food processing, textiles, and other industries that utilize agricultural raw materials. Through processes like milling, canning, and refining, agroprocessing transforms crops into value-added products, extending their shelf life, improving quality, and enhancing their market value. This not only benefits farmers by creating additional revenue streams but also contributes to food security, reduces postharvest losses, and fosters economic growth by generating employment and supporting downstream industries.

1.6.6 Irrigation and water management

In Pakistan, where agriculture heavily depends on irrigation, effective water resource management is of paramount importance. The country's prosperity hinges on its ability to efficiently allocate and manage water for agricultural purposes. Complex irrigation systems play a pivotal role in ensuring equitable water distribution among crops, facilitating their optimal growth and productivity. Proper irrigation and water management practices are vital not only for sustaining agricultural output but also for supporting rural livelihoods, food security, and the overall economic well-being of the nation.

1.6.7 Agricultural machinery and technology

The use of modern agricultural machinery, tools, and technology has become essential to enhance productivity and efficiency in farming. In the modern agricultural landscape, the incorporation of advanced agricultural machinery, tools, and technology has become indispensable for elevating productivity and operational efficiency within the sector. Mechanization has revolutionized traditional farming practices, enabling tasks like planting, harvesting, and soil preparation to be accomplished with greater speed and precision. Through innovations like GPS-guided tractors, automated irrigation systems, and data-driven decision-making, technology optimizes resource utilization, reduces manual labor, and enhances crop yields. Moreover, the integration of machinery and technology empowers farmers to navigate challenges posed by evolving environmental conditions and market demands. By harnessing these tools, agriculture is positioned to embrace sustainable practices, increase profitability, and secure the sustenance of future generations.

1.6.8 Agricultural research and extension

Research institutions and extension services play a key role in developing new crop varieties, techniques, and best practices to improve agricultural outcomes. In Pakistan, the synergy between agricultural research institutions and extension services plays a pivotal role in propelling the agricultural sector forward. These institutions serve as bastions of innovation, dedicating their efforts to the development of novel crop varieties, cutting-edge techniques, and best practices that hold the potential to elevate agricultural outcomes. By conducting research to address challenges like changing climate patterns, pest and disease management, and soil health, they equip farmers with the knowledge and tools required to optimize their yields and sustainably manage resources. The extension services then bridge the gap between research and implementation, disseminating valuable insights to farmers across the country. This symbiotic relationship between research and extension serves as the cornerstone of progress, facilitating the evolution of Pakistan's agriculture toward enhanced resilience, productivity, and food security.

1.6.9 Agricultural markets and trade

In Pakistan, the intricate network of agricultural markets and trade plays a pivotal role in connecting producers with consumers, both within the nation's borders and across international boundaries. This component forms a vital link in the agricultural supply chain, facilitating the efficient movement of diverse products from farms to tables. The dynamic nature of these markets involves processes such as pricing, storage, transportation, and distribution, each contributing to the accessibility and availability of agricultural goods. Furthermore, international trade of agricultural products enhances economic growth by generating export earnings and diversifying revenue sources. By ensuring the smooth flow of products and fostering competitive markets, Pakistan's agricultural trade ecosystem contributes to food security, economic stability, and the overall prosperity of the nation.

1.6.10 Rural development

Rural development stands as a crucial pillar in Pakistan, aimed at nurturing the well-being of its rural communities and fostering sustainable progress. With the majority of the population residing in rural areas, initiatives dedicated to rural development play an integral role in elevating living standards and promoting equitable growth. By providing access to basic services such as education, healthcare, clean water, and sanitation, these efforts address critical human needs and pave the way for improved quality of life. Moreover, the development of rural infrastructure, including roads, electricity, and communication networks, enhances connectivity and opens doors to economic opportunities. As rural development endeavors continue to unfold, they play a pivotal role in bridging the urban-rural divide, bolstering the agricultural sector, and empowering individuals and communities to contribute actively to Pakistan's broader socioeconomic advancement.

1.6.11 Government policies and support

The role of the government in shaping the agricultural landscape of Pakistan is undeniable, as it wields its influence through a spectrum of policies and support mechanisms. The formulation of effective agricultural policies sets the tone for sustainable growth, addressing challenges such as food security, environmental conservation, and economic stability. These policies guide subsidy programs that provide financial assistance to farmers, thereby easing their burden and enhancing their access to critical inputs such as seeds, fertilizers, and machinery. Moreover, price support mechanisms ensure that farmers receive fair compensation for their produce, safeguarding their livelihoods in an often unpredictable market environment.

Institutions dedicated to agriculture, both at the federal and provincial levels, serve as vital conduits for government support. Entities like the Ministry of National Food Security and Research collaborate with provincial agriculture departments to design and implement policies tailored to the unique needs of different regions. Research institutes contribute by developing improved crop varieties, sustainable farming practices, and solutions to combat emerging challenges like climate change and pests. Extension services extend the government's arm to rural areas, offering farmers advisory services, disseminating best practices, and conducting capacity-building workshops that empower farmers with the knowledge and skills necessary for modern, efficient agriculture.

The government's commitment to agriculture is not just confined to policy; it resonates through the creation of a supportive ecosystem that empowers farmers, fosters innovation, and aligns with the nation's broader development goals. By championing policies and institutions that cater to the needs of the agricultural sector, Pakistan paves the way for a resilient and vibrant farming community that contributes significantly to the country's progress and sustenance.

1.6.12 Climate and environmental factors

In the context of Pakistan, the intricate interplay between climate and agriculture underscores the paramount importance of climate and environmental factors in shaping the sector's dynamics. The country's diverse geography and varying climatic zones expose agriculture to a range of conditions, from arid deserts to temperate valleys and high-altitude plateaus. Rainfall patterns and temperature fluctuations wield considerable influence over planting seasons, crop growth, and overall agricultural productivity. With a predominantly agrarian economy, shifts in these climatic elements can ripple through the entire socioeconomic fabric, affecting food production, livelihoods, and even rural-urban migration patterns.

Moreover, the ongoing challenges posed by climate change intensify the need for adaptation strategies within agriculture. Erratic weather events, extended droughts, and increased instances of extreme weather phenomena disrupt established growing cycles and necessitate the adoption of resilient agricultural practices. As Pakistan grapples with these environmental intricacies, innovative approaches such as drought-resistant crops, water-efficient irrigation, and climate-resilient farming techniques gain significance. The collaboration between scientific research institutions, governmental bodies, and local communities becomes pivotal in devising strategies that not only mitigate the adverse impacts of climate change but also harness its potential for sustainable growth.

Recognizing the intricate dance between climate and agriculture, Pakistan's commitment to climate-smart policies and practices is integral to ensuring the sector's adaptability and long-term viability. As climate continues to exert its influence, a proactive response that integrates scientific insights, technology, and community participation is essential for nurturing a resilient agricultural landscape that can withstand the challenges of today and tomorrow.

1.6.13 Natural resource utilization

In Pakistan, the intricate relationship between agriculture and natural resource utilization underscores the pivotal role of these resources in sustaining the nation's agrarian economy. Water, being a paramount resource, occupies a central position due to the country's reliance on irrigation to support crop production. A combination of canal irrigation systems, tube wells, and advanced sprinkler systems is employed to ensure a consistent water supply to fields, particularly in regions with irregular rainfall patterns. Efficient water management practices are imperative to combat water scarcity, enhance water use efficiency, and mitigate the impacts of changing climate conditions.

Land utilization is another cornerstone of Pakistan's agricultural landscape. Vast expanses of arable land are cultivated using a variety of techniques, from traditional plowing and tilling to modern conservation practices that minimize soil erosion and maintain soil health. The judicious use of land resources is critical for maximizing productivity while preserving the long-term fertility of the soil.

In this intricate interplay between agriculture and natural resources, soil management assumes paramount importance. Sustainable farming practices, including crop rotation, cover cropping, and organic matter incorporation, ensure that soil quality remains optimal for long-term productivity. Moreover, biodiversity conservation emerges as a critical consideration, with the need to strike a balance between agricultural expansion and safeguarding natural habitats.

As Pakistan navigates the challenges of population growth, climate change, and environmental degradation, the sustainable utilization of natural resources remains a top priority. Embracing innovative water-saving technologies, adopting soil-conserving practices, and promoting biodiversity-friendly approaches are central to the nation's efforts to ensure food security, economic stability, and environmental preservation. The harmonious coexistence of agriculture and nature becomes pivotal, fostering a resilient and prosperous agricultural sector that is well-equipped to meet the demands of the present and future generations.

1.6.14 Agricultural value chains

In Pakistan, the concept of agricultural value chains encompasses the intricate web of interconnected activities that transform raw agricultural products into finished goods, with each stage adding value to the final product. The nation's agriculture sector encompasses a diverse array of participants, ranging from input suppliers and farmers at the production end to processors, wholesalers, retailers, and exporters further down the line.

This multifaceted value chain is characterized by its dynamic nature, as each stage contributes to the quality, quantity, and marketability of the products. Input suppliers provide essential resources such as seeds, fertilizers, and machinery to farmers, enabling them to cultivate their crops effectively. The farmers, who are at the heart of the chain, utilize these resources to produce crops that eventually find their way to processors.

The processing stage is particularly vital, as it transforms raw agricultural produce into processed goods, such as packaged foods, textiles, dairy products, and leather goods. These agro-based industries rely on agricultural raw materials, creating a symbiotic relationship that generates economic value and employment opportunities. Processors collaborate with wholesalers and retailers to distribute the products throughout the country, ensuring their availability to consumers.

Additionally, the value chain extends to international markets through exporters, who facilitate the trade of Pakistan's agricultural products worldwide. The intricate connections between these stages underline the sector's importance, not only for food security and rural livelihoods but also for contributing to the overall economy. By fostering collaboration and innovation along the value chain, Pakistan's agricultural sector holds the potential to drive economic growth, enhance export revenue, and create a sustainable framework that benefits both producers and consumers.

1.6.15 International trade

International trade forms a vital dimension of Pakistan's agriculture sector, enabling the country to engage in the global market by exporting a diverse range of agricultural products. Key exports include commodities such as rice, textiles, fruits, and seafood, which not only contribute to the nation's economy but also strengthen its position on the international stage.

Participation in international trade offers Pakistan an avenue for expanding its market reach beyond domestic borders. By tapping into diverse global markets, the agriculture sector can leverage its competitive advantages, thereby enhancing its revenue potential and creating additional income streams for farmers and stakeholders.

To succeed in the international trade arena, adherence to international quality and safety standards is crucial. Meeting these standards ensures that Pakistani agricultural products are well-received in international markets, fostering consumer trust and loyalty. Compliance with these standards bolsters the reputation of Pakistani exports, positioning them as reliable and high-quality offerings.

Moreover, international trade significantly contributes to foreign exchange earnings. By capitalizing on its strengths in agriculture, Pakistan can bolster its balance of payments through the export of agricultural goods. This foreign exchange influx can play a critical role in supporting economic growth, financing developmental projects, and strengthening the nation's overall economic resilience.

In a globalized world, international trade opens doors to economic prosperity, knowledge exchange, and cultural interactions. By strategically aligning its agricultural exports with international demand and quality standards, Pakistan's agriculture sector stands poised to seize new opportunities, diversify its revenue sources, and contribute to the nation's sustainable development trajectory.

1.6.16 Agricultural development programs

Agricultural development programs in Pakistan encompass a range of concerted efforts aimed at propelling the agriculture sector toward sustainable growth, resilience, and improved well-being for farmers and rural communities. These programs are typically a collaborative endeavor involving the government, international organizations, development partners, and local stakeholders.

With the overarching goal of bolstering agricultural productivity and livelihoods, these programs address multifaceted dimensions of the sector. They play a critical role in enhancing the efficiency and accessibility of irrigation systems, ensuring that water reaches crops efficiently and minimizing water wastage. Additionally, these initiatives promote the development of high-quality seeds and improved agricultural practices, both of which contribute to elevated yields and crop quality.

Research and development are core components of these programs, as they foster innovation and the adoption of advanced technologies. By focusing on areas such as improved crop varieties, pest and disease management, and sustainable farming practices, these initiatives contribute to overall agricultural advancement.

Capacity building forms a vital pillar, as these programs equip farmers with the knowledge and skills necessary to optimize their farming practices. Through training, workshops, and extension services, farmers gain insights into modern techniques, risk management, and effective resource utilization.

Furthermore, these programs emphasize forging linkages between farmers and markets. By enhancing market access and providing relevant information, farmers can make informed decisions about what to produce and where to sell their products. This not only ensures better returns for farmers but also contributes to food security by facilitating the efficient distribution of agricultural goods.

Collectively, agricultural development programs act as catalysts for positive transformation within the sector. By addressing a comprehensive spectrum of challenges, from production to market linkages, these initiatives contribute to enhanced agricultural sustainability, economic growth, and the well-being of rural communities across Pakistan.

At the culmination of this exploration, the intricate components of Pakistan's agricultural system converge to create a dynamic and multifaceted sector. These components – ranging from crops and livestock to agroprocessing, irrigation, policy support, and more – collectively form the backbone of the nation's agricultural landscape. This interconnectedness necessitates a holistic approach to address challenges and foster sustainability within each facet.

Intricacies notwithstanding, Pakistan's agriculture assumes a role of paramount significance, extending its influence across economic, social, and environmental domains. A cohesive strategy to confront challenges, promote sustainable practices, and ensure the well-being of farmers is pivotal in safeguarding food security, fostering economic development, and preserving the environment. By forging alliances between stakeholders, implementing innovative technologies, and endorsing policies that nurture resilience, Pakistan's agricultural sector can further fortify its contribution to the nation's prosperity.

As the bedrock of the nation's economy, agriculture in Pakistan serves as the cornerstone of progress and sustenance. It generates employment opportunities, supports livelihoods, and ensures a stable supply of essential food items. This sector's impact resonates beyond economic realms, encompassing cultural and social dimensions, as rural communities remain deeply intertwined with agricultural traditions. Thus, the agriculture sector becomes the linchpin that propels rural development and drives economic growth, influencing the trajectory of the entire nation.

The tapestry of Pakistan's agriculture is woven with intricate threads, connecting various components and challenges in a way that underscores the sector's centrality. Addressing these challenges while embracing sustainable practices and innovations is pivotal for elevating the sector's contributions to food security, economic prosperity, and environmental stewardship. As the backbone of Pakistan's economy, agriculture embodies both the resilience of its people and the promise of a flourishing future.

In the heart of Pakistan's landscape lies the thriving nucleus of its economy and sustenance – agriculture. This chapter has illuminated the pivotal role that agriculture plays in shaping the country's destiny. By delving into its history, contributions, and challenges, we have gained an appreciation for its profound impact on livelihoods, food security, and rural development. As we move forward, it is essential to acknowledge the formidable challenges facing this sector and embrace innovative solutions to ensure the continuity of this essential backbone.

While agriculture is the cornerstone of Pakistan's economy, it is not without its hurdles. **Chapter 2** elucidates the manifold challenges confronting the sector. From climate change to water scarcity and market access issues, these challenges stand as significant impediments to sustainable agricultural growth. Understanding these challenges is crucial as we navigate the path toward a resilient and productive agricultural landscape.

References

Altieri, M. A., & Nicholls, C. I. (2018). Agroecology and the reconstruction of a post-COVID-19 agriculture. Agronomy for Sustainable Development, 40(5), 1–4.

Amanullah. (2024). Harvesting Success: A Comprehensive Guide to Agronomy Courses. Punjab Books, Lahore.

Amanullah & Khan, U. (2024). Enhancing public health through modern agronomy: Sustainable and nutrient-rich practices. Journal of Agriculture, Food Systems, and Community Development, 13(2), 11–18.

Amanullah., Ondrasek, G., Al –Tawaha, A. R. M. S., eds. (2023). Integrated nutrients management: An approach for sustainable crop production and food security in changing climates. Lausanne: Frontiers Media SA. doi: 10.3389/978-2-8325-3169-3

Amanullah, & Khalid, S. (2020). Agronomy – Climate Change & Food Security. IntechOpen, London, United Kingdom.

Amanullah, & Fahad, S. (2018). Nitrogen in Agriculture-Updates. InTechOpen, London, UK.

Amanullah, & Fahad. (2018). Corn – Production and Human Health in Changing Climate. InTech, London, UK.

Bashir, M. K., Shabbir, M. S., & Bashir, S. (2014). Impact of agricultural growth on economic development of Pakistan: Time series analysis from 1970 to 2012. Journal of Economics and Sustainable Development, 5(20), 123–131.

Bünemann, E. K., Bongiorno, G., Bai, Z., Creamer, R. E., De Deyn, G., de Goede, R., . . . & Brussaard, L. (2018). Soil quality – A critical review. Soil Biology and Biochemistry, 120, 105–125.

Dorosh, P. A., & Rashid, S. (2015). Food and Agriculture in Pakistan: Progress and Priorities. IFPRI Discussion Paper 01445.

FAO. (2018). Climate-smart Agriculture Sourcebook. Food and Agriculture Organization of the United Nations, Rome, Italy.

Francis, C. A., & Porter, P. (2011). Sustainable agriculture in temperate zones. In: Handbook of Sustainable Agriculture (pp. 1–30). CRC Press, London, UK.
Gliessman, S. R. (Ed.). (2014). Agroecology: The Ecology of Sustainable Food Systems (3rd ed.). CRC Press, London, UK.
Government of Pakistan. (2019). Pakistan Economic Survey 2018–19. Ministry of Finance.
Hussain, I. (2004). Agriculture sector of Pakistan: An analysis of growth, instability and determinants. The Pakistan Development Review, 43(4), 769–786.
Hussain, M. (2016). Agriculture in Pakistan: A comprehensive overview. International Journal of Economics and Financial Issues, 6(3), 1452–1457.
Jabeen, A., Ijaz, M. S., & Iqbal, M. J. (2017). The role of agriculture in economic growth of Pakistan: An empirical investigation. Journal of Agriculture and Environment for International Development, 111(1), 139–154.
Jadoon, I. A., & Haider, M. I. (2016). Agricultural development in Pakistan: Growth, constraints and future strategies. Journal of Agriculture and Environmental Sciences, 5(2), 129–141.
Klerkx, L., Aarts, N., & Leeuwis, C. (2010). Adaptive management in agricultural innovation systems: The interactions between innovation networks and their environment. Agricultural Systems, 103(6), 390–400.
Mustafa, G., Saboor, A., & Maqbool, R. (2015). Role of agriculture in economic growth of Pakistan. Journal of Economic Info, 1(1), 15–22.
Pretty, J. (2008). Agricultural sustainability: Concepts, principles, and evidence. Philosophical Transactions of the Royal Society B: Biological Sciences, 363(1491), 447–465.
Pretty, J., & Bharucha, Z. P. (2014). Sustainable intensification in agricultural systems. Annals of Botany, 114(8), 1571–1596.
Pretty, J., Toulmin, C., & Williams, S. (2011). Sustainable intensification in African agriculture. International Journal of Agricultural Sustainability, 9(1), 5–24.
Qasim, M., & Nazir, M. S. (2018). The role of agriculture in economic development of Pakistan. Research Journal of Finance and Accounting, 9(8), 61–66.
Qasim, M., Mahmood, N., & Nazir, M. S. (2019). Contribution of agriculture sector in the economic development of Pakistan. International Journal of Economics and Financial Issues, 9(3), 102–110.
Vanlauwe, B., Bationo, A., Chianu, J., Giller, K. E., Merckx, R., Mokwunye, U., . . . & Sanginga, N. (2010). Integrated soil fertility management: Operational definition and consequences for implementation and dissemination. Outlook on Agriculture, 39(1), 17–24.

Chapter 2
Global challenges in agriculture

Introduction

In today's interconnected world, agricultural systems face a myriad of challenges that transcend national borders and require collective action to address. From the impacts of climate change to the pressures of resource scarcity and the complexities of market dynamics, agricultural systems worldwide are confronted with a host of interconnected challenges that threaten food security, environmental sustainability, and rural livelihoods. This chapter delves into the common challenges faced by agricultural systems across the globe (Figure 2.1), examining their implications and exploring innovative and sustainable solutions to mitigate their impacts.

Figure 2.1: The three major global challenges of agriculture.

2.1 Climate change

Climate change poses one of the most pressing challenges to global agriculture, altering weather patterns, increasing the frequency and intensity of extreme weather events, and exacerbating environmental stresses. Rising temperatures, changing precipitation patterns, and shifts in growing seasons disrupt agricultural production, reduce crop

https://doi.org/10.1515/9783111448077-002

yields, and threaten food security for millions of people worldwide. Moreover, climate change amplifies existing vulnerabilities in marginalized communities, exacerbating inequalities in access to food, water, and resources. Addressing the impacts of climate change requires implementing adaptation and mitigation strategies that enhance the resilience of agricultural systems, promote sustainable land management practices, and reduce greenhouse gas emissions from agricultural activities.

Climate change presents an existential threat to global agriculture, manifesting in altered weather patterns, heightened frequency and severity of extreme weather events, and exacerbated environmental stresses. These impacts are already being felt across the globe, posing significant challenges to agricultural production, food security, and rural livelihoods.

Here are some examples illustrating the impacts of climate change on global agriculture:

1. Increased frequency and intensity of droughts in regions like California (USA) and Australia's Murray-Darling Basin have reduced water availability for crop irrigation, leading to yield losses and economic hardship for farmers.
2. Changing rainfall patterns and prolonged dry spells in East Africa have affected crop production, particularly for staple crops like maize and beans, threatening food security and exacerbating malnutrition in vulnerable communities.
3. Rising temperatures in the Arctic region have accelerated permafrost thawing, affecting traditional farming practices of indigenous communities and disrupting reindeer herding, a vital aspect of their livelihoods.
4. Extreme weather events such as hurricanes and typhoons have devastated agricultural landscapes in countries like Bangladesh and the Philippines, causing crop damage, livestock losses, and infrastructure destruction, undermining food production and rural economies.
5. Shifts in pest and disease patterns, driven by warmer temperatures and changing climatic conditions, have affected crop yields and quality globally. For instance, the spread of pests like the fall armyworm in Africa has resulted in significant losses for maize farmers, threatening food security in the region.

Rising temperatures are disrupting traditional growing seasons, leading to shifts in planting and harvesting schedules and affecting the suitability of crops in certain regions. Additionally, changing precipitation patterns are causing droughts, floods, and erratic rainfall, further destabilizing agricultural systems and diminishing crop yields. Extreme weather events, such as hurricanes, cyclones, and wildfires, are becoming more frequent and severe, causing widespread devastation to crops, infrastructure, and agricultural livelihoods.

The consequences of climate change are particularly dire for vulnerable and marginalized communities, exacerbating existing inequalities in access to food, water, and resources. Smallholder farmers, indigenous peoples, and rural communities are disproportionately affected by the impacts of climate change, as they often lack the resources, infrastructure, and support systems to adapt to rapidly changing environmental conditions. Moreover, marginalized populations are more likely to rely on agriculture for their livelihoods and sustenance, making them highly vulnerable to fluctuations in weather patterns and agricultural productivity.

Addressing the impacts of climate change on agriculture requires a multifaceted approach that encompasses both adaptation and mitigation strategies. Adaptation measures aim to enhance the resilience of agricultural systems to climate-related risks, such as implementing drought-tolerant crop varieties, improving water management practices, and adopting climate-smart agricultural techniques. These measures can help farmers cope with the impacts of climate change and minimize losses in crop yields and income.

Mitigation efforts, on the other hand, seek to reduce greenhouse gas emissions from agricultural activities and mitigate the drivers of climate change. Agriculture is a significant contributor to greenhouse gas emissions, accounting for approximately a quarter of global emissions, primarily from methane and nitrous oxide emissions from livestock, rice paddies, and fertilizer use. Implementing sustainable land management practices, such as agroforestry, conservation agriculture, and soil carbon sequestration, can help mitigate emissions and enhance carbon storage in agricultural soils.

Promoting sustainable land management practices is essential for building resilience to climate change and reducing greenhouse gas emissions from agriculture. These practices not only help to mitigate climate change but also contribute to improved soil health, water quality, and biodiversity conservation. Investing in agricultural research, extension services, and farmer education programs can facilitate the adoption of sustainable farming practices and promote the transition to more climate-resilient and environmentally friendly agricultural systems.

Addressing the impacts of climate change on agriculture is essential for ensuring global food security, environmental sustainability, and rural livelihoods. By implementing adaptation and mitigation strategies that enhance the resilience of agricultural systems, promote sustainable land management practices, and reduce greenhouse gas emissions, we can build more resilient, adaptive, and sustainable food systems that can withstand the challenges of a changing climate and provide for the needs of present and future generations.

2.2 Resource scarcity

Resource scarcity, including water scarcity, land degradation, and depletion of natural resources, presents significant challenges to agricultural sustainability and productivity. With growing demand for food, water, and energy, coupled with competing land uses and environmental pressures, agricultural systems are under increasing strain to meet the needs of a growing global population. In many regions, unsustainable water use, soil erosion, and deforestation threaten the long-term viability of agricultural production, exacerbating food insecurity and environmental degradation.

Here are a few examples illustrating the threats to agricultural viability due to unsustainable practices (Figure 2.2):

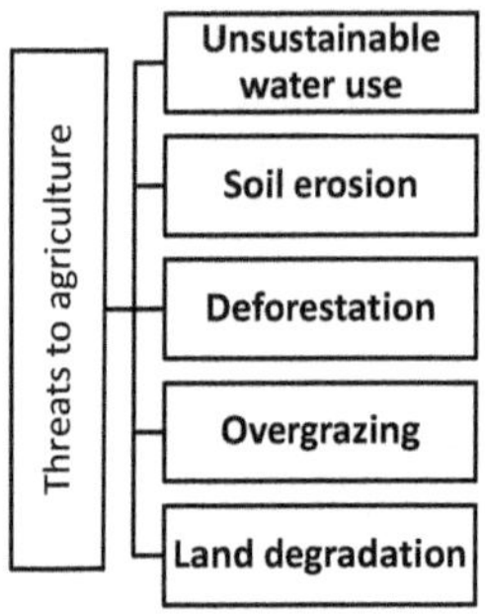

Figure 2.2: Threats to agriculture due to unsustainable practices.

1. **Unsustainable water use** in the Ogallala Aquifer region in the United States has led to declining water levels, jeopardizing irrigation for crops such as corn, wheat, and cotton. This unsustainable water extraction puts the future of agriculture in the region at risk.
2. **Soil erosion** in the Loess Plateau of China, caused by intensive farming practices and deforestation, has resulted in the loss of fertile topsoil. This degradation of soil quality threatens agricultural productivity and livelihoods for millions of farmers in the region.
3. **Deforestation** in the Amazon rainforest in Brazil has led to the loss of valuable biodiversity and ecosystem services essential for agriculture, such as pollination and soil fertility. The conversion of forests to agricultural land contributes to soil degradation, water scarcity, and increased greenhouse gas emissions, posing long-term challenges for sustainable food production.
4. **Overgrazing** in semiarid regions like the Sahel in Africa has led to land degradation and desertification, reducing the productivity of rangelands and threatening the livelihoods of pastoral communities. Without sustainable management practices, such as rotational grazing and reforestation, these areas face continued degradation and food insecurity.
5. **Land degradation** due to intensive monoculture farming practices, such as soybean cultivation in the Cerrado region of Brazil, has depleted soil nutrients and increased vulnerability to pests and diseases. This unsustainable agricultural expansion threatens both local ecosystems and global food security.

Addressing resource scarcity requires promoting efficient and sustainable resource management practices, investing in water-saving technologies, soil conservation measures, and agroforestry systems, and fostering integrated approaches that balance competing demands for land, water, and resources.

Resource scarcity is a critical challenge confronting global agriculture, encompassing issues such as water scarcity, land degradation, and depletion of natural resources. As the demand for food, water, and energy continues to escalate, agricultural systems worldwide are under mounting pressure to sustainably meet the needs of a growing global population. However, the intensification of competing land uses, environmental pressures, and unsustainable resource management practices exacerbate the strain on agricultural sustainability and productivity.

Water scarcity poses a significant threat to agricultural sustainability, particularly in arid and semiarid regions where water resources are limited. Unsustainable water use practices, such as excessive irrigation, groundwater depletion, and inefficient water management, contribute to water stress, jeopardizing the availability of water for agricultural production. Moreover, competition for water resources among various sectors, including agriculture, industry, and urbanization, further exacerbates water scarcity and undermines agricultural resilience.

Land degradation, including soil erosion, loss of soil fertility, and desertification, poses another formidable challenge to agricultural sustainability. Unsustainable land management practices, such as deforestation, overgrazing, and monoculture farming, degrade soil quality, diminish agricultural productivity, and threaten the long-term viability of agricultural production systems. Soil erosion, in particular, undermines soil health, reduces water retention capacity, and increases susceptibility to droughts and floods, exacerbating food insecurity and environmental degradation. Furthermore, the depletion of natural resources, such as fossil fuels, minerals, and biodiversity, poses significant risks to agricultural sustainability and resilience.

Here are a few examples illustrating these risks:

1. **Depletion of fossil fuels**: The reliance on fossil fuels for mechanized agriculture, transportation, and agrochemical production contributes to greenhouse gas emissions and climate change. As fossil fuel reserves decline, the cost of energy-intensive farming practices increases, potentially affecting the affordability and accessibility of food production inputs.
2. **Depletion of minerals**: Intensive agricultural practices, such as excessive fertilizer use, can deplete soil fertility and essential minerals over time. For example, phosphorus mining, a critical component of fertilizer production, is becoming increasingly unsustainable as global reserves dwindle. Without sustainable soil management practices and alternative nutrient sources, agricultural productivity may decline, impacting global food security.
3. **Loss of biodiversity**: Agricultural expansion and intensification have led to the loss of biodiversity, including the decline of pollinators, beneficial insects, and native plant species. This loss of biodiversity can disrupt ecosystem services essential for agriculture, such as pollination, pest control, and soil fertility. For instance, the

decline of bee populations due to habitat loss and pesticide use threatens crop yields and the stability of food production systems.
4. **Overexploitation of water resources**: Unsustainable water use for irrigation, livestock production, and agroindustrial activities depletes freshwater resources and exacerbates water scarcity in many regions. For example, the overextraction of groundwater for irrigation in the Punjab region of India has led to declining water tables and reduced agricultural productivity. Without improved water management practices and conservation efforts, agricultural systems reliant on irrigation face increasing risks of water shortages and crop failures.
5. **Loss of arable land**: Urbanization, soil erosion, and land degradation contribute to the loss of arable land suitable for agriculture. In regions like the Nile Delta in Egypt, land subsidence due to groundwater extraction and sea-level rise threatens fertile agricultural areas, displacing farming communities and reducing food production capacity. Protecting and restoring arable land through sustainable land management practices is essential for ensuring long-term agricultural sustainability and resilience.

The reliance on finite resources for agricultural inputs, such as synthetic fertilizers, pesticides, and mechanization, exacerbates environmental degradation and contributes to greenhouse gas emissions. Additionally, biodiversity loss and habitat destruction undermine ecosystem services essential for agricultural productivity, such as pollination, pest control, and soil fertility.

Here are a few examples that illustrate the points mentioned:

1. **Synthetic fertilizers**: The widespread use of synthetic fertilizers in modern agriculture contributes to environmental degradation through nutrient runoff and soil pollution. For instance, excess nitrogen and phosphorus from fertilizers can leach into water bodies, leading to eutrophication, algal blooms, and aquatic ecosystem degradation. In regions like the Gulf of Mexico, agricultural runoff containing nitrogen and phosphorus from fertilizer application has created a large dead zone where marine life cannot survive.
2. **Pesticides**: Chemical pesticides used to control pests and diseases in crops can have unintended environmental consequences, such as harming nontarget organisms and disrupting ecosystems. For example, neonicotinoid pesticides, which are commonly used in agriculture, have been implicated in the decline of bee populations worldwide. Without bees for pollination, agricultural productivity and crop yields could suffer, impacting food security and ecosystem stability.
3. **Mechanization**: The intensive use of machinery and equipment in modern agriculture contributes to greenhouse gas emissions and fossil fuel consumption. Tractors, harvesters, and other agricultural machinery typically run on diesel or gasoline, emitting carbon dioxide (CO_2) and other greenhouse gases into the atmosphere. In countries like the United States, agricultural machinery accounts for a significant portion of total greenhouse gas emissions from the agricultural sector.

4. **Biodiversity loss**: The conversion of natural habitats for agriculture, such as forests, wetlands, and grasslands, results in biodiversity loss and habitat destruction. This loss of biodiversity undermines ecosystem resilience and the provision of essential ecosystem services. For example, the conversion of tropical rainforests to agricultural land in countries like Brazil and Indonesia not only reduces biodiversity but also contributes to carbon emissions and climate change.
5. **Habitat destruction**: Agricultural expansion often involves clearing land and draining wetlands, leading to habitat destruction and fragmentation. This habitat loss threatens wildlife populations and disrupts ecological balance. For instance, the conversion of wetlands for rice cultivation in Southeast Asia has led to the loss of critical habitats for migratory birds and aquatic species, affecting biodiversity and ecosystem functioning.

Addressing resource scarcity requires adopting efficient and sustainable resource management practices that prioritize the conservation and stewardship of natural resources. Investing in water-saving technologies, such as drip irrigation, rainwater harvesting, and water recycling systems, can enhance water efficiency and reduce water consumption in agriculture. Implementing soil conservation measures, such as terracing, contour plowing, and cover cropping, can mitigate soil erosion, improve soil health, and enhance agricultural productivity.

Agroforestry systems, which integrate trees and shrubs into agricultural landscapes, offer multifaceted benefits for resource conservation, biodiversity conservation, and climate change mitigation. Agroforestry practices, such as alley cropping, windbreaks, and silvopasture, provide additional sources of income, improve soil fertility, and enhance ecosystem resilience, contributing to sustainable agricultural development.

Moreover, fostering integrated approaches that balance competing demands for land, water, and resources is essential for achieving agricultural sustainability and resilience. Integrated land and water management strategies, such as watershed management, integrated pest management, and ecosystem-based adaptation, promote synergies between agricultural production, environmental conservation, and community development, fostering resilient and sustainable food systems.

Addressing resource scarcity is imperative for ensuring the long-term viability and resilience of agricultural systems worldwide. By promoting efficient and sustainable resource management practices, investing in innovative technologies and approaches, and fostering integrated and holistic solutions, we can build more resilient, adaptive, and sustainable agricultural systems that can meet the needs of present and future generations while safeguarding the health of the planet.

2.3 Market dynamics

Market dynamics, including price volatility, trade barriers, and market distortions, present challenges for farmers, particularly smallholder farmers in developing countries. Fluctuations in global commodity prices, trade policies, and market uncertainties can affect farmers' incomes, livelihoods, and food security, exacerbating poverty and inequality in rural communities. Moreover, market distortions, such as subsidies and trade barriers, can undermine local food systems, discourage sustainable farming practices, and perpetuate inequities in global food supply chains. Addressing market dynamics requires promoting fair and transparent market mechanisms, supporting value-added agriculture and diversification, and empowering smallholder farmers with access to markets, finance, and market information.

Market dynamics play a pivotal role in shaping the livelihoods and well-being of farmers, especially smallholder farmers in developing countries. These dynamics, characterized by price volatility, trade barriers, and market distortions, present formidable challenges to agricultural sustainability, economic prosperity, and food security. Understanding and addressing these market dynamics are essential for fostering equitable and resilient agricultural systems worldwide.

Price volatility is a significant concern for farmers, as fluctuating prices for agricultural commodities can lead to unpredictable incomes and financial instability. Sudden spikes or drops in commodity prices can have profound impacts on farmers' livelihoods, affecting their ability to invest in inputs, adopt new technologies, and access essential services. Moreover, price volatility can exacerbate food insecurity and poverty, particularly among vulnerable populations who rely on agriculture for their livelihoods and food security.

Trade barriers, including tariffs, quotas, and nontariff barriers, hinder farmers' access to markets and limit their ability to compete in global markets. Protectionist trade policies and trade disputes can disrupt agricultural trade flows, leading to market uncertainty and reducing farmers' market access and export opportunities. Smallholder farmers in developing countries are particularly vulnerable to trade barriers, as they often lack the resources and capacity to navigate complex trade regulations and compliance requirements.

Market distortions, such as subsidies, price controls, and market monopolies, can undermine the efficiency and fairness of agricultural markets, distorting price signals and inhibiting market competition. Subsidies for agricultural inputs, such as fertilizers and pesticides, can artificially lower production costs for large-scale agribusinesses, creating unfair advantages and crowding out smallholder farmers. Price controls and market monopolies can restrict farmers' access to fair prices and limit their ability to negotiate favorable terms of trade, perpetuating poverty and inequality in rural communities.

Addressing market dynamics requires promoting fair and transparent market mechanisms that ensure equitable access to markets, facilitate price discovery, and foster competition. Supporting value-added agriculture and diversification can help farmers capture higher returns for their products and reduce their reliance on volatile commodity markets. Empowering smallholder farmers with access to markets, finance, and market information is essential for enhancing their resilience and enabling them to adapt to changing market conditions.

Furthermore, strengthening farmers' organizations, cooperatives, and collective marketing initiatives can enhance their bargaining power and enable them to negotiate fair prices and better terms of trade. Investing in market infrastructure, such as storage facilities, transportation networks, and market information systems, can improve market efficiency and reduce postharvest losses, increasing farmers' incomes and food security.

Addressing market dynamics is crucial for promoting inclusive and sustainable agricultural development, reducing poverty and inequality, and enhancing food security and rural livelihoods. By promoting fair and transparent market mechanisms, supporting value-added agriculture and diversification, and empowering smallholder farmers with access to markets, finance, and market information, we can build more resilient, equitable, and prosperous agricultural systems that benefit farmers, consumers, and the broader economy.

Addressing the global challenges facing agriculture requires collective action and collaboration among stakeholders across sectors and regions. From the impacts of climate change and resource scarcity to the complexities of market dynamics, agricultural systems worldwide are confronted with interconnected challenges that demand innovative and sustainable solutions. By recognizing the common challenges faced by agricultural systems and working together to implement adaptive and transformative measures, we can build more resilient, inclusive, and sustainable food systems that ensure food security, protect the environment, and promote the well-being of present and future generations.

2.4 Major challenges faced by Pakistan's agricultural sector

The major challenges faced by Pakistan's agricultural sector are given in Figure 2.3:

Pakistan's agricultural sector stands at a critical juncture, facing a multitude of challenges that hinder its growth and sustainability. From persistent water scarcity and land degradation to climate variability and inadequate infrastructure, the sector grapples with a complex web of obstacles. These challenges not only threaten food security and rural livelihoods but also impede the sector's ability to contribute effectively to the nation's economic development. In this context, it becomes imperative to delve into the major challenges confronting Pakistan's agricultural sector, understanding their underlying causes and exploring potential strategies for overcoming them.

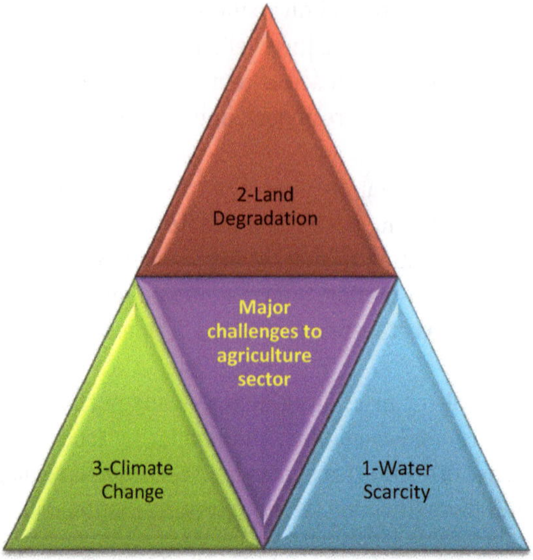

Figure 2.3: Major challenges to agriculture sector.

2.4.1 Water scarcity

Pakistan's agriculture heavily relies on irrigation water, but the country faces a water scarcity issue. Limited water availability, inefficient irrigation systems, and climate change-related impacts lead to water shortages, affecting crop yields and agricultural productivity. Water scarcity is a significant challenge in Pakistan's agriculture. The country heavily relies on irrigation for agricultural production, but its water resources are limited and depleting. Inefficient water management practices, inadequate infrastructure, and competing demands from other sectors exacerbate the problem. Climate change impacts, such as irregular rainfall patterns and glacial melting, further aggravate water scarcity issues.

2.4.2 Land degradation

Land degradation, including soil erosion, salinization, and waterlogging, poses a significant challenge to agriculture in Pakistan. Unsustainable farming practices, such as excessive use of chemical fertilizers, poor irrigation methods, and deforestation, contribute to soil degradation. This results in reduced soil fertility, decreased crop yields, and loss of arable land.

2.4.3 Climate change

Pakistan is vulnerable to climate change impacts, including increased frequency and intensity of droughts, floods, and extreme weather events. Erratic rainfall patterns and rising temperatures affect agricultural productivity, crop yields, and livestock health. Climate change also disrupts cropping patterns, water availability, and overall agricultural planning, making adaptation strategies critical for the sector.

2.4.4 Outdated farming techniques

Pakistan's agriculture sector struggles with low productivity levels, partly due to limited adoption of modern farming technologies and practices. Lack of access to improved seeds, mechanization, advanced irrigation systems, and agricultural machinery hinders productivity growth. Many farmers in Pakistan still use traditional and outdated farming techniques that limit productivity and efficiency. Limited access to modern machinery, quality seeds, and improved farming practices hinders progress. Lack of knowledge about sustainable agriculture techniques and low adoption of technology further contribute to suboptimal yields and farm incomes.

2.4.5 Fragmented landholdings and small-scale farming

Fragmented landholdings and small-scale farming practices are common in Pakistan, limiting economies of scale, mechanization, and investment in modern agricultural techniques. Small farmers often face challenges accessing credit, advanced machinery, quality inputs, and technical knowledge. The fragmentation of land also hampers the implementation of integrated agriculture practices and efficient farming systems.

2.4.6 Inadequate infrastructure

Insufficient infrastructure, including irrigation systems, storage facilities, and transportation networks, affects agricultural productivity and market access. Limited irrigation infrastructure results in water wastage and uneven distribution, while inadequate storage facilities lead to postharvest losses. Poor transportation networks increase logistic costs and restrict farmers' access to markets. Weak agricultural value chains, inadequate storage and transportation facilities, and postharvest losses contribute to inefficient marketing and distribution of agricultural produce. These challenges result in reduced profitability for farmers and limited access to markets.

2.4.7 Inadequate agricultural extension services

Insufficient investment in agricultural extension services limits the sector's progress. Inadequate funding for agriculture extension institutions, limited collaboration between academia and farmers, and a lack of technological innovation hinder the adoption of improved agricultural practices. The absence of tailored research and extension services prevents the development and dissemination of technologies and knowledge necessary for addressing agricultural challenges.

2.4.8 Lack of research and development

Inadequate investment in agricultural research and development limits the sector's progress. Insufficient funding for research institutions, limited collaboration between academia and farmers, and a lack of technological innovation impede the adoption of improved agricultural practices. The absence of tailored research and development hampers the development of crop varieties resilient to pests, diseases, and changing climatic conditions. Insufficient investment in agricultural research and development limits the sector's ability to innovate and address emerging challenges. Limited access to improved crop varieties, pest and disease management strategies, and sustainable agricultural practices hinder productivity and resilience.

2.4.9 Market constraints and price volatility

Agricultural markets in Pakistan face challenges related to limited market infrastructure, middlemen dominance, and price volatility. Farmers often struggle to access fair prices for their produce due to market inefficiencies and lack of direct market linkages. Inadequate storage and cold chain facilities contribute to postharvest losses, while weak value chains limit income generation opportunities for farmers.

2.4.10 Limited access to finance and agricultural inputs

Limited access to finance and agricultural inputs, such as high-quality seeds, fertilizers, and machinery, poses challenges to farmers. Small farmers, in particular, struggle to access credit facilities, making it difficult to invest in modern farming technologies and inputs. The lack of affordable and timely access to inputs affects agricultural productivity and inhibits the adoption of improved farming practices.

2.4.11 Policy and governance issues

Inconsistent policies, inadequate implementation of agricultural reforms, and weak governance structures impact the agricultural sector. Lack of coordination among different government departments, inadequate farmer support mechanisms, and delays in policy implementation hinder agricultural development.

2.4.12 Rural poverty and socioeconomic inequality

Agriculture is a significant source of employment and livelihoods in rural areas. However, widespread rural poverty, limited access to education and healthcare, and unequal distribution of resources contribute to socioeconomic disparities, impacting agricultural development and farmer well-being.

Addressing these challenges requires a multifaceted approach, including investments in water management, soil conservation, climate-resilient agriculture, and modernizing farming practices. Strengthening agricultural research and extension services, improving access to credit and markets, enhancing rural infrastructure, and promoting farmer education and capacity building are also crucial for overcoming the challenges and unlocking the potential of Pakistan's agriculture sector.

2.5 Land degradation, water scarcity, and climate change

Land degradation, water scarcity, and climate change represent a triad of interconnected challenges confronting Pakistan's agricultural sector. Land degradation, resulting from unsustainable land management practices such as deforestation and soil erosion, threatens the fertility and productivity of agricultural lands. Simultaneously, water scarcity exacerbates the problem, with dwindling water resources due to overextraction, inefficient irrigation methods, and erratic precipitation patterns.

Climate change further compounds these issues, manifesting in rising temperatures, altered rainfall patterns, and increased frequency of extreme weather events. These challenges collectively undermine agricultural productivity, exacerbate food insecurity, and pose significant hurdles to the sector's resilience and sustainability. Addressing these interrelated challenges requires integrated approaches that prioritize sustainable land management, efficient water use, and climate-resilient agricultural practices.

2.5.1 Land degradation

Land degradation refers to the deterioration of the physical, chemical, and biological properties of land, resulting in reduced soil fertility, loss of vegetation cover, and decreased agricultural productivity.

The following issues contribute to land degradation.

2.5.1.1 Soil erosion

Unsustainable agricultural practices, such as improper tillage methods, overgrazing, and deforestation, accelerate soil erosion. Wind and water erosion lead to the loss of topsoil, which is rich in organic matter and nutrients necessary for plant growth.

2.5.1.2 Salinization and waterlogging

Excessive irrigation and poor drainage systems contribute to salinization and waterlogging in agricultural areas. The accumulation of salts in the soil reduces crop yields and renders land unsuitable for cultivation.

2.5.1.3 Deforestation

Deforestation for timber, agriculture, and urbanization causes the loss of forest cover, leading to soil erosion, reduced water infiltration, and loss of biodiversity.

2.5.2 Water scarcity

Water scarcity is a significant challenge in Pakistan, driven by various factors such as population growth, inefficient water management, and climate change.

The following issues contribute to water scarcity.

2.5.2.1 Limited water resources

Pakistan relies heavily on the Indus River system for irrigation, but the availability of water is limited, especially during the dry season. The growing demand from various sectors, including agriculture, industry, and domestic use, exacerbates the water scarcity issue.

2.5.2.2 Inefficient water management

Inefficient water management practices, including outdated irrigation infrastructure, water losses during distribution, and low water-use efficiency in agriculture, contribute to water scarcity. Inadequate water storage facilities and the absence of effective water conservation measures further compound the problem.

2.5.2.3 Climate change impacts

Climate change exacerbates water scarcity by altering rainfall patterns and affecting the timing and availability of water resources. Increasing temperatures and changes in precipitation patterns pose challenges to water availability for agriculture and exacerbate water stress in arid and semiarid regions.

2.5.3 Climate change

Climate change poses significant challenges to agriculture in Pakistan, impacting agricultural productivity, crop yields, and livestock health.

The following issues are associated with climate change.

2.5.3.1 Erratic rainfall patterns

Changes in rainfall patterns, including irregular and unpredictable rainfall, affect agricultural planning, crop growth cycles, and water availability for irrigation. Droughts and dry spells increase water stress for crops and livestock, leading to reduced productivity.

2.5.3.2 Temperature extremes

Rising temperatures pose challenges to agricultural systems, especially heat-sensitive crops. High temperatures can negatively impact crop development, flowering, and fruiting, leading to reduced yields. Heat stress also affects livestock health and productivity.

2.5.3.3 Increased frequency of extreme weather events

Climate change leads to an increased frequency and intensity of extreme weather events, such as floods and droughts. These events cause crop losses, damage infrastructure, and disrupt agricultural operations, affecting food security and livelihoods.

2.5.3.4 Glacier melting and water availability

Glacial melt in the Himalayas affects water availability in river systems, including the Indus River, which is vital for agriculture in Pakistan. Changes in glacier dynamics can disrupt water availability, affecting irrigation, hydroelectric power generation, and overall water security.

Addressing these issues requires comprehensive strategies and actions. Implementing sustainable agricultural practices, such as conservation agriculture, agroforestry, and efficient irrigation techniques, can mitigate land degradation and water scarcity. Adaptation measures, including climate-resilient crop varieties, improved water management, and weather-based advisory systems, can help address climate change challenges in agriculture. Integrated approaches that consider the interconnectedness of land, water, and climate are crucial for promoting sustainable agricultural development in Pakistan.

2.6 Pakistan's socioeconomic challenges

Pakistan faces various socioeconomic challenges that have a significant impact on the agricultural sector. These challenges stem from factors such as rapid population growth, poverty, limited access to resources and services, and social inequality.

2.6.1 Population growth

Pakistan has a rapidly growing population, which puts immense pressure on the agricultural sector to meet the increasing food demands. The challenge lies in ensuring food security for a growing population while maintaining sustainable agricultural practices. The need to produce more food leads to intensified agricultural activities, which can contribute to environmental degradation, land fragmentation, and increased competition for resources.

2.6.2 Poverty and rural livelihoods

A significant proportion of Pakistan's population resides in rural areas and relies on agriculture for their livelihoods. However, poverty and limited economic opportunities in rural areas perpetuate the cycle of poverty. Many farmers lack access to credit, markets, and quality inputs, making it difficult to invest in modern farming practices. Insufficient income and limited access to social services hinder agricultural development and perpetuate socioeconomic inequalities.

2.6.3 Limited access to resources

Unequal access to resources such as land, water, credit, and technology is a major challenge in Pakistan's agricultural sector. Large landholdings are concentrated in the hands of a few, while small-scale farmers face land fragmentation and limited access to land. Similarly, water scarcity and inefficient irrigation systems restrict farmers' access to water, affecting crop yields and productivity. Limited access to credit, modern inputs, and technology further hampers agricultural development.

2.6.4 Gender inequality

Gender inequality is prevalent in Pakistan's agricultural sector. Women make substantial contributions to agricultural production, but their access to resources, credit, and decision-making power is often limited. Lack of gender parity in land ownership, access to markets, and agricultural extension services hampers women's empowerment and constrains agricultural productivity and overall development.

2.6.5 Education and knowledge gap

Low literacy rates and limited access to quality education in rural areas contribute to an information and knowledge gap among farmers. Lack of awareness about modern farming techniques, climate-resilient practices, and market dynamics inhibits the adoption of improved agricultural practices. Limited access to agricultural extension services and inadequate dissemination of research findings further restrict farmers' ability to enhance productivity and adapt to changing conditions.

2.6.6 Infrastructure and market access

Inadequate infrastructure, including road networks, storage facilities, and market linkages, poses challenges to agricultural development. Poor transportation and market infrastructure result in high transaction costs, post-harvest losses, and limited market access for farmers. Limited value addition and weak market linkages also restrict opportunities for farmers to obtain fair prices and maximize their income.

2.6.7 Environmental degradation and climate vulnerability

Environmental challenges, such as land degradation, water scarcity, and climate change, have socioeconomic implications for the agricultural sector. Environmental degradation reduces agricultural productivity, exacerbates resource constraints, and increases farmers' vulnerability to climate-related risks. Climate change impacts, including erratic rainfall patterns and extreme weather events, disrupt agricultural production, exacerbating food insecurity and economic vulnerabilities.

Addressing these socioeconomic challenges requires a multifaceted approach. It includes implementing policies and programs that promote rural development, poverty alleviation, and inclusive growth. Investments in education, training, and agricultural extension services are essential to enhance farmers' knowledge and skills. Enhancing access to credit, resources, and markets can improve farmers' economic opportunities. Gender mainstreaming and empowering women in agriculture are crucial for promoting gender equality and overall sector development. Additionally, investments in rural infrastructure, irrigation systems, and sustainable agricultural practices are essential to address environmental and climate-related challenges and promote sustainable agricultural development in Pakistan.

2.7 Implications of population growth and food security concerns

Population growth and food security concerns have significant implications for Pakistan, particularly in terms of meeting the food needs of a growing population.

2.7.1 Increased food demand

As the population continues to grow, the demand for food increases. Pakistan's population is projected to reach 400 million by 2050, placing immense pressure on the agricultural sector to produce sufficient food to meet the rising demand. The challenge lies in ensuring food security and adequate nutrition for the growing population.

2.7.2 Agricultural productivity

To meet the increasing food demand, agricultural productivity must be improved. However, population growth often leads to land fragmentation, reducing the average farm size. Small landholdings pose challenges for mechanization, economies of scale, and investment in modern farming techniques, affecting productivity. Enhancing agricultural productivity becomes crucial to meet the growing food demand.

2.7.3 Pressure on natural resources

Population growth places additional strain on natural resources, including land, water, and forests. The expansion of agriculture to meet food demand can lead to deforestation, encroachment on ecologically sensitive areas, and unsustainable water extraction. This puts pressure on ecosystems and exacerbates issues such as land degradation, water scarcity, and biodiversity loss.

2.7.4 Food price inflation

Rapid population growth and high food demand can result in food price inflation. If food production does not keep pace with population growth, the prices of essential food items can increase, making it difficult for low-income households to afford an adequate and nutritious diet. Food price inflation can exacerbate poverty and increase food insecurity among vulnerable populations.

2.7.5 Food security and nutrition

Population growth poses challenges to food security and nutrition. Ensuring access to sufficient, safe, and nutritious food for all is crucial. Food security concerns arise when the production, availability, and affordability of food are compromised. It is essential to address issues such as postharvest losses, inefficient supply chains, and unequal distribution to ensure food security and improve nutritional outcomes.

2.7.6 Poverty and inequality

Population growth can exacerbate poverty and socio-economic inequalities. The inability of the agricultural sector to generate sufficient employment opportunities and income sources for the growing population can perpetuate poverty. Limited access to resources, market constraints, and lack of opportunities for marginalized groups can further widen the income gap and exacerbate social disparities.

2.7.7 Urbanization and land use changes

Population growth often leads to rapid urbanization, resulting in the conversion of agricultural land into residential or industrial areas. This conversion reduces the agricultural land base, leading to decreased agricultural production and increased dependence on imports to meet food needs. Balancing urbanization with sustainable land use practices becomes crucial for ensuring long-term food security.

Addressing the implications of population growth and food security concerns requires a comprehensive approach. This includes investing in sustainable agriculture practices, improving agricultural productivity, enhancing access to resources and credit for small farmers, promoting research and development in agriculture, and strengthening value chains and market linkages. Implementing policies that focus on poverty alleviation, social safety nets, and nutrition programs is essential to address food security concerns. Moreover, promoting sustainable urban planning and land management can help reconcile urbanization with the preservation of agricultural land and food production capacity.

2.8 An integrated approach to overcome the challenges in agriculture

Overcoming the multifaceted challenges faced in agriculture demands a comprehensive and integrated approach that addresses each obstacle with targeted strategies.

Strategies that can be employed to tackle the challenges and foster sustainable agricultural development are listed further.

2.8.1 Water resource management

- Invest in modernizing irrigation infrastructure to enhance water efficiency and reduce wastage.
- Promote the adoption of water-efficient technologies such as drip and sprinkler irrigation.
- Implement water conservation practices like rainwater harvesting and water storage facilities.
- Enforce equitable water distribution systems and encourage crop rotation patterns that require less water.

2.8.2 Climate change resilience

- Develop climate-smart agricultural practices that adapt to changing weather patterns and extreme events.
- Promote the cultivation of climate-resilient crop varieties that can withstand droughts, floods, and temperature fluctuations.
- Invest in research and extension services that educate farmers on climate-responsive farming techniques.
- Encourage agroforestry and afforestation practices to mitigate climate impacts and promote carbon sequestration.

2.8.3 Low productivity

- Increase accessibility to high-quality seeds and improved crop varieties that exhibit higher yields.
- Provide farmers with training in modern agricultural techniques and best practices.
- Promote integrated pest management to control pests and diseases effectively.
- Facilitate the use of fertilizers and soil amendments based on soil nutrient assessments.

2.8.4 Land degradation and soil health

- Encourage conservation tillage practices to minimize soil erosion and maintain soil structure.
- Promote organic farming methods to enhance soil fertility and reduce reliance on synthetic inputs.
- Implement agroforestry systems to prevent soil degradation and improve land quality.
- Conduct soil testing and offer tailored recommendations for nutrient management.

2.8.5 Access to finance and technology

- Establish agricultural credit programs that provide farmers with accessible and affordable loans.
- Facilitate partnerships with technology providers to ensure farmers have access to modern tools and equipment.
- Create platforms for knowledge sharing and technology dissemination through mobile apps, workshops, and training programs.

2.8.6 Market access and value addition

- Develop efficient agricultural markets that ensure fair prices and reduce postharvest losses.
- Promote the establishment of agroprocessing industries to add value to agricultural products.
- Strengthen farmer cooperatives and associations to enhance bargaining power and collective marketing.

2.8.7 Policy reforms and institutional support

- Advocate for policies that prioritize agricultural development, sustainability, and farmer welfare.
- Strengthen agricultural extension services to provide accurate information, training, and technical support.
- Establish regulatory frameworks that discourage deforestation, overgrazing, and unsustainable land use practices.

2.8.8 Community engagement and capacity building

- Empower farming communities through education and training on modern techniques.
- Encourage participatory approaches in decision-making, involving farmers in the planning of development programs.
- Support women's involvement in agriculture and provide gender-sensitive training programs.
- By embracing a multifaceted approach that combines technological innovations, policy reforms, capacity building, and community involvement, Pakistan's agriculture can navigate the challenges it faces and pave the way for a resilient, sustainable, and thriving sector.

As we conclude this exploration, it becomes evident that the path toward agricultural sustainability and prosperity is not devoid of obstacles. From water scarcity and land degradation to climate change impacts and low productivity, the challenges confronting Pakistan's agriculture are both diverse and formidable.

However, within these challenges also reside opportunities for transformation and growth. The trials of water scarcity prompt innovation in water management techniques, while climate change impacts drive the adoption of resilient farming practices. Low productivity becomes an impetus for embracing modern technologies and enhancing knowledge dissemination among farmers. The mosaic of challenges, when viewed through the lens of proactive approaches, underscores the sector's potential for evolution and adaptation.

The journey ahead requires concerted efforts, a blend of policies, research, technology, and community engagement. The collaboration of government institutions, research organizations, international partners, and local communities is vital in developing and implementing solutions that address these challenges holistically.

The resilience of Pakistan's agricultural sector hinges on the ability to navigate these challenges with determination and foresight. By leveraging innovative approaches, embracing sustainable practices, and integrating lessons from both successes and setbacks, the sector can transcend its current limitations.

As we bid farewell to this chapter, the lens shifts from challenges to possibilities, from obstacles to solutions. It is a call to action, an invitation to collectively steer the course of Pakistan's agriculture toward a sustainable, resilient, and prosperous future. The following chapters will delve into strategies, innovations, and practices that hold promise in transforming these challenges into opportunities, casting a spotlight on the potential that lies within the complexities of Pakistan's agricultural landscape.

The journey through Pakistan's agricultural landscape reveals the intricate web of challenges that this sector grapples with. As we confront these challenges head-on, the need for strategic policy formulations and institutional support becomes apparent. With a clear understanding of the hurdles at hand, we set the stage for a comprehensive approach that integrates knowledge from various disciplines to forge a sustainable path forward.

As we transition to the realm of Chapter 3, we recognize the pivotal role it plays in addressing the challenges laid out in the preceding chapter. With its focus on optimizing crop production and soil management, agronomy offers tangible solutions to improve yields and enhance food security. This integration of scientific knowledge and practical application underscores the significance of this discipline as a bedrock for sustainable agriculture.

References

Ahmed, M., Mahmood, S., & Bano, S. (2014). Challenges of Pakistani Agriculture. International Journal of Information Technology and Management, 13(1), 79–90.

Ahmed, Z. U., & Shah, S. A. (2016). Major challenges for agriculture sector in Pakistan. International Journal of Social Sciences and Humanities Reviews, 6(1), 39–42.

Ali, H., Zulfiqar, F., & Siddiqui, A. (2018). Challenges in agriculture in Pakistan and their solutions. International Journal of Environmental & Agriculture Research, 4(3), 87–93.

Altieri, M. A., & Nicholls, C. I. (2018). Agroecology and the reconstruction of a post-COVID-19 agriculture. Agronomy for Sustainable Development, 40(5), 1–4.

Amanullah. (2024). Harvesting Success: A Comprehensive Guide to Agronomy Courses Punjab Books, Lahore, PK.

Amanullah., Ondrasek, G., Al –Tawaha, A. R. M. S., eds. (2023). Integrated nutrients management: An approach for sustainable crop production and food security in changing climates. Lausanne: Frontiers Media SA. doi: 10.3389/978-2-8325-3169-3

Amanullah, Khalid, S., Muhammad, A., Yar, M., et al. (2021). Integrated use of biofertilizers with organic and inorganic phosphorus sources improve dry matter partitioning and yield of hybrid maize. Communications in Soil Science and Plant Analysis, 52(21), 2732–2747.

Amanullah, Adil Khan, Khalid, S., et al. (2019). Integrated management of phosphorus, organic sources, and beneficial microbes improve dry matter partitioning of maize. Communications in Soil Science and Plant Analysis, 50(20), 2544–2569.

Amanullah, Asif, M., & Almas, L. K. (2012). Agronomic efficiency and profitability of P-fertilizers applied at different planting densities of maize in Northwest Pakistan. Journal of Plant Nutrition, 35, 331–341.

Bünemann, E. K., Bongiorno, G., Bai, Z., Creamer, R. E., De Deyn, G., De Goede, R., . . . & Brussaard, L. (2018). Soil quality – A critical review. Soil Biology and Biochemistry, 120, 105–125.

Chaudhry, I. S., Khan, I., & Shahbaz, B. (2019). Agriculture challenges and their impact on rural development in Pakistan. European Online Journal of Natural and Social Sciences: Proceedings, 8(1), 185–191.

FAO. (2018). Climate-smart agriculture sourcebook. Food and Agriculture Organization of the United Nations.

Francis, C. A., & Porter, P. (2011). Sustainable agriculture in temperate zones. In: Handbook of Sustainable Agriculture (pp. 1–30). CRC Press, London, UK.

Gliessman, S. R. (Ed.). (2014). Agroecology: The Ecology of Sustainable Food Systems (3rd ed.). CRC Press, London, UK.

Haq, A. U., Khan, M. N., & Hussain, S. (2013). Challenges and opportunities in agriculture and rural development in Pakistan: An overview. Pakistan Economic and Social Review, 51(1), 1–34.

Khan, I., Amanullah, & Jamal, A. (2022). Partial substitution of chemical fertilizers with organic supplements increased wheat productivity and profitability under limited and assured irrigation regimes. Agriculture, 12, 1754.

Klerkx, L., Aarts, N., & Leeuwis, C. (2010). Adaptive management in agricultural innovation systems: The interactions between innovation networks and their environment. Agricultural Systems, 103(6), 390–400.

Krasilnikov, P., Miguel A. Taboada, & Amanullah. (2021). Fertilizer Use, Soil Health and Agricultural Sustainability. Agriculture (SI). Multidisciplinary Digital Publishing Institute (MDPI), Basel, Switzerland.

Nadia, Amanullah, et al. (2023). Improvement in wheat productivity with integrated management of beneficial microbes along with organic and inorganic phosphorus sources. Agriculture, 13, 1118.

Nazli, H. (2015). Agriculture and poverty in Pakistan: drivers, constraints and opportunities. Pakistan Institute of Development Economics.

Pretty, J. (2008). Agricultural sustainability: Concepts, principles, and evidence. Philosophical Transactions of the Royal Society B: Biological Sciences, 363(1491), 447–465.

Pretty, J., & Bharucha, Z. P. (2014). Sustainable intensification in agricultural systems. Annals of Botany, 114(8), 1571–1596.

Pretty, J., Toulmin, C., & Williams, S. (2011). Sustainable intensification in African agriculture. International Journal of Agricultural Sustainability, 9(1), 5–24.

Riaz, M. A., & Rao, S. A. (2013). Major challenges to agriculture and rural development in Pakistan: A review. Sarhad Journal of Agriculture, 29(3), 357–364.

Shabbir, R. M., & Ahmad, A. (2017). Challenges faced by agriculture sector in Pakistan: A critical analysis. International Journal of Agricultural and Environmental Sciences, 3(2), 96–100.

Shahzad, A., & Rasool, F. (2015). Challenges to agriculture in Pakistan and solution for development. International Journal of Economics, Commerce and Management, 3(5), 49–60.

Siddiqui, M. A., & Ghaus, A. (2012). Challenges of agriculture sector in Pakistan: A critical analysis. Journal of Economics and Sustainable Development, 3(1), 99–105.

Vanlauwe, B., Bationo, A., Chianu, J., Giller, K. E., Merckx, R., Mokwunye, U., . . . & Sanginga, N. (2010). Integrated soil fertility management: Operational definition and consequences for implementation and dissemination. Outlook on Agriculture, 39(1), 17–24.

Chapter 3
Agronomy: the foundation of modern agriculture

Introduction

This chapter encapsulates the fundamental role that agronomy plays in the realm of agriculture. Agronomy serves as the cornerstone of agricultural practices by delving into the scientific principles governing crop cultivation, soil management, and sustainable agricultural systems. It forms the basis upon which various agricultural disciplines build, enabling effective and efficient food production. Just as a mother nurtures and guides the growth of her children, agronomy nurtures the growth of crops and the agricultural sector as a whole. Through research, innovation, and practical application, agronomy ensures that the foundation of agriculture remains strong, resilient, and capable of meeting the challenges of feeding a growing global population while safeguarding our natural resources.

Agronomy is the scientific field of study that focuses on improving and optimizing the production of crops and plants within agricultural systems. It involves a multidisciplinary approach that combines knowledge from various fields such as biology, chemistry, soil science, and meteorology. Agronomists work to understand the interactions between plants, soils, climate, and management practices to enhance agricultural productivity, sustainability, and profitability. They research and develop strategies for crop cultivation, soil management, irrigation, pest control, and crop rotation, among other aspects. Agronomy plays a critical role in advancing agricultural practices that ensure food security, promote environmental stewardship, and support the livelihoods of farming communities.

3.1 Importance of agronomy

Agronomy holds great importance globally due to its significant contributions to agricultural productivity, sustainability, and food security (Figure 3.1).

3.1.1 Enhanced crop productivity

Agronomy focuses on the science and practice of crop production. By studying factors such as soil fertility, crop nutrition, crop physiology, pest and disease management, and agronomic practices, agronomist's help optimize crop productivity. This is particularly crucial for countries, where agriculture is a primary source of food and livelihood for millions of people.

https://doi.org/10.1515/9783111448077-003

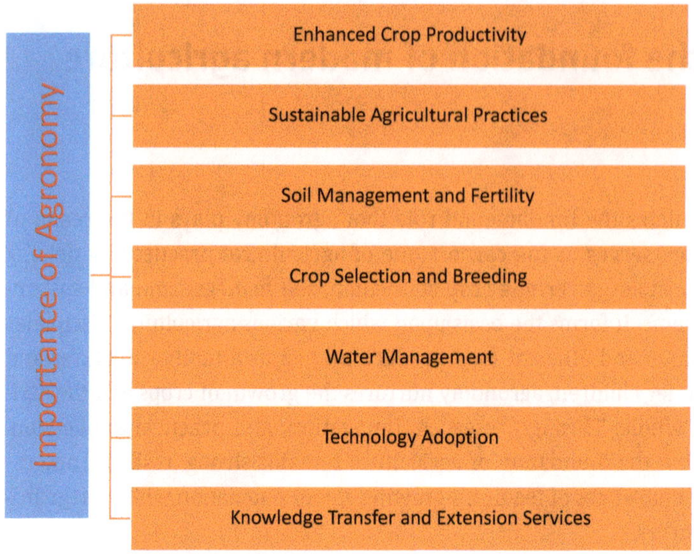

Figure 3.1: Importance of agronomy.

3.1.2 Sustainable agricultural practices

Agronomy plays a vital role in promoting sustainable agricultural practices. Agronomists develop and disseminate knowledge about conservation agriculture, soil and water conservation, integrated pest management (IPM), precision farming, and crop rotation. By adopting these practices, farmers can reduce soil erosion, improve soil health, conserve water, and minimize the use of chemical inputs, thus promoting long-term sustainability of agriculture.

3.1.3 Soil management and fertility

Agronomy emphasizes the study of soil properties, soil fertility, and soil management practices. Countries around the world face significant challenges related to soil degradation, erosion, and nutrient deficiencies. By understanding soil characteristics, agronomists can provide recommendations for proper soil management, including appropriate nutrient application, soil amendments, and erosion control measures, to improve soil fertility and crop productivity.

3.1.4 Crop selection and breeding

Agronomists contribute to crop selection and breeding programs, helping farmers identify suitable crop varieties for specific agroclimatic conditions. They play a critical role in evaluating and recommending high-yielding, disease-resistant, and climate-resilient crop varieties. This facilitates farmers in making informed decisions regarding crop selection, leading to improved productivity and profitability.

3.1.5 Water management

Efficient water management is crucial in countries, where water resources are limited. Agronomists study irrigation techniques, water requirements of different crops, and water-use efficiency. They provide guidance on optimal irrigation schedules, water conservation practices, and the adoption of precision irrigation technologies, enabling farmers to maximize water-use efficiency and cope with water scarcity challenges.

3.1.6 Technology adoption

Agronomy contributes to the adoption of advanced agricultural technologies. By evaluating and promoting the use of improved seeds, efficient fertilization techniques, mechanization, and precision farming technologies, agronomists help farmers increase productivity, reduce input costs, and improve resource-use efficiency. This is particularly important in a country like Pakistan, where smallholder farmers can benefit from technology adoption to enhance their agricultural practices.

3.1.7 Knowledge transfer and extension services

Agronomists play a key role in knowledge transfer and extension services. They disseminate research findings, best practices, and innovative techniques to farmers through extension programs, farmer field schools, and advisory services. This helps bridge the gap between scientific knowledge and on-ground farming practices, empowering farmers with the latest information and enabling them to make informed decisions.

Agronomy plays a crucial role in improving agricultural productivity, sustainability, and farmer livelihoods globally. By integrating scientific knowledge, technology, and best practices, agronomists contribute to the development of a more efficient, resilient, and profitable agriculture sector worldwide.

3.2 Integrated crop management

Integrated crop management (ICM) is a holistic and innovative approach developed by **agronomists** to optimize crop production while minimizing negative environmental impacts. It involves combining various agricultural practices, technologies, and strategies to create a balanced and sustainable farming system.

ICM integrates different aspects of crop management, including soil health, pest and disease control, water management, and nutrient optimization. It emphasizes the use of practices that are both economically viable and environmentally friendly. Instead of relying solely on chemical inputs, ICM promotes the use of IPM, precision agriculture techniques, crop rotation, cover cropping, and other sustainable methods to improve overall crop health and yield.

By considering the entire ecosystem and promoting a systems-based approach, ICM aims to achieve long-term agricultural sustainability while addressing challenges such as soil degradation, water scarcity, and biodiversity loss. It reflects a more proactive and comprehensive way of managing crops that benefits both farmers and the environment.

ICM is an approach to crop production that emphasizes the integration of various practices and strategies to achieve sustainable and efficient agricultural systems. It involves the holistic management of crops, taking into account factors such as crop nutrition, pest and disease control, water management, soil health, and biodiversity conservation.

3.2.1 Benefits of integrated crop management

ICM offers a multitude of advantages, revolutionizing the approach to agricultural practices and sustainability. The following are the seven key benefits of ICM (Figure 3.2).

3.2.1.1 Enhanced productivity and higher yields

ICM focuses on optimizing crop productivity by employing best management practices tailored to specific crop and environmental conditions. By adopting integrated approaches, such as balanced nutrient management, IPM, efficient irrigation, and improved seed selection, farmers can enhance crop yields and overall productivity.

3.2.1.2 Risk reduction and improved food security

ICM helps mitigate risks associated with crop production. By diversifying farming practices, implementing pest and disease control measures, and adopting climate-smart agricultural techniques, farmers are better prepared to manage unexpected challenges such as pest outbreaks, extreme weather events, or changes in market con-

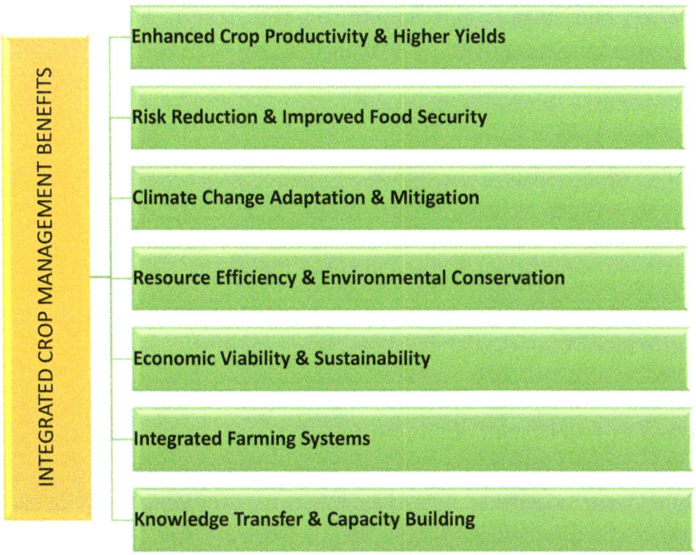

Figure 3.2: Benefits of integrated crop management.

ditions. This reduces the vulnerability of agricultural systems and contributes to improved food security.

3.2.1.3 Change adaptation and mitigation

ICM plays a crucial role in adapting to and mitigating the impacts of climate change. Through the integration of climate-smart practices, such as conservation agriculture, agroforestry, and efficient water management, ICM helps enhance the resilience of agricultural systems to changing climatic conditions. It also contributes to the reduction of greenhouse gas emissions, thus aiding in global efforts to combat climate change and reduce global warming.

3.2.1.4 Efficiency and environmental conservation

ICM promotes the efficient use of resources, including water, fertilizers, and pesticides. By employing precision agriculture techniques, optimizing nutrient application, and reducing chemical inputs through biological control methods, ICM minimizes resource wastage and decreases environmental pollution. It also helps conserve soil health, protect water quality, and preserve biodiversity.

3.2.1.5 Economic viability and sustainability

Integrating ICM into agricultural systems can lead to improved economic viability and long-term sustainability. By maximizing yields, reducing input costs, minimizing pest and disease damage, and optimizing resource use, farmers can enhance their profitability and ensure the long-term viability of their farming operations.

3.2.1.6 Integrated farming systems

ICM aligns well with integrated farming systems, which combine multiple agricultural activities within a single farming system. By integrating crop cultivation with livestock rearing, fish farming, agroforestry, or other complementary activities, farmers can optimize resource utilization, nutrient recycling, and diversify income sources. This integrated approach promotes resilience, enhances productivity, and supports sustainable rural development.

3.2.1.7 Knowledge transfer and capacity building

Implementing ICM requires knowledge transfer and capacity building among farmers and agricultural stakeholders. By disseminating best practices, providing training and extension services, and fostering farmer-to-farmer knowledge sharing, ICM contributes to the continuous improvement of agricultural practices and the adoption of sustainable farming techniques.

Integrating ICM into agriculture offers numerous benefits, including enhanced productivity, risk reduction, climate change adaptation, environmental conservation, economic viability, and sustainable agricultural systems. By adopting an integrated approach, farmers can achieve better yields, ensure food security, contribute to climate change mitigation, and promote a more sustainable and resilient agricultural sector.

3.3 Crop production and crop productivity

Crop production and crop productivity are related terms in agriculture, but they refer to slightly different aspects of the agricultural process:

3.3.1 Crop production

Crop production refers to the total quantity of crops that are grown and harvested within a specific area during a given time period. It represents the overall output of crops from a field, farm, or region. Crop production encompasses the entire process of cultivating crops, including preparing the land, planting seeds or seedlings, managing the crops through their growth stages, controlling pests and diseases, and finally harvesting the mature crops.

3.3.2 Crop productivity

Crop productivity, on the other hand, focuses on the efficiency of crop production in terms of the yield achieved relative to the resources (such as land, labor, water, and fertilizers) invested in the process. It measures how effectively inputs are converted into outputs. Crop productivity is usually expressed as yield per unit of resource input. For example, it could be measured as bushels of grain per acre of land, kilograms of fruit per liter of water used for irrigation, or tons of vegetables per unit of fertilizer applied.

The key distinction lies in the scope of measurement. Crop production quantifies the total quantity of crops harvested, while crop productivity assesses how efficiently resources are utilized to achieve that output. A farm might have high crop production due to cultivating a large area, but if the inputs used are also high, the crop productivity might be relatively low. On the other hand, a farm with moderate crop production but efficient use of resources could have a higher crop productivity. Balancing both high production and high productivity is the goal of sustainable and efficient agricultural practices.

3.4 Strategies to increase crop productivity

Enhancing crop productivity is a central focus of agricultural development, driving efforts to ensure food security and economic prosperity. Various strategies are employed to maximize yields and optimize agricultural outputs.

The following are the 10 key strategies to increase crop productivity (Figure 3.3).

3.4.1 Improved seed selection

Using high-quality and genetically improved seeds that are well-adapted to local conditions can significantly enhance crop productivity. Selecting seeds with traits such as high yield potential, disease resistance, drought tolerance, and nutrient-use efficiency can improve crop performance.

3.4.2 Balanced nutrient management

Implementing appropriate nutrient management practices is essential to optimize crop growth and productivity. Conducting soil tests to assess nutrient levels and applying fertilizers based on crop nutrient requirements can help ensure that crops receive adequate nutrition. Efficient fertilizer application techniques, such as split

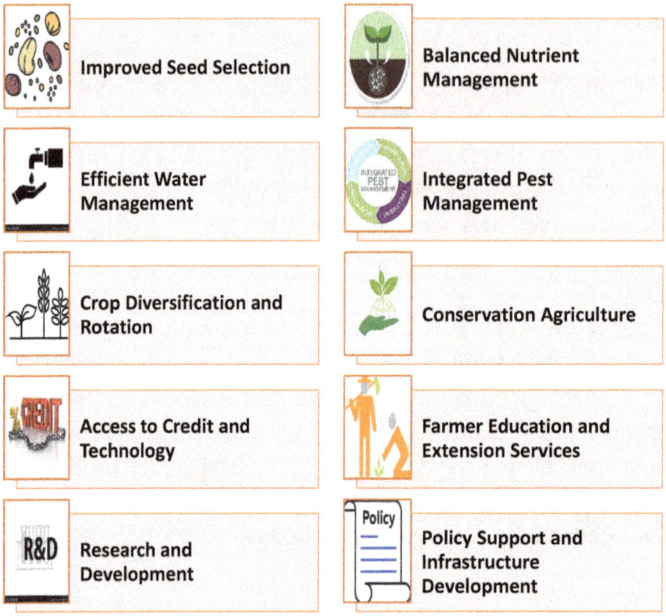

Figure 3.3: Key strategies to increase crop productivity.

application and site-specific nutrient management, should be adopted to minimize wastage and environmental impacts.

3.4.3 Efficient water management

Given the water scarcity challenges in Pakistan, efficient water management techniques are crucial for increasing crop productivity. Implementing practices such as drip irrigation, sprinkler irrigation, and laser leveling can help conserve water, improve irrigation efficiency, and ensure crops receive optimal water supply.

3.4.4 Integrated pest management (IPM)

Adopting IPM practices helps manage pests, diseases, and weeds effectively while minimizing reliance on chemical pesticides. Integrated approaches include the use of biological control agents, crop rotation, cultural practices, and resistant crop varieties. Regular monitoring of pests and diseases allows for timely intervention and targeted control measures.

3.4.5 Crop diversification and rotation

Crop diversification and rotation can enhance soil fertility, reduce pest and disease pressure, and improve overall productivity. Introducing crop rotation systems and incorporating legumes, cover crops, or cash crops with different growth requirements can help break pest cycles, improve nutrient cycling, and reduce soil-borne diseases.

3.4.6 Conservation agriculture

Conservation agriculture practices, including minimum tillage, crop residue management, and soil cover through mulching or cover crops, can improve soil health, moisture retention, and nutrient availability. This approach reduces soil erosion, enhances organic matter content, and promotes sustainable crop production.

3.4.7 Access to credit and technology

Providing farmers with access to affordable credit, improved farm machinery, and modern agricultural technologies can facilitate increased productivity. Access to financial resources allows farmers to invest in inputs, machinery, and technologies that can positively impact crop production and overall efficiency.

3.4.8 Farmer education and extension services

Effective farmer education and extension services are crucial for disseminating knowledge, best practices, and innovations to farmers. Extension programs can provide training on modern farming techniques, pest and disease management, water conservation, and efficient use of inputs, enabling farmers to make informed decisions and adopt practices that enhance productivity.

3.4.9 Research and development

Continuous research and development efforts in crop science and agronomy are essential for developing improved crop varieties, best management practices, and climate-resilient farming systems. Collaborative efforts between research institutions, universities, and farmers can lead to the development and dissemination of technologies and practices that increase crop productivity.

3.4.10 Policy support and infrastructure development

Enabling policies and investments in agricultural infrastructure, including irrigation systems, rural roads, storage facilities, and market access, can create an enabling environment for increased crop productivity. These measures support efficient input supply, reduce postharvest losses, and facilitate the marketing of agricultural products.

3.5 The role of agronomy in food security

Agronomy plays a crucial role in ensuring food security by focusing on the science and practice of sustainable crop production. It encompasses a range of disciplines such as soil science, crop physiology, plant breeding, and pest management, all aimed at optimizing agricultural practices to maximize crop yields and quality.

The following are key ways in which agronomy contributes to food security:

1. **Optimizing crop production**: Agronomists work to develop and promote efficient and sustainable farming practices. This includes selecting the right crop varieties for specific agroclimatic conditions, determining optimal planting times, and recommending appropriate fertilization and irrigation techniques. By optimizing these factors, agronomy helps farmers achieve higher yields per hectare, increasing overall food production.
2. **Improving soil health**: Soil is a critical component of agriculture, and agronomy focuses on soil management practices that enhance fertility and health. This includes techniques such as conservation tillage, crop rotation, and cover cropping, which help to maintain soil structure, minimize erosion, and increase nutrient availability. Healthy soils are essential for robust plant growth and sustainable crop production.
3. **Pest and disease management**: Agronomy plays a role in developing IPM strategies that minimize the use of chemical pesticides while effectively controlling pests and diseases. This approach includes monitoring pest populations, promoting natural predators, using resistant crop varieties, and implementing cultural practices that reduce pest pressure. By managing pests and diseases effectively, agronomy protects crop yields and reduces postharvest losses.
4. **Water management**: With increasing water scarcity and climate variability, agronomy helps farmers optimize water-use efficiency through practices such as drip irrigation, rainwater harvesting, and mulching. By improving water management, agronomy ensures that crops have sufficient moisture for growth, particularly in regions prone to drought.
5. **Adapting to climate change**: Agronomy is at the forefront of developing climate-smart agricultural practices that help farmers adapt to changing climatic conditions. This includes promoting climate-resilient crop varieties, agroforestry systems, and soil conservation techniques. These strategies enhance the resilience of

agricultural systems, ensuring food production in the face of climate-related challenges.

6. **Promoting sustainable practices**: Agronomy emphasizes sustainable agriculture, which involves balancing the need for increased food production with environmental conservation and social equity. By promoting practices that minimize environmental impact, such as organic farming, agroecology, and precision agriculture, agronomy contributes to long-term food security by preserving natural resources for future generations.

Agronomy is indispensable for achieving global food security goals. Through its focus on sustainable crop production, soil health, pest management, water efficiency, climate resilience, and sustainability, agronomy ensures that farmers have the knowledge and tools to produce more food while safeguarding the environment and livelihoods of communities around the world.

3.6 The role of agronomy in sustainability

Agronomy plays a pivotal role in global sustainability by promoting practices and technologies that balance agricultural productivity with environmental conservation and social equity. As the science and practice of sustainable crop production, agronomy contributes to several aspects of global sustainability:

1. **Sustainable food production**: Agronomy focuses on maximizing crop yields while minimizing negative environmental impacts. Through advancements in plant breeding, soil management, and crop protection, agronomy helps farmers produce more food on existing agricultural land. This is crucial for meeting the growing global demand for food while reducing the need for deforestation and expansion of agricultural land into natural habitats.
2. **Soil health and conservation**: Healthy soils are essential for sustainable agriculture, and agronomy emphasizes practices that improve soil fertility, structure, and biodiversity. Techniques such as conservation tillage, cover cropping, and agroforestry promote soil health, reduce erosion, and enhance carbon sequestration. By maintaining soil health, agronomy contributes to the long-term productivity and sustainability of agricultural systems.
3. **Water management**: Agronomy plays a key role in optimizing water-use efficiency in agriculture. With water scarcity becoming a major concern in many regions, agronomists develop and promote irrigation technologies, drought-tolerant crop varieties, and water-saving practices such as drip irrigation and rainwater harvesting. These efforts help conserve water resources and ensure sustainable agricultural production in water-stressed areas.
4. **Biodiversity conservation**: Agronomy seeks to balance agricultural production with the conservation of biodiversity. By promoting diverse cropping systems,

crop rotations, and IPM, agronomy helps maintain biodiversity on farms. This is crucial for supporting pollinators, natural pest predators, and other beneficial organisms that contribute to ecosystem resilience and sustainable food production.

5. **Climate change mitigation and adaptation**: Agronomy is central to climate-smart agriculture, which aims to mitigate greenhouse gas emissions from agriculture while adapting to climate change impacts. This includes promoting climate-resilient crop varieties, agroforestry systems, and practices that enhance carbon sequestration in soils. Agronomy helps farmers adapt to changing climate conditions while contributing to global efforts to reduce agriculture's carbon footprint.
6. **Promoting sustainable livelihoods**: Sustainable agronomy practices not only benefit the environment but also support the livelihoods of farmers and rural communities. By improving agricultural productivity, reducing input costs, and enhancing market access, agronomy helps increase farmers' incomes and improve their resilience to economic shocks.

Agronomy is essential for achieving global sustainability goals by promoting sustainable food production, soil health, water management, biodiversity conservation, climate change mitigation, and sustainable livelihoods. Through research, education, and the adoption of best practices, agronomy contributes to a more sustainable and resilient agricultural sector that can meet the needs of current and future generations while protecting the planet's resources.

Agronomy serves as the foundation of modern agriculture due to its pivotal role in understanding and optimizing crop production systems. As a scientific discipline, agronomy encompasses various aspects of crop cultivation, including soil management, crop nutrition, plant genetics, pest and weed control, and crop physiology. By applying principles from agronomy, farmers can make informed decisions to maximize yields, improve crop quality, and enhance resource efficiency. Furthermore, agronomy integrates advancements in technology, such as precision agriculture and genetic engineering, to continually innovate and address the challenges faced by modern agriculture, ensuring its sustainability and productivity. Overall, agronomy provides the scientific framework and practical knowledge essential for the success and advancement of agricultural practices worldwide.

By implementing these strategies and adopting sustainable and science-based practices, crop productivity globally can be enhanced, contributing to increased food security, improved farmer livelihoods, and the overall development of the agricultural sector worldwide. From the principles of agronomy, we step into the realm of integrated agriculture: sustainable agriculture and decarbonizing practices in **Chapter 4**. Here, the convergence of diverse practices, from crop rotation to conservation techniques, offers a holistic approach to harmonizing productivity and environmental health. As we explore this chapter, the bridges between individual components of agriculture come to life, forming a cohesive framework that aims to propel us toward a more balanced and resilient agricultural system.

References

Altieri, M. A., & Nicholls, C. I. (2018). Agroecology and the reconstruction of a post-COVID-19 agriculture. Agronomy for Sustainable Development, 40(5), 1–4.

Amanullah. (2024). Harvesting Success: A Comprehensive Guide to Agronomy Courses. Punjab Books, Lahore, PK.

Amanullah., Ondrasek, G., Al –Tawaha, A. R. M. S., eds. (2023). Integrated nutrients management: An approach for sustainable crop production and food security in changing climates. Lausanne: Frontiers Media SA. doi: 10.3389/978-2-8325-3169-3

Amanullah, Khalid, S., Muhammad, A., Yar, M., et al. (2021). Integrated use of biofertilizers with organic and inorganic phosphorus sources improve dry matter partitioning and yield of hybrid maize. Communications in Soil Science and Plant Analysis, 52(21), 2732–2747.

Amanullah, & Fahad, S. (2018). Integrated nutrient management in corn production: Symbiosis for food security and grower's income in arid and semiarid climates. In: Amanullah & S. Fahad (Eds.). Corn – Production and Human Health in Changing Climate (pp. 03–12). InTech, London, UK.

Amanullah, Khan, A., Khalid, S., et al. (2019). Integrated management of phosphorus, organic sources, and beneficial microbes improve dry matter partitioning of maize. Communications in Soil Science and Plant Analysis, 50(20), 2544–2569.

Amanullah, & Khalid, S. (2016). Integrated use of phosphorus, animal manures and biofertilizers improve maize productivity under semiarid condition. In: L. L. Marcelo & S. Soloneski (Eds.). Organic Fertilizers – From Basic Concepts to Applied Outcomes (pp. 137–155). InTech, Rijeka, Croatia.

Ardwaj, A. K., & Bhatia, A. (2013). Importance of agronomy in sustainable agriculture. Journal of Soil and Water Conservation, 12(2), 127–132.

Bremner, P. M. (2009). Agronomy: Science of sustainable agriculture. In: Proceedings of the 22nd International Conference of Agricultural Economists, Foz do Iguaçu, Brazil.

Bünemann, E. K., Bongiorno, G., Bai, Z., Creamer, R. E., De Deyn, G., de Goede, R., . . . & Brussaard, L. (2018). Soil quality – A critical review. Soil Biology and Biochemistry, 120, 105–125.

Cavigelli, M. A., & Lengnick, L. (2008). Integrated farming systems: A framework for developing sustainable agricultural systems. International Journal of Agricultural Sustainability, 6(3), 185–202.

Clark, A. J., & Ferris, H. (2011). The role of the soil food web in sustainable agriculture. In: Soil Ecology and Ecosystem Services (pp. 57–83). Oxford University Press. Oxford, UK.

FAO. (2018). Climate-smart Agriculture Sourcebook. Food and Agriculture Organization of the United Nations. Rome, Italy.

Francis, C. A., & Porter, P. (2011). Sustainable agriculture in temperate zones. In: Handbook of Sustainable Agriculture (pp. 1–30). CRC Press. London, UK.

Gliessman, S. R. (2014). Agroecology: The Ecology of Sustainable Food Systems. CRC Press. London, UK.

Gliessman, S. R. (Ed.). (2014). Agroecology: The Ecology of Sustainable Food Systems (3rd ed.). CRC Press.

Hillel, D. (2004). Introduction to Environmental Soil Physics. Elsevier Academic Press. Amsterdam, The Netherlands.

Jackson, M. L. (1973). Soil Chemical Analysis. Prentice-Hall. Upper Saddle River, New Jersey, US.

Klerkx, L., Aarts, N., & Leeuwis, C. (2010). Adaptive management in agricultural innovation systems: The interactions between innovation networks and their environment. Agricultural Systems, 103(6), 390–400.

Lal, R. (2013). Sustainable crop production for food security. In: Achieving Sustainable Cultivation of Maize (Vol. 1, pp. 1–27). Burleigh Dodds Science Publishing. Castle Hill, Cambridge, UK.

Pretty, J. (2008). Agricultural sustainability: Concepts, principles, and evidence. Philosophical Transactions of the Royal Society B: Biological Sciences, 363(1491), 447–465.

Pretty, J., & Bharucha, Z. P. (2014). Sustainable intensification in agricultural systems. Annals of Botany, 114(8), 1571–1596.

Pretty, J., Toulmin, C., & Williams, S. (2011). Sustainable intensification in African agriculture. International Journal of Agricultural Sustainability, 9(1), 5–24.

Sanchez, P. A., Shepherd, K. D., Soule, M. J., Place, F. M., Buresh, R. J., Izac, A. N., . . . & Mokwunye, A. U. (1997). Soil fertility replenishment in Africa: An investment in natural resource capital. In: Replenishing Soil Fertility in Africa (pp. 1–46). Soil Science Society of America, Madison, United States.

Snapp, S. S., & Gentry, L. E. (2017). Sustaining soils and soul: The promise of agroecological farming. In: Agroecology (pp. 179–208). CRC Press, London, UK.

Vanlauwe, B., Bationo, A., Chianu, J., Giller, K. E., Merckx, R., Mokwunye, U., . . . & Sanginga, N. (2010). Integrated soil fertility management: Operational definition and consequences for implementation and dissemination. Outlook on Agriculture, 39(1), 17–24.

Chapter 4
Integrated agriculture: sustainable agriculture and decarbonizing practices

Introduction

Integrated agriculture (integrated farming) is a sustainable agricultural system that harmoniously integrates various components such as livestock, crop production, fish, poultry, and trees. This holistic approach fosters biodiversity and enhances ecosystem resilience, creating a balanced and self-sustaining environment. By combining different farming activities, integrated farming maximizes resource efficiency, minimizes waste, and promotes synergistic interactions between different elements of the farm ecosystem. Through the integration of diverse agricultural practices, integrated farming not only boosts productivity and profitability but also contributes to environmental conservation and the long-term sustainability of agricultural systems.

Integrated and sustainable agriculture represents a transformative approach to farming that seeks to enhance productivity, resilience, and environmental sustainability. In the face of climate change and its associated challenges, decarbonizing agriculture has become an imperative for mitigating greenhouse gas emissions, reducing carbon footprints, and promoting climate resilience. This chapter explores the principles, practices, and strategies of integrated and sustainable agriculture with a focus on decarbonization in response to changing climates.

4.1 Understanding integrated and sustainable agriculture

Integrated and sustainable agriculture encompasses a holistic approach to farming that integrates ecological, economic, and social dimensions. It emphasizes the optimization of resource use, the enhancement of ecosystem services, and the promotion of agroecological principles to achieve long-term viability and resilience. By adopting integrated farming systems that combine crop production, livestock management, agroforestry, and soil conservation practices, farmers can minimize environmental impacts, increase productivity, and improve livelihoods.

Integrated and sustainable agriculture represents a holistic approach to farming that recognizes the interconnectedness of ecological, economic, and social factors in agricultural systems. This approach seeks to optimize resource use while enhancing ecosystem services and promoting agroecological principles to ensure the long-term viability and resilience of farming practices (Figure 4.1).

https://doi.org/10.1515/9783111448077-004

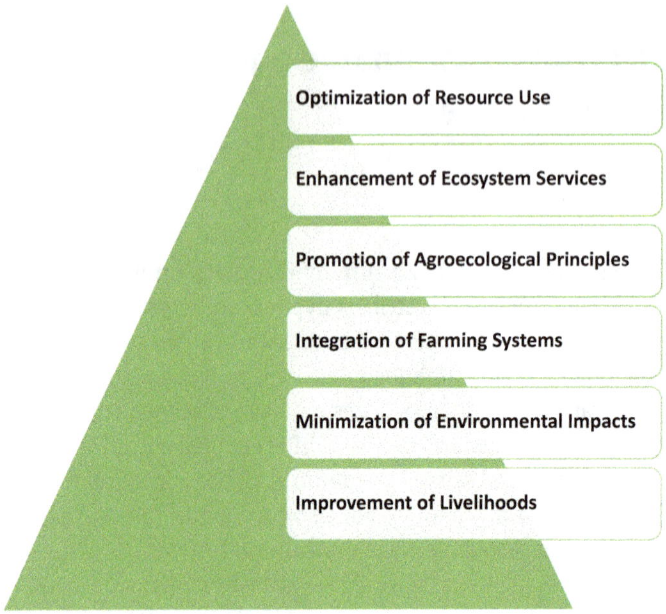

Figure 4.1: Agroecological principles to ensure the long-term viability and resilience of farming practices.

4.1.1 Optimization of resource use

Integrated and sustainable agriculture prioritizes the efficient utilization of resources such as water, soil, energy, and nutrients. Rather than relying solely on external inputs, farmers aim to harness natural processes and cycles to meet their agricultural needs. This includes practices such as crop rotation, intercropping, and cover cropping, which enhance soil fertility, reduce the need for synthetic fertilizers, and minimize nutrient runoff, thereby preserving soil health and promoting sustainable land use.

4.1.2 Enhancement of ecosystem services

A key tenet of integrated and sustainable agriculture is the recognition of the valuable ecosystem services provided by agricultural landscapes. These services include pollination, pest regulation, soil formation, and water purification, among others. By adopting practices that support biodiversity, such as maintaining hedgerows, planting native vegetation, and creating wildlife habitats, farmers can enhance ecosystem resilience and contribute to the overall health and functioning of agricultural ecosystems.

4.1.3 Promotion of agroecological principles

Agroecology lies at the heart of integrated and sustainable agriculture, emphasizing the integration of ecological principles into agricultural systems. This involves representing natural ecosystems and harnessing ecological processes to enhance productivity, resilience, and sustainability. Agroecological practices such as agroforestry, polyculture, and biological pest control promote biodiversity, improve soil health, and reduce the reliance on external inputs, thereby fostering resilient and self-sustaining farming systems.

4.1.4 Integration of farming systems

Integrated farming systems lie at the intersection of crop production, livestock management, agroforestry, and soil conservation practices. By combining diverse components within a single farming operation, farmers can capitalize on synergies and optimize resource use. For example, integrating livestock into cropping systems can provide natural fertilizer, reduce weed pressure, and improve soil structure, while agroforestry systems can enhance biodiversity, provide shade and shelter, and diversify farm income streams.

4.1.5 Minimization of environmental impacts

One of the primary goals of integrated and sustainable agriculture is to minimize environmental impacts associated with agricultural practices. This includes reducing greenhouse gas emissions, conserving water resources, preventing soil erosion, and preserving biodiversity. By adopting practices such as conservation tillage, precision irrigation, and agroecological zoning, farmers can mitigate their environmental footprint and contribute to broader conservation efforts.

4.1.6 Improvement of livelihoods

Ultimately, integrated and sustainable agriculture seeks to improve the livelihoods of farmers and rural communities. By enhancing productivity, resilience, and sustainability, farmers can secure stable incomes, reduce vulnerability to external shocks, and improve overall quality of life. Moreover, by promoting local food systems, value-added processing, and direct marketing initiatives, integrated agriculture can create economic opportunities and strengthen community ties, fostering vibrant and resilient rural economies.

Integrated and sustainable agriculture offers a holistic approach to farming that prioritizes the optimization of resource use, enhancement of ecosystem services, promotion of agroecological principles, integration of farming systems, minimization of environmental impacts, and improvement of livelihoods. By embracing these principles and practices, farmers can build resilient and sustainable agricultural systems that benefit both people and the planet.

4.2 Fostering agroecological approaches

Agroecological approaches emphasize the integration of ecological principles into agricultural systems to promote sustainability and resilience (Figure 4.2). By mimicking natural ecosystems and harnessing ecological processes, agroecological practices can reduce the need for external inputs, minimize environmental impacts, and enhance biodiversity. Strategies such as agroforestry, polyculture, and biological pest control promote ecosystem health, increase resilience to climate change, and contribute to decarbonizing agriculture.

Agroecological approaches represent a paradigm shift in agriculture, emphasizing the integration of ecological principles into farming systems to promote sustainability, resilience, and biodiversity conservation. By mimicking natural ecosystems and harnessing ecological processes, agroecological practices strive to reduce reliance on external inputs, minimize environmental impacts, and enhance ecosystem health.

Figure 4.2: Agroecological approaches emphasizing the integration of ecological principles systems to promote sustainability and resilience.

4.2.1 Agroforestry

Agroforestry integrates trees and shrubs into agricultural landscapes, creating diverse and multifunctional farming systems. By combining agricultural crops with trees or shrubs, agroforestry systems provide numerous ecological benefits, including enhanced carbon sequestration, improved soil fertility, and increased biodiversity. Agroforestry practices such as alley cropping, windbreaks, and silvopasture not only contribute to decarbonizing agriculture by sequestering carbon in biomass and soils but also provide additional ecosystem services such as erosion control, water regulation, and habitat provision.

4.2.2 Polyculture

Polyculture involves cultivating multiple crop species within the same field, mimicking natural ecosystems' diversity and complexity. By diversifying crop species, polyculture systems enhance pest and disease resistance, improve soil health, and promote nutrient cycling, reducing the need for chemical inputs and enhancing agricultural sustainability. Polyculture also fosters ecological resilience, as diverse cropping systems are more resilient to climate variability and extreme weather events than monoculture systems. Moreover, polyculture promotes habitat diversity and supports beneficial insects and pollinators, contributing to overall ecosystem health and biodiversity conservation.

4.2.3 Biological pest control

Biological pest control involves using natural enemies such as predators, parasites, and pathogens to manage pest populations in agricultural systems. By harnessing natural ecological processes, biological pest control reduces reliance on synthetic pesticides, minimizes environmental impacts, and promotes biodiversity conservation. Strategies such as conservation biological control, habitat manipulation, and augmentative biological control enhance natural pest control services, contributing to sustainable pest management and reducing chemical inputs in agriculture. Moreover, by promoting the presence of natural enemies and enhancing habitat diversity, biological pest control supports ecological balance and resilience in agricultural ecosystems.

4.2.4 Soil health management

Agroecological approaches prioritize soil health management through practices such as organic matter addition, crop diversification, and minimal soil disturbance. By enhancing soil organic matter content, improving soil structure, and promoting beneficial soil organisms, agroecological practices improve soil fertility, water retention, and nutrient cycling, reducing the need for synthetic fertilizers and enhancing carbon sequestration in soils. Moreover, by maintaining healthy soils, agroecological approaches support plant health, productivity, and resilience to environmental stresses, contributing to sustainable agricultural production and climate resilience.

Fostering agroecological approaches is essential for promoting sustainable and resilient agriculture while mitigating climate change impacts and enhancing biodiversity conservation. By integrating ecological principles into farming systems through practices such as agroforestry, polyculture, biological pest control, and soil health management, farmers can reduce reliance on external inputs, minimize environmental impacts, and enhance ecosystem services. These approaches not only contribute to decarbonizing agriculture but also support broader goals of sustainable development, food security, and environmental conservation.

4.2.5 Empowering farmers and communities

Empowering farmers and communities is essential for promoting integrated and sustainable agriculture and fostering climate resilience. Providing access to knowledge, resources, and support services can help farmers adopt sustainable practices, improve productivity, and enhance resilience to climate change. Moreover, investing in rural infrastructure, market access, and social capital can strengthen local economies, reduce poverty, and build community resilience in the face of climate-related challenges.

Empowering farmers and communities is not only essential for promoting integrated and sustainable agriculture but also crucial for fostering climate resilience and achieving broader goals of sustainable development. By providing farmers and communities with access to knowledge, resources, and support services, stakeholders can facilitate the adoption of sustainable practices, enhance productivity, and build resilience to climate change impacts. Moreover, investing in rural infrastructure, market access, and social capital can strengthen local economies, alleviate poverty, and foster community resilience in the face of climate-related challenges.

4.2.6 Access to knowledge and resources

Access to knowledge and resources is fundamental for enabling farmers to adopt sustainable agricultural practices. Providing farmers with training, extension services, and technical support can enhance their capacity to implement agroecological approaches, improve soil health, and reduce reliance on external inputs. Moreover, promoting peer-to-peer learning, farmer field schools, and participatory research can facilitate knowledge sharing and innovation within farming communities, fostering the adoption of context-specific and locally adapted practices.

4.2.7 Support services and financial assistance

In addition to knowledge and training, farmers require access to support services and financial assistance to implement sustainable practices effectively. Providing farmers with access to credit, subsidies, and incentives for adopting agroecological approaches can help overcome financial barriers and incentivize sustainable investments. Moreover, establishing insurance schemes, risk-sharing mechanisms, and disaster preparedness programs can help farmers cope with climate-related risks and uncertainties, safeguarding their livelihoods and promoting long-term resilience.

4.2.8 Rural infrastructure development

Investing in rural infrastructure is essential for improving access to markets, inputs, and essential services in agricultural communities. Developing transportation networks, storage facilities, and market linkages can facilitate the efficient movement of agricultural produce, reduce postharvest losses, and enhance market access for smallholder farmers. Moreover, investing in irrigation systems, renewable energy technologies, and climate-resilient infrastructure can enhance agricultural productivity, water management, and energy access in rural areas, contributing to sustainable development and poverty alleviation.

4.2.9 Market access and value chains

Facilitating market access and strengthening value chains are critical for enhancing the economic viability of sustainable agriculture and improving farmers' livelihoods. Supporting farmers' cooperatives, producer organizations, and collective marketing initiatives can help smallholder farmers access markets, negotiate better prices, and capture more value from their agricultural produce. Moreover, promoting inclusive value chains, fair trade practices, and market transparency can ensure equitable distribution of benefits and empower marginalized farmers, particularly women and youth, to participate in agricultural markets and value-added activities.

4.2.10 Social capital and community resilience

Building social capital and fostering community resilience are essential for strengthening the social fabric and collective action within farming communities. Investing in social cohesion, community organizations, and participatory decision-making processes can promote solidarity, cooperation, and mutual support among farmers, enabling them to address common challenges and capitalize on opportunities collectively. Moreover, strengthening local institutions, governance structures, and social networks can enhance community resilience and adaptive capacity in the face of climate-related shocks and stresses, ensuring that farmers are better prepared to cope with and recover from adverse events.

Empowering farmers and communities is essential for promoting integrated and sustainable agriculture, fostering climate resilience, and achieving inclusive and sustainable development. By providing farmers with access to knowledge, resources, and support services, stakeholders can enable the adoption of sustainable practices and enhance productivity while building community resilience and reducing vulnerability to climate change impacts. Moreover, investing in rural infrastructure, market access, and social capital can strengthen local economies, alleviate poverty, and promote inclusive growth, ensuring that agricultural development is equitable, resilient, and sustainable.

Integrated and sustainable agriculture offers a pathway to decarbonizing agriculture and promoting climate resilience in changing climates. By adopting holistic approaches that integrate ecological, economic, and social dimensions, farmers can enhance productivity, reduce environmental impacts, and adapt to climate change. By investing in sustainable land management practices, fostering agroecological approaches, and empowering farmers and communities, we can build more resilient, equitable, and sustainable agricultural systems that benefit both people and the planet.

4.3 Sustainable agriculture

Sustainable agriculture is an approach to farming that aims to meet the current and future needs of agriculture while ensuring environmental stewardship, economic viability, and social well-being. It involves practices and systems that optimize resource use, minimize environmental impacts, promote biodiversity, and support the long-term viability of agricultural production (Figure 4.3).

4.3.1 Conservation of soil resources

Sustainable agriculture practices, such as conservation tillage, cover cropping, and crop rotation, help protect soil from erosion, maintain soil structure, and enhance organic matter content. By preserving soil fertility and health, sustainable agriculture ensures the long-term productivity of agricultural land and prevents soil degradation.

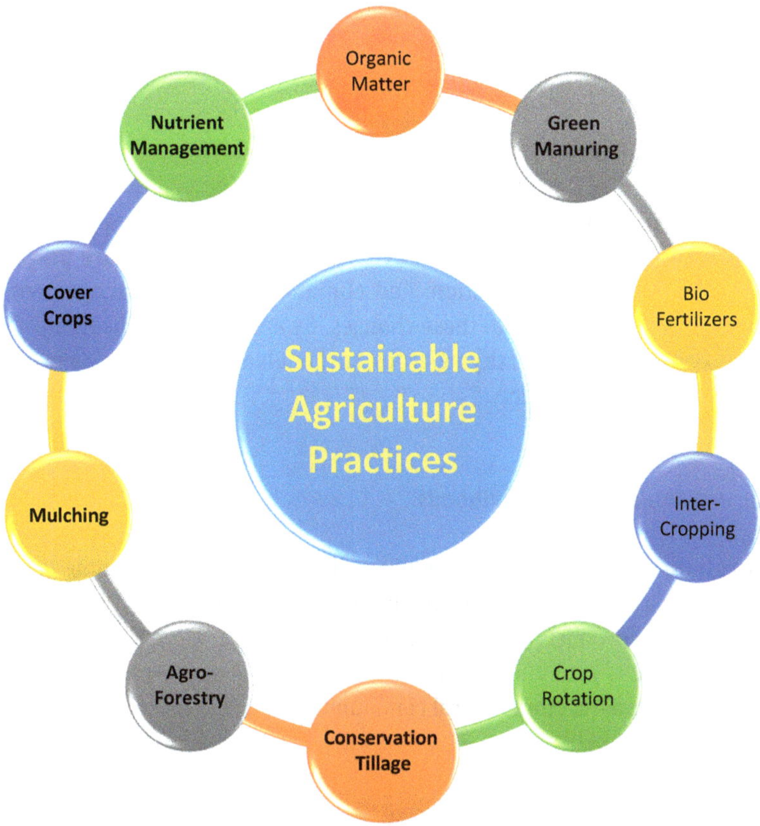

Figure 4.3: Sustainable agriculture practices.

4.3.2 Water resource management

Pakistan faces water scarcity challenges, making efficient water management crucial. Sustainable agriculture practices, including precision irrigation, water-saving techniques, and proper drainage systems, help optimize water use and conserve this valuable resource. By reducing water waste and improving irrigation efficiency, sustainable agriculture contributes to sustainable water resource management.

4.3.3 Biodiversity conservation

Sustainable agriculture practices prioritize biodiversity conservation by promoting habitat preservation, ecological balance, and the use of integrated pest management (IPM) strategies. By preserving natural habitats, protecting pollinators, and promoting beneficial insects, sustainable agriculture contributes to the conservation of biodiversity and ecosystem services.

4.3.4 Climate change resilience

Pakistan is vulnerable to climate change impacts, including increased temperatures, erratic rainfall patterns, and extreme weather events. Sustainable agriculture practices, such as agroforestry, soil conservation, and climate-smart crop management, help build resilience and adaptability to these changes. By enhancing soil health, promoting water conservation, and diversifying crops, sustainable agriculture aids in climate change adaptation and mitigation.

4.3.5 Economic viability and rural livelihoods

Sustainable agriculture practices can improve the economic viability of farming communities. By optimizing resource use, reducing input costs, and enhancing market opportunities for sustainable products, farmers can improve their income and livelihoods. Sustainable agriculture also promotes equitable distribution of resources, enhances rural employment, and supports vibrant rural communities.

4.3.6 Food security and nutrition

Sustainable agriculture plays a critical role in ensuring food security and nutrition. By promoting diverse cropping systems, improving soil fertility, and reducing post-harvest losses, sustainable agriculture enhances agricultural productivity and the availability of nutritious food. It also encourages sustainable farming practices among smallholder farmers, who form the backbone of agriculture in Pakistan.

4.3.7 Environmental protection

Sustainable agriculture helps protect the environment by minimizing the use of synthetic fertilizers and pesticides, reducing chemical runoff into water bodies, and preserving soil and water quality. It promotes the use of organic fertilizers, IPM, and ecological approaches to pest and disease control. By reducing chemical inputs and preserving natural resources, sustainable agriculture contributes to environmental conservation.

4.3.8 Long-term sustainability

Sustainable agriculture focuses on long-term sustainability by considering the economic, environmental, and social dimensions of agricultural systems. It emphasizes the prudent use of resources, the protection of natural ecosystems, and the equitable distribution of benefits. By adopting sustainable practices, Pakistan can ensure the long-term viability of its agricultural sector, enhance resilience to environmental challenges, and contribute to sustainable development goals.

Promoting sustainable agriculture requires a combination of farmer education, policy support, research and development, and market incentives. By embracing sustainable agricultural practices, we can achieve natural resource conservation, enhance food security, and promote the sustainable development of its agricultural sector.

4.4 Integrated agriculture

Integrated agriculture, also known as integrated farming or integrated crop-livestock systems, is an approach to farming that aims to achieve sustainable and holistic agricultural production by optimizing the interactions and synergies between different components of a farming system. It involves integrating various agricultural practices and enterprises to create a more balanced and resilient system that enhances productivity, minimizes negative environmental impacts, and promotes economic viability. This approach embodies the principles of sustainable agriculture by emphasizing efficiency, resource conservation, and environmental stewardship. It involves the integration of different agricultural practices, such as crop cultivation, livestock rearing, fisheries, agroforestry, and cottage industry, to maximize productivity, optimize resource utilization, and minimize negative environmental impacts.

4.4.1 Concept of integrated agriculture

Integrated agriculture, also known as integrated farming systems or integrated crop-livestock systems, is an agricultural approach that aims to optimize the interactions and synergies between different components of the farming system. It involves the integration of multiple agricultural activities, such as crop production, livestock rearing, aquaculture, agroforestry, and cottage industry, in a harmonious and sustainable manner.

The concept of integrated agriculture recognizes that various elements of the agricultural system are interconnected and interdependent. Instead of viewing crops, livestock, fisheries, and other activities as separate entities, integrated agriculture seeks to create a synergistic relationship between them. By leveraging the interactions between different components, integrated agriculture aims to achieve multiple benefits, including increased productivity, efficient resource utilization, risk reduction, and environmental sustainability.

4.4.2 Principles of integrated agriculture

Integrated agriculture encompasses principles that emphasize the holistic management of agricultural systems to optimize productivity, sustainability, and resilience. At its core, integrated agriculture seeks to harness synergies among various components of the farming ecosystem, including crops, livestock, trees, and other elements. Key principles include diversification, which involves growing multiple crops and integrating livestock to enhance resource-use efficiency and reduce risks associated with monocropping (Figure 4.4).

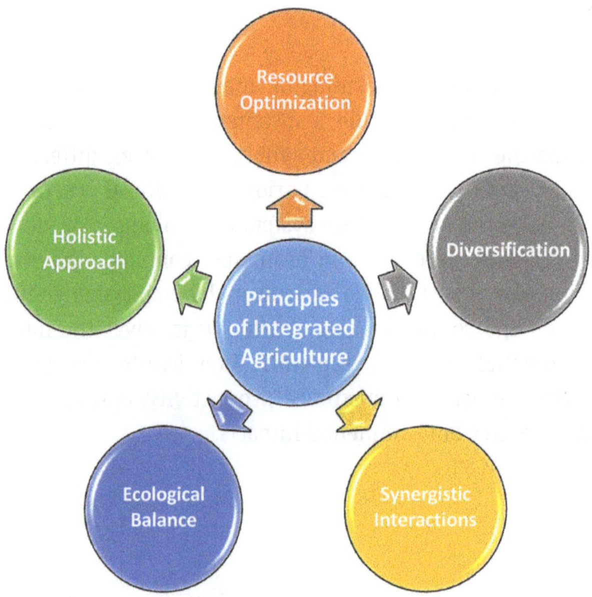

Figure 4.4: Principles of integrated agriculture.

Another principle is resource recycling, where waste products from one component, such as crop residues or animal manure, are utilized as inputs for another, promoting nutrient cycling and minimizing waste. Additionally, integrated agriculture prioritizes ecological balance by promoting biodiversity, conserving natural resources, and minimizing environmental impacts. By adhering to these principles, integrated agriculture systems aim to achieve sustainable production while fostering ecosystem health and resilience.

4.4.2.1 Resource optimization

Integrated agriculture focuses on optimizing the use of available resources, including land, water, nutrients, and energy. By integrating different agricultural activities, farmers can make efficient use of resources and reduce wastage. For example, livestock waste can be utilized as organic fertilizer for crops, minimizing the need for external inputs and improving nutrient cycling.

4.4.2.2 Diversification

Integrated agriculture encourages the diversification of farming activities. By integrating multiple components, farmers can reduce their dependence on a single commodity and spread their risks. Diversification can include growing multiple crops, raising different livestock species, and incorporating nonagricultural activities such as agroforestry or beekeeping. This diversification enhances resilience and stability in the farming system.

4.4.2.3 Synergistic interactions

Integrated agriculture seeks to create positive interactions between different components of the farming system. For example, livestock can provide manure for crop fertilization, while crop residues can be used as feed for animals. These interactions can improve resource efficiency, nutrient cycling, and overall productivity. Integrated aquaculture systems can utilize nutrients from fishponds to fertilize crops, and crop residues can serve as feed for fish.

4.4.2.4 Ecological balance

Integrated agriculture emphasizes the importance of maintaining ecological balance within the farming system. It promotes practices that conserve biodiversity, protect natural habitats, and minimize the use of agrochemicals. By adopting sustainable pest management strategies, enhancing soil health, and preserving ecosystem services, integrated agriculture aims to minimize negative environmental impacts.

4.4.2.5 Holistic approach

Integrated agriculture takes a holistic approach to farming by considering the entire production system and its surrounding environment. It goes beyond individual components and looks at the interactions and feedback loops within the system. This holistic perspective allows for a more comprehensive understanding of the complex relationships and dynamics involved in agricultural production.

The concept of integrated agriculture is applicable across different scales, from small-scale farming systems to large commercial operations. It can be adapted to various agroecological conditions and is flexible in its implementation. Integrated agriculture is not a one-size-fits-all approach, but rather a framework that can be customized based on local conditions, farmer preferences, and available resources.

Integrated agriculture is an approach that integrates multiple components of the farming system to achieve sustainable and efficient agricultural production. It emphasizes resource optimization, diversification, synergistic interactions, ecological balance, and a holistic perspective. By adopting integrated agriculture practices, farmers can enhance productivity, profitability, and environmental sustainability in their farming operations.

Integrated agriculture addresses global agricultural challenges by enhancing productivity, optimizing resource utilization, promoting climate change adaptation, conserving biodiversity, improving soil health, implementing sustainable pest management, ensuring food security and nutrition, and supporting sustainable livelihoods. By embracing integrated approaches, the agricultural sector can move toward a more resilient, environmentally sustainable, and socially inclusive future.

4.4.3 Key aspects of integrated agriculture

Key aspects of integrated agriculture (Figure 4.5) are discussed further.

4.4.3.1 Diversification

Integrated agriculture encourages diversifying the farm's activities beyond monoculture cropping. This might involve combining crop cultivation with livestock rearing, agroforestry, aquaculture, or other complementary enterprises. Diversification helps spread risks, enhance resource-use efficiency, and increase overall system stability.

4.4.3.2 Nutrient cycling

Nutrient cycling is a fundamental concept in integrated agriculture. Crop residues, animal manure, and organic matter are recycled within the system to replenish soil fertility. Livestock waste can be used as organic fertilizer for crops, reducing the need for external inputs and promoting sustainable soil health.

4.4.3.3 Crop-livestock integration

Combining crop and livestock components allows for synergistic interactions. For instance, crop residues can be used as feed for livestock, and animal waste can enrich the soil. This integration can improve nutrient cycling, reduce waste, and increase overall farm productivity.

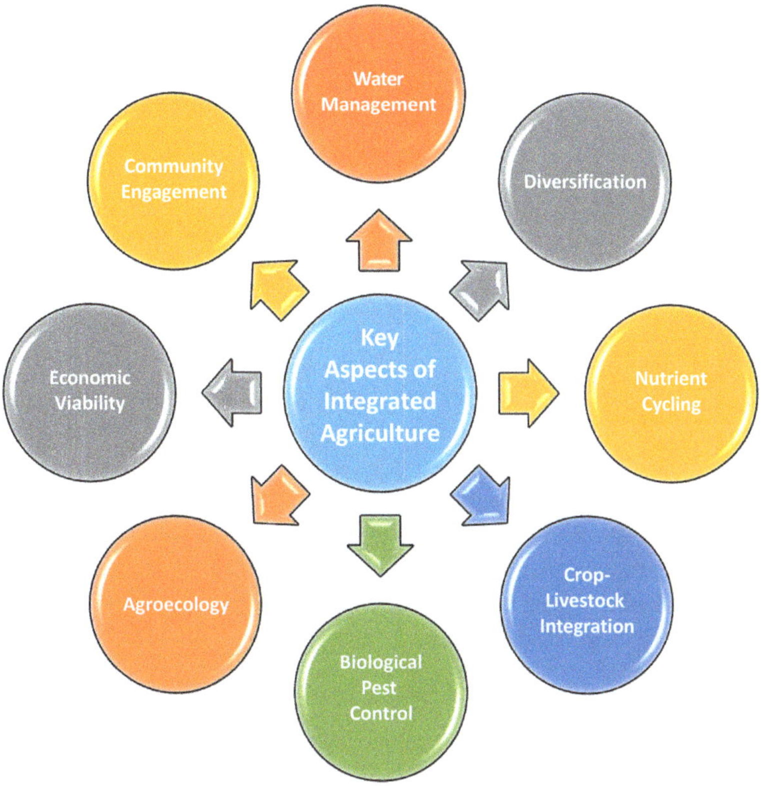

Figure 4.5: Key aspects of integrated agriculture.

4.4.3.4 Biological pest control

Integrated agriculture promotes natural mechanisms of pest control, such as using beneficial insects, trap crops, and crop rotation to reduce the reliance on chemical pesticides. This approach helps maintain ecosystem balance and reduces negative impacts on the environment and human health.

4.4.3.5 Water management

Efficient water use is crucial in integrated agriculture. Practices like agroforestry, cover cropping, and mulching help retain soil moisture and reduce water runoff, improving water-use efficiency and mitigating the impacts of water scarcity.

4.4.3.6 Agroecology

Integrated agriculture is aligned with agroecological principles that focus on adapting farming practices to local ecological conditions. It takes into account factors like climate, soil type, and local biodiversity to create a harmonious and sustainable farming system.

4.4.3.7 Economic viability

Integrated agriculture seeks to improve farm income and livelihoods through diversified income streams and reduced production costs. By integrating different enterprises, farmers can tap into multiple markets and reduce dependency on a single crop or product.

4.4.3.8 Community engagement

Integrated agriculture often involves active community participation and knowledge sharing. Farmers exchange experiences, best practices, and innovations, contributing to the overall development of the farming community.

The benefits of integrated agriculture include enhanced productivity, improved resilience to climate change, reduced environmental degradation, and better livelihoods for farmers. By embracing this approach, farmers can contribute to the broader goal of achieving sustainable agriculture that meets the needs of the present without compromising the ability of future generations to meet their own needs.

4.4.4 Role of integrated agriculture and global agricultural challenges

The role of integrated agriculture in addressing global agricultural challenges is significant, as it offers holistic and sustainable solutions to various issues faced by the agricultural sector worldwide.

Here are key ways in which integrated agriculture addresses global agricultural challenges (Figure 4.6).

4.4.4.1 Enhancing productivity

Integrated agriculture promotes increased agricultural productivity by leveraging synergies between different components of the farming system. For example, the integration of livestock and crop production allows for efficient nutrient cycling and resource utilization. This integration can result in higher crop yields, improved soil fertility, and enhanced overall productivity, which is crucial for meeting the growing global demand for food.

4.4.4.2 Resource efficiency and conservation

Integrated agriculture optimizes the use of resources, such as land, water, and nutrients, leading to improved resource efficiency. By diversifying production and implementing sustainable practices, integrated systems minimize resource wastage,

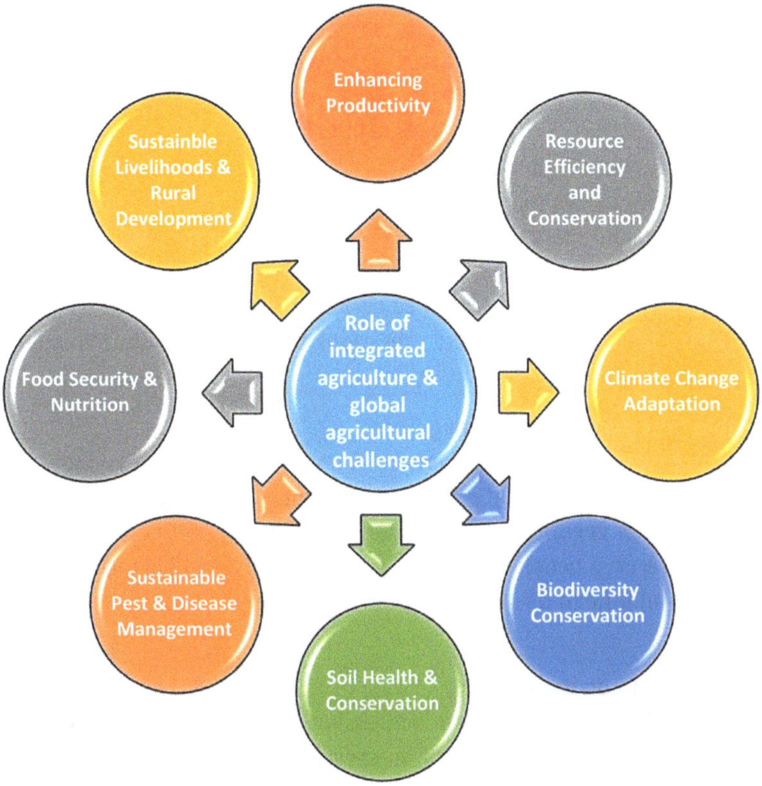

Figure 4.6: Role of integrated agriculture and global agricultural challenges.

reduce environmental degradation, and enhance the long-term sustainability of agriculture. This is particularly important in the face of limited land availability, water scarcity, and concerns about climate change.

4.4.4.3 Climate change adaptation

Integrated agriculture plays a vital role in addressing the challenges posed by climate change. Through diversification, integrated systems can enhance resilience and mitigate risks associated with changing climate patterns. For instance, agroforestry practices, which involve the integration of trees with crops and livestock, can provide shade, conserve water, sequester carbon, and reduce vulnerability to extreme weather events.

4.4.4.4 Biodiversity conservation

Integrated agriculture contributes to biodiversity conservation by creating diverse and ecologically balanced farming systems. By integrating different components, such as crops, livestock, trees, and water bodies, integrated systems provide habitats for a wide range of plant and animal species. This biodiversity helps to maintain ecological balance, promote pollination, enhance pest control, and support overall ecosystem health.

4.4.4.5 Soil health and conservation

Integrated agriculture prioritizes soil health and conservation. Practices such as crop rotation, cover cropping, and the use of organic amendments improve soil structure, fertility, and microbial activity. Healthy soils, with improved water-holding capacity and nutrient availability, contribute to sustained agricultural productivity and reduce soil erosion, mitigating land degradation and preserving valuable agricultural land.

4.4.4.6 Sustainable pest and disease management

Integrated agriculture employs holistic and environmentally friendly pest and disease management strategies. By diversifying crops and implementing IPM practices, farmers can minimize the use of chemical pesticides and reduce the risk of pest outbreaks. This approach helps preserve beneficial insects, maintain ecological balance, and reduce negative impacts on human health and the environment.

4.4.4.7 Food security and nutrition

Integrated agriculture plays a crucial role in addressing global food security and nutrition challenges. By diversifying production, integrated systems increase the availability and variety of food sources, contributing to a more balanced and nutritious diet. The integration of livestock, aquaculture, and horticulture can provide diverse sources of protein, essential nutrients, and micronutrients, reducing malnutrition and improving overall food security.

4.4.4.8 Sustainable livelihoods and rural development

Integrated agriculture contributes to sustainable livelihoods and rural development by creating income-generating opportunities and promoting entrepreneurship. Integrated systems often offer multiple revenue streams, such as crop sales, livestock products, and value-added activities, thus increasing income stability and improving rural livelihoods. This helps reduce poverty, create employment, and foster sustainable economic growth in rural communities.

4.4.5 Importance and benefits of integrated agriculture

Integrated agriculture plays a crucial role in promoting sustainable development by addressing multiple challenges faced by the agricultural sector and fostering a harmonious relationship between agriculture, the environment, and society.

Integrated agriculture, with its holistic approach, offers a multitude of benefits and holds significant importance in the realm of sustainable development. One of the primary advantages is its ability to tackle various challenges simultaneously, creating a more resilient and efficient agricultural system. By integrating diverse practices such as crop rotation, agroforestry, and IPM, farmers can enhance soil fertility, reduce pests and diseases, and optimize water usage. This not only improves crop yields but also ensures the long-term health of the land.

Furthermore, integrated agriculture fosters a harmonious relationship between agriculture, the environment, and society. It promotes biodiversity conservation by creating habitats for beneficial organisms, thereby enhancing natural pest control and reducing the reliance on chemical pesticides. This has positive ripple effects on the environment, such as improved water quality and reduced pollution from agricultural runoff.

Another crucial aspect of integrated agriculture is its role in mitigating climate change. Practices like agroforestry, which combines trees with crops or livestock, sequester carbon dioxide from the atmosphere, helping to offset greenhouse gas emissions. This contributes to climate change resilience by building more robust agricultural systems capable of withstanding extreme weather events.

Moreover, integrated agriculture benefits farmers economically by diversifying income sources. For example, integrating livestock into crop farming allows farmers to sell meat, dairy, or other animal products in addition to crops, creating multiple revenue streams. This can buffer against market fluctuations and improve overall financial stability for farming households.

The importance of integrated agriculture cannot be overstated. Its benefits extend far beyond individual farms to encompass environmental sustainability, food security, climate resilience, and economic prosperity. Embracing integrated agricultural practices is not just a choice for farmers; it is a necessity for building a more sustainable and resilient agricultural sector for generations to come.

4.4.6 Benefits of integrated agriculture in sustainable development

Key benefits of integrated agriculture in sustainable development are described further (Figure 4.7).

4.4.6.1 Sustainable resource management

Integrated agriculture promotes sustainable use of natural resources by minimizing waste, optimizing resource utilization, and reducing environmental pollution. For example, livestock waste can be recycled as organic fertilizer for crops, reducing the need for synthetic fertilizers. This approach helps to maintain soil fertility, conserve water, and protect biodiversity.

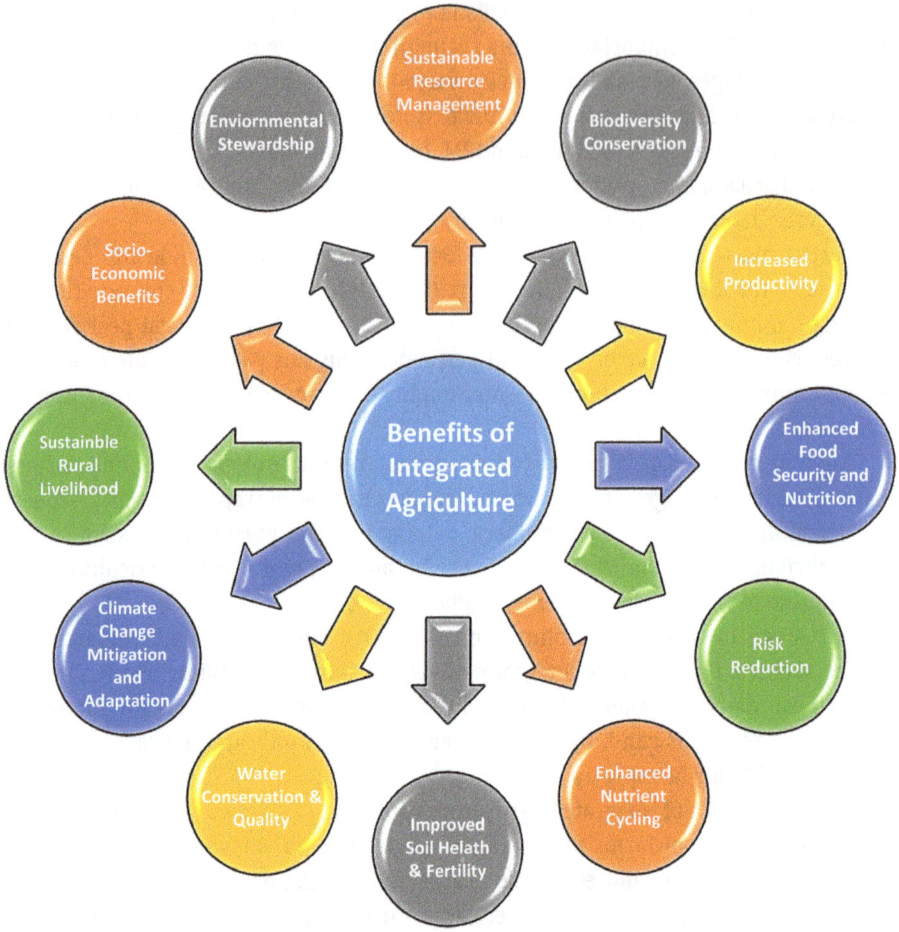

Figure 4.7: Benefits of integrated agriculture in sustainable development.

Integrated agriculture optimizes the use of resources such as land, water, nutrients, and energy. By integrating different components, farmers can utilize resources more efficiently. For example, livestock waste can be recycled as organic fertilizer for crops, reducing the need for synthetic fertilizers. This resource efficiency minimizes waste, conserves natural resources, and reduces the environmental footprint of agriculture.

4.4.6.2 Biodiversity conservation

Integrated agriculture promotes biodiversity conservation by providing a diverse range of habitats, food sources, and ecological niches. Diverse farming systems support a variety of plant and animal species, including beneficial insects, pollinators, and natural predators that contribute to pest control. The preservation of biodiversity enhances ecological resilience, promotes ecosystem stability, and supports long-term sustainability.

4.4.6.3 Increased productivity

Integration of different components allows for the efficient use of available resources. For instance, crop residues can be used as feed for livestock, optimizing resource utilization and reducing feed costs. The diversity of agricultural activities within an integrated system helps to maximize productivity and income potential for farmers.

4.4.6.4 Enhanced food security and nutrition

Integrated agriculture contributes to food security by diversifying production and improving access to nutritious food. The integration of crops, livestock, and aquaculture systems provides a variety of food sources and increases the availability of diverse, nutritious products. Additionally, integrated farming systems often have higher productivity and stability, reducing the vulnerability of farmers to food production fluctuations.

4.4.6.5 Risk reduction

Integrated agriculture provides a buffer against risks and uncertainties. By diversifying their activities, farmers are less vulnerable to the negative impacts of pests, diseases, or adverse weather conditions. For instance, if a crop fails due to a pest infestation, the income from livestock or fish farming can help offset the losses.

4.4.6.6 Enhanced nutrient cycling

Integrated systems facilitate the cycling of nutrients within the agricultural system. For example, livestock manure can be used as fertilizer for crops, reducing the reliance on synthetic fertilizers. This not only saves costs but also reduces nutrient runoff and water pollution.

4.4.6.7 Improved soil health and fertility

Integrated agriculture practices, such as crop rotation, cover cropping, and the integration of livestock, contribute to improving soil health and fertility. These practices enhance soil structure, increase organic matter content, promote beneficial microbial activity, and reduce soil erosion. Healthy soils are essential for sustained agricultural productivity, nutrient cycling, and carbon sequestration.

4.4.6.8 Water conservation and quality

Integrated agriculture promotes efficient water management strategies, reducing water consumption and preserving water quality. Practices such as drip irrigation, precision agriculture, and integrated water management systems minimize water wastage and nutrient runoff, thus protecting water resources and preventing water pollution. This is particularly crucial in regions facing water scarcity and where agriculture is a significant water user.

4.4.6.9 Climate change mitigation and adaptation

Integrated agriculture has the potential to mitigate climate change by reducing greenhouse gas emissions and sequestering carbon in soil and biomass. Agroforestry practices, for instance, involve planting trees on agricultural lands, which can sequester carbon, provide shade, and mitigate the impacts of extreme weather events. Additionally, integrated systems are more resilient to climate variability, as diversification reduces vulnerability to climate-related risks and shocks.

4.4.6.10 Sustainable rural livelihoods

Integrated agriculture supports the development of sustainable rural livelihoods by providing opportunities for small-scale farmers to engage in diverse agricultural activities. This reduces their dependency on single commodities and creates a more secure and resilient livelihood base. Integrated agriculture can also facilitate the growth of cottage industries and rural entrepreneurship, leading to increased income and improved living standards in rural communities.

4.4.6.11 Socioeconomic benefits

Integrated agriculture can improve the economic viability of farming operations. By diversifying income sources and value-added activities, farmers can generate additional revenue streams and reduce their dependence on a single commodity. This enhances the financial stability and sustainability of farming enterprises. Integrated agriculture can have positive socioeconomic impacts by generating additional income streams and creating employment opportunities. Integrated systems can support value-added activities such as processing and marketing, which can increase the profitability of farming enterprises and contribute to rural development. Moreover, the diversification of income sources enhances the economic resilience of farming communities.

4.4.6.12 Environmental stewardship

Integrated agriculture promotes environmentally friendly practices by minimizing the use of agrochemicals, conserving water resources, preserving biodiversity, and reducing greenhouse gas emissions. It aligns with the principles of sustainable agriculture and contributes to the conservation of ecosystems and natural resources.

Integrated agriculture is of utmost importance in promoting sustainable development in agriculture. Its benefits include resource efficiency, biodiversity conservation, improved soil health, climate change mitigation and adaptation, water conservation, enhanced food security and nutrition, socioeconomic benefits, and sustainable rural livelihoods. By adopting integrated agricultural practices, societies can achieve a more sustainable and resilient agricultural system that balances environmental, social, and economic needs.

Integrated agriculture is a holistic approach that integrates different components of agricultural production to achieve sustainable and efficient farming systems. By considering the interactions between crops, livestock, fisheries, and other elements, integrated agriculture offers numerous benefits such as sustainable resource management, increased productivity, risk reduction, and environmental stewardship. It is an important approach for addressing the challenges faced by agriculture in Pakistan and globally, and holds great potential for improving agricultural systems and rural livelihoods.

4.4.7 Potential areas of innovation of integrated agriculture

Integrated agriculture continues to evolve and adapt to various global challenges and advancements. Some of the potential areas of innovation and trends that could have emerged in integrated agriculture are described below.

4.4.7.1 Climate-smart agriculture

Given the increasing impacts of climate change, there has likely been a growing emphasis on climate-smart agricultural practices within the realm of integrated agriculture. These practices aim to enhance resilience to climate variability while minimizing environmental impact.

As our world faces the escalating challenges of climate change, the integration of climate-smart agriculture practices within integrated farming systems serves as a beacon of hope. By adapting to changing climatic conditions and bolstering the resilience of crops and ecosystems, these practices exemplify the dynamic synergy between human ingenuity and the natural world. As we embark on a journey toward a more sustainable future, the cultivation of climate-smart agriculture underscores our commitment to safeguarding food security, enhancing ecosystem health, and leaving behind a legacy of adaptation and innovation.

4.4.7.2 Digital agriculture and precision farming

The integration of digital technologies, such as remote sensing, Internet of things devices, and data analytics, into agricultural systems has likely expanded. These technologies enable precise monitoring and management of crops, livestock, and resources, enhancing efficiency and sustainability.

In the digital age, the marriage of technology and agriculture has given birth to a revolution known as digital agriculture and precision farming. Through the seamless integration of sensors, data analytics, and smart devices, our approach to cultivation has transformed into a realm of precision, efficiency, and informed decision-making. As we navigate the complexities of agricultural challenges, the embrace of digital tools not only optimizes resource utilization but also propels us into an era where sustainability and productivity harmonize in ways previously unimaginable.

4.4.7.3 Agroecology and regenerative agriculture

Agroecological approaches that emphasize the integration of ecological principles into agriculture have gained traction. Regenerative agriculture practices focus on restoring soil health, enhancing biodiversity, and building resilient ecosystems.

Within the embrace of agroecology and regenerative agriculture lies a profound realization: that the harmonious interaction between nature and farming can lead to a future of abundance. This commitment to cultivating in alignment with ecological principles revitalizes soil health, nurtures biodiversity, and rejuvenates ecosystems. As we embark on this journey, we acknowledge that the soil beneath our feet is not just a medium for growth but a living testament to our role as stewards of the land, nurturing it for generations yet to come.

4.4.7.4 Circular agriculture

Circular economy concepts, where waste from one part of the system becomes a resource for another, have likely been integrated into agricultural systems. This could involve recycling organic waste into compost, generating energy from agricultural residues, or using by-products for value-added products.

Circular agriculture integrates principles of the circular economy into agricultural systems, fostering resource efficiency and sustainability. By adopting practices where waste from one aspect of farming serves as a valuable resource for another, such as recycling organic waste into compost or generating energy from agricultural residues, circular agriculture promotes a closed-loop approach to farming. This approach minimizes waste, conserves resources, and enhances the overall resilience of agricultural systems, contributing to both environmental sustainability and economic viability.

4.4.7.5 Ecosystem services and biodiversity conservation

There might be an increased emphasis on harnessing ecosystem services and conserving biodiversity within integrated agriculture. These practices help improve pollination, natural pest control, and overall ecological balance.

In the intricate tapestry of nature's interconnectedness, the integration of ecosystem services and biodiversity conservation resonates as a symphony of harmony and balance. As we recognize the irreplaceable roles played by pollinators, natural predators, and myriad organisms, we honor our commitment to a world where coexistence with nature isn't just a choice but a necessity. Through the preservation of biodiversity and the nurturing of ecosystem services, we cultivate a legacy where life thrives in harmony, and our fields yield not just crops, but a celebration of life itself.

4.4.7.6 Social and economic dimensions

Integrated agriculture is not only about ecological integration but also about socioeconomic integration. This could involve strengthening rural livelihoods, empowering local communities, and ensuring equitable access to resources and benefits.

Amidst the green fields and rolling landscapes of integrated agriculture lies a profound realization – that agriculture is not just about the harvest but about the lives it touches. Within its embrace, rural communities thrive, livelihoods are secured, and social bonds are strengthened. As we navigate the intersections of agriculture with society, we acknowledge that our commitment to sustainable practices extends beyond the land, enriching lives, weaving stories, and transforming futures.

4.4.7.7 Cross-sector collaborations

Integrated agriculture often involves collaboration across sectors, including agriculture, forestry, and fisheries. Partnerships and integrated approaches across different sectors could have evolved to address complex challenges more effectively.

In the tapestry of modern agriculture, the threads of cross-sector collaborations intricately weave together diverse disciplines into a unified fabric of progress. As we embrace the interplay between agriculture, forestry, fisheries, and more, we acknowledge that our challenges are shared, and our solutions can be amplified through collective action. In these partnerships, the power to shape a more sustainable and resilient future lies not just within individual hands, but in the harmonious symphony of collaborative minds.

4.4.7.8 Policy support and incentives

Governments and international organizations might have implemented policies and incentives to promote integrated agricultural practices, recognizing their potential for addressing multiple sustainability goals.

Within the realm of policy support and incentives, we glimpse the potential to transform integrated agriculture from aspiration to reality. As governments and institutions lend their weight to sustainable practices, we witness the dawn of a new era where economic growth aligns with ecological responsibility. These policies not only nurture fertile fields but cultivate a landscape where sustainability flourishes, communities thrive, and the promise of a greener future becomes an undeniable truth.

4.4.7.9 Value chains and market access

Integrated agriculture could have evolved to better align with market demands for sustainably produced products. This might include practices that enhance traceability, quality, and access to markets for integrated agricultural products.

In the intricate dance of supply chains and market access, integrated agriculture steps into the spotlight as a beacon of responsible production. From field to fork, these value chains not only bring nourishment but also amplify the voices of responsible growers and stewards of the land. As we embrace transparency, traceability, and ethical sourcing, we cultivate a future where every bite tells a story of sustainability, and every purchase supports a web of positive impact.

4.4.7.10 Education and knowledge sharing

Efforts to educate farmers, practitioners, and the public about integrated agriculture's benefits might have expanded, promoting its adoption and fostering a culture of sustainability.

Amidst the fields of integrated agriculture, the seeds of education and knowledge sharing germinate into a legacy of empowerment. As we equip farmers, practitioners, and communities with the tools to make informed decisions, we acknowledge that the greatest harvest isn't just that of crops but of wisdom. Through education, we cultivate not only yields but a culture of sustainability, where the knowledge sown today blooms into a resilient, vibrant future.

In the crucible of integrated and sustainable agriculture, concepts from agronomy fuse with ecological wisdom to create a harmonious dance of productivity and conservation. The nurturing of healthy soils, the art of precision farming, and the adoption of agroforestry stand as testaments to the innovative spirit of this approach. Through the integration of crop and livestock systems, this chapter illuminates the path to rejuvenating our landscapes, safeguarding biodiversity, and ensuring a legacy of nourishment for generations to come.

Integrated agriculture stands as a beacon of hope for achieving sustainable agricultural practices. Through its holistic approach, which combines ecological, economic, and social dimensions, integrated agriculture fosters resilience, resource efficiency, and environmental stewardship. By integrating diverse farming systems, optimizing resource use, and promoting ecosystem health, integrated agriculture not only enhances productivity and profitability but also safeguards the well-being of future generations and the planet. As we navigate the complexities of modern agriculture, embracing integrated approaches is essential to ensure a sustainable and prosperous future for agriculture and society as a whole.

4.4.8 The role integrated agriculture in decarbonizing agriculture

Integrated agriculture plays a crucial role in decarbonizing agriculture by implementing practices that reduce greenhouse gas emissions and enhance carbon sequestration. This approach involves combining various farming activities in a synergistic manner to optimize resource use, minimize environmental impact, and mitigate climate change.

One way integrated agriculture contributes to decarbonization is through the adoption of conservation agriculture practices such as minimal tillage, cover cropping, and crop rotation. These practices help to maintain soil organic matter, improve soil structure, and reduce the need for synthetic fertilizers, thereby lowering emissions from soil degradation and fertilizer use.

Furthermore, integrated agriculture often incorporates agroforestry systems, which involve planting trees alongside crops or livestock. Trees act as carbon sinks, sequestering atmospheric carbon dioxide through photosynthesis and storing it in biomass and soil organic matter. By integrating trees into agricultural landscapes, integrated farming systems enhance carbon storage capacity and mitigate greenhouse gas emissions.

Livestock management practices within integrated agriculture also play a role in decarbonization. For example, rotational grazing systems can improve soil health and carbon sequestration while reducing methane emissions from enteric fermentation in ruminant animals. Additionally, the use of manure as organic fertilizer contributes to soil carbon sequestration and reduces the need for synthetic fertilizers.

Overall, integrated agriculture offers a multifaceted approach to decarbonizing agriculture by promoting sustainable land management, enhancing carbon sequestration, and minimizing greenhouse gas emissions across various farming activities. By adopting integrated farming systems, farmers can contribute to mitigating climate change while maintaining agricultural productivity and resilience.

4.5 Decarbonizing agriculture

Decarbonizing agriculture involves reducing carbon emissions from agricultural activities and enhancing carbon sequestration in soils and vegetation. As agriculture is a significant contributor to greenhouse gas emissions, decarbonization strategies are essential for mitigating climate change and promoting climate-smart agriculture. Practices such as conservation agriculture, agroforestry, cover cropping, and organic farming can help reduce emissions from soil degradation, deforestation, and fertilizer use, while enhancing carbon storage in soils and biomass.

Decarbonizing agriculture is a critical component of efforts to mitigate climate change and promote climate-smart agriculture. It entails reducing carbon emissions generated by agricultural activities while simultaneously enhancing carbon sequestration in soils and vegetation. Given agriculture's substantial contribution to greenhouse gas emissions, particularly from activities such as soil degradation, deforestation, and fertilizer use, decarbonization strategies are essential for fostering sustainable agricultural practices and mitigating the impacts of climate change.

4.5.1 Reducing carbon emissions

Decarbonization involves implementing practices and technologies that minimize carbon emissions associated with agricultural activities. One approach is conservation agriculture, which emphasizes minimal soil disturbance, permanent soil cover, and diversified crop rotations. By reducing tillage, conserving soil moisture, and enhancing soil organic matter, conservation agriculture mitigates emissions from soil degradation and promotes carbon sequestration in agricultural soils.

Additionally, agroforestry practices integrate trees and shrubs into agricultural landscapes, providing multiple benefits such as carbon sequestration, biodiversity conservation, and soil erosion control. By planting trees alongside crops or livestock, agroforestry systems enhance carbon storage in biomass and soil organic matter while diversifying farm income streams and improving ecosystem resilience.

Cover cropping involves planting nonharvested crops during fallow periods or between cash crop rotations to protect and enrich the soil. Cover crops contribute to carbon sequestration by capturing atmospheric carbon dioxide through photosynthesis and incorporating it into soil organic matter. Moreover, cover crops help suppress weeds, improve soil structure, and enhance nutrient cycling, thereby reducing the need for synthetic fertilizers and mitigating emissions associated with their production and application.

Organic farming practices, which exclude synthetic inputs such as pesticides and fertilizers, promote soil health and biodiversity while reducing greenhouse gas emissions. By avoiding the use of synthetic nitrogen fertilizers, organic farming mitigates emissions of nitrous oxide, a potent greenhouse gas, while enhancing soil carbon storage and microbial activity. Additionally, organic farming systems often incorporate practices such as crop rotations, composting, and biological pest control, which further contribute to carbon sequestration and emissions reduction.

4.5.2 Enhancing carbon sequestration

Decarbonizing agriculture also involves enhancing carbon sequestration in soils and vegetation to offset emissions from agricultural activities. Improving soil health and increasing soil organic matter content are key strategies for enhancing carbon sequestration in agricultural soils. Practices such as no-till farming, cover cropping, and crop rotation promote soil carbon storage by reducing soil disturbance, increasing plant biomass inputs, and enhancing microbial activity.

Agroforestry systems play a crucial role in enhancing carbon sequestration by storing carbon in both aboveground biomass and belowground roots and soil organic matter. The presence of trees in agroforestry systems contributes to long-term carbon storage while providing additional benefits such as shade, wind protection, and habitat for wildlife.

Additionally, restoring degraded ecosystems such as grasslands, wetlands, and degraded forests can significantly enhance carbon sequestration and biodiversity conservation. Land restoration practices such as reforestation, afforestation, and revegetation promote carbon storage in biomass and soils while restoring ecosystem services and enhancing resilience to climate change.

Decarbonizing agriculture involves implementing a suite of practices and strategies to reduce carbon emissions from agricultural activities and enhance carbon sequestration in soils and vegetation. By adopting practices such as conservation agriculture, agroforestry, cover cropping, and organic farming, farmers can mitigate emissions while promoting soil health, biodiversity, and resilience to climate change. Additionally, enhancing carbon sequestration through improved soil management, agroforestry, and ecosystem restoration can contribute to broader climate mitigation efforts and foster sustainable agricultural systems.

4.5.3 Promoting sustainable land management

Sustainable land management practices play a crucial role in decarbonizing agriculture and promoting climate resilience. Practices such as no-till farming, crop rotation, and agroecological approaches improve soil health, enhance water retention, and reduce erosion, thereby increasing carbon sequestration and mitigating climate change impacts. By investing in sustainable land management practices, farmers can enhance ecosystem services, increase agricultural productivity, and adapt to changing climatic conditions.

Sustainable land management practices are fundamental to the decarbonization of agriculture and the enhancement of climate resilience in agricultural systems. These practices aim to improve soil health, enhance water retention, reduce erosion, and ultimately increase carbon sequestration in soils, thereby mitigating the impacts of climate change on agricultural productivity and ecosystem stability.

4.5.3.1 No-till farming

No-till farming is a sustainable land management practice that involves planting crops without disturbing the soil through tillage. By leaving crop residues on the soil surface and minimizing soil disturbance, no-till farming helps to preserve soil structure, reduce erosion, and enhance soil organic matter content. This promotes carbon sequestration in soils while improving soil water retention and nutrient cycling, thereby increasing agricultural productivity and resilience to climate variability.

4.5.3.2 Crop rotation

Crop rotation is another essential practice for sustainable land management, involving the sequential planting of different crops on the same land over time. Crop rotation helps to break pest and disease cycles, improve soil fertility, and reduce reliance on chemical inputs. By diversifying crop species and root structures, crop rotation enhances soil health and structure, increases carbon inputs into the soil, and promotes microbial activity, thus enhancing carbon sequestration and mitigating climate change impacts.

4.5.3.3 Agroecological approaches

Agroecological approaches integrate ecological principles into agricultural systems to enhance sustainability, resilience, and productivity. These approaches focus on mimicking natural ecosystems, enhancing biodiversity, and optimizing resource use. Practices such as intercropping, polyculture, and agroforestry promote species diversity, improve soil health, and increase carbon sequestration in soils and vegetation. By fostering ecological balance and reducing reliance on external inputs, agroecological approaches contribute to sustainable land management and climate resilience in agricultural systems.

4.5.3.4 Soil conservation measures

Implementing soil conservation measures such as contour plowing, terracing, and windbreaks helps to prevent soil erosion and degradation, thereby preserving soil fertility and enhancing carbon sequestration. These measures reduce soil loss, improve water infiltration, and protect against extreme weather events, mitigating the impacts of climate change on soil productivity and agricultural sustainability.

4.5.3.5 Cover cropping

Cover cropping involves planting nonharvested crops during fallow periods or between cash crop rotations to protect and enrich the soil. Cover crops capture atmospheric carbon dioxide through photosynthesis and incorporate it into soil organic matter, thereby increasing carbon sequestration in agricultural soils. Additionally, cover crops improve soil structure, suppress weeds, and enhance nutrient cycling, thus promoting soil health and resilience to climate variability.

By investing in sustainable land management practices such as no-till farming, crop rotation, agroecological approaches, soil conservation measures, and cover cropping, farmers can enhance ecosystem services, increase agricultural productivity, and adapt to changing climatic conditions. These practices not only promote carbon sequestration and mitigate climate change impacts but also contribute to broader goals of sustainable agriculture, environmental conservation, and food security.

Within the fertile landscape of integrated and sustainable agriculture, **Chapter 5** embarks on a journey deeper into the intricate tapestry – the realm of integrated farming systems. This chapter encapsulates the harmonious symphony of crops, livestock, and the environment, intertwining their roles into a seamless whole. The stage is set for a profound exploration of the interplay between various components, painting a vivid picture of resilience and resourcefulness.

References

Altieri, M. A. (1999). The ecological role of biodiversity in agroecosystems. Agriculture, Ecosystems & Environment, 74(1–3), 19–31.

Amanullah,. (2024). Harvesting Success: A Comprehensive Guide to Agronomy Courses. Punjab Books, Lahore, Pakistan.

Amanullah., Ondrasek, G., Al –Tawaha, A. R. M. S., eds. (2023). Integrated nutrients management: An approach for sustainable crop production and food security in changing climates. Lausanne: Frontiers Media SA. doi: 10.3389/978-2-8325-3169-3

Amanullah, Ullah, H., Soliman Elshikh, M., Alwahibi, M. S., Alkahtani, J., Muhammad, A., & Khalid, S. (2020). Nitrogen contents in soil, grains, and straw of hybrid rice differ when applied with different organic nitrogen sources. Agriculture, 10, 386.

Amanullah, & Khalid, S. (2020). Agronomy – Climate Change & Food Security. IntechOpen, London, United Kingdom.

Amanullah, Inamullah, Alkahtani, J., Elshikh, M. S., Alwahibi, M. S., Muhammad, A., Ahmad, M., & Khalid, S. (2020). Phosphorus and zinc fertilization influence crop growth rates and total biomass of coarse vs. fine types rice cultivars. Agronomy, 10, 1356.

Amanullah, Inamullah, Alwahibi, M. S., Elshikh, M. S., Alkahtani, J., Muhammad, A., Khalid, S., Ahmad, M., Khan, N., Ullah, S., & Ali, I. (2020). Phosphorus and zinc fertilization improve zinc biofortification in grains and straw of coarse vs. fine rice genotypes. Agronomy, 10, 1155.

Amanullah, Hidayatullah, A. J., Shah, Z., Khan, M. J., Parmar, B., & Fahad, S. (2019). Organic carbon sources and nitrogen management improve biomass of hybrid rice (Oryza sativa L.) under nitrogen deficient condition. In: M. Hasanuzzaman, M. Fujita, K. Nahar & J. K. Biswas (Eds.) Advances in Rice Research for Abiotic Stress Tolerance (pp. 447–468). Woodhead Publishing, Chennai.

Amanullah, & Fahad, S. (2018). Nitrogen in Agriculture – Updates. InTechOpen, London, UK.

Amanullah and Fahad. (2018). Corn – Production and Human Health in Changing Climate InTech, London, UK.

Bàrberi, P., & Lo Cascio, B. (2011). Long-term tillage and crop rotation effects on weed seedbank size and composition. Weed Research, 51(5), 573–581.

Gurr, G. M., Wratten, S. D., & Landis, D. A. (2017). Habitat management to suppress pest populations: Progress and prospects. Annual Review of Entomology, 62, 91–109.

Kromp, B. (1999). Carabid beetles in sustainable agriculture: A review on pest control efficacy, cultivation impacts and enhancement. Agriculture, Ecosystems & Environment, 74(1–3), 187–228.

Leakey, R. R. (2019). Integrated crop and resource management: What can it do for sustainable intensification of tropical agriculture?. Global Food Security, 20, 98–106.

Mangan, F. X., & Adler, L. S. (2017). Management-intensive grazing increases native forb richness and diversity and decreases weed abundance. Agriculture, Ecosystems & Environment, 240, 1–7.
Norsworthy, J. K., Oliveira, M. J., Jha, P., Malik, M. S., & Buckelew, J. K. (2017). Integrating herbicide programs with harvest weed seed control and impact on Amaranthus palmeri seedbank depletion over six years of sorghum cropping. Pest Management Science, 73(10), 2160–2166.
Oerke, E. C. (2006). Crop losses to pests. The Journal of Agricultural Science, 144(1), 31–43.
Pickett, J. A., & Khan, Z. R. (2016). Plant volatile-mediated signalling and its application in agriculture: Successes and challenges. New Phytologist, 212(1), 856–870.
Power, A. G. (2010). Ecosystem services and agriculture: Tradeoffs and synergies. Philosophical Transactions of the Royal Society B: Biological Sciences, 365(1554), 2959–2971.
Pretty, J. (2008). Agricultural sustainability: Concepts, principles and evidence. Philosophical Transactions of the Royal Society B: Biological Sciences, 363(1491), 447–465.
Pretty, J. (2018). Agroecology: What it is and what it has to offer. In: Agroecology (pp. 1–28). Springer, Heidelberg, Germany.
Reynolds, H. L., Packer, A., Bever, J. D., & Clay, K. (2003). Grassroots ecology: Plant-microbe-soil interactions as drivers of plant community structure and dynamics. Ecology, 84(9), 2281–2291.
Soto, G., & Caswell, M. (2019). Is integrated pest management sustainable? An economic analysis. Journal of Environmental Management, 233, 11–21.
Tilman, D., Cassman, K. G., Matson, P. A., Naylor, R., & Polasky, S. (2002). Agricultural sustainability and intensive production practices. Nature, 418(6898), 671–677.
Tuck, S. L., Winqvist, C., Mota, F., Ahnström, J., Turnbull, L. A., & Bengtsson, J. (2014). Land-use intensity and the effects of organic farming on biodiversity: A hierarchical meta-analysis. Journal of Applied Ecology, 51(3), 746–755.
Uphoff, N. (2002). Agroecological Innovations: Increasing Food Production with Participatory Development Earthscan, Earthscan, London, UK.
Vandermeer, J. H., & Perfecto, I. (2007). The agricultural matrix and a future paradigm for conservation. Conservation Biology, 21(6), 1389–1391.
Wezel, A., Bellon, S., Doré, T., & Francis, C. (2009). Agroecology as a science, a movement and a practice. A review. Agronomy for Sustainable Development, 29(4), 503–515.
Zou, Y., & Qiang, S. (2005). The balance of nutrient and energy of the integrated agricultural system. Chinese Journal of Ecology, 24(1), 81–86.

Chapter 5
Integrated farming systems: a global approach

Introduction

In the face of mounting challenges such as climate change, population growth, and environmental degradation, the need for sustainable and resilient agricultural systems has never been more urgent. Integrated farming system (IFS) emerges as a holistic and forward-thinking approach that transcends traditional agricultural practices, offering a multifaceted solution to address the complex and interconnected issues facing global food production.

IFS represents a paradigm shift in agricultural thinking, emphasizing the integration of diverse components within farming systems to optimize resource use, enhance productivity, and promote environmental sustainability. Unlike conventional monoculture farming, which often relies on external inputs and leads to soil degradation and biodiversity loss, IFS adopts a more holistic and diversified approach that harnesses synergies between different components of the farming system.

At its core, IFS seeks to mimic the complexity and resilience of natural ecosystems, where diverse species interact and support each other in a symbiotic relationship. By integrating crops, livestock, agroforestry, aquaculture, and other agricultural activities within a single farming operation, IFS promotes resource efficiency, ecosystem services, and resilience to environmental stresses.

The global adoption of IFS represents a paradigm shift in agricultural thinking, as farmers and policymakers recognize the limitations of conventional farming practices and the urgent need for more sustainable alternatives. In regions around the world, from smallholder farms in developing countries to large-scale commercial operations in industrialized nations, interest in IFS is growing as farmers seek to improve productivity, reduce environmental impacts, and adapt to changing climatic conditions.

In this context, this chapter explores the concept of IFS as a global approach to agriculture, examining its principles, benefits, and challenges across diverse agroecological contexts. Drawing on case studies and examples from around the world, we highlight innovative practices and success stories that demonstrate the transformative potential of IFS in promoting sustainable and resilient food production systems.

Through a comprehensive exploration of IFS, this chapter aims to inspire and empower farmers, policymakers, and stakeholders to embrace a more holistic and integrated approach to agriculture that prioritizes sustainability, resilience, and food security for present and future generations. As we confront the unprecedented challenges of the twenty-first century, IFS emerges as a beacon of hope and innovation, offering a pathway toward a more sustainable and equitable food system for all.

https://doi.org/10.1515/9783111448077-005

5.1 Farming system

A farming system encompasses the entire set of activities and practices carried out on a farm to produce various agricultural products. This includes crops, livestock, and other components such as agroforestry, fisheries, and nonagricultural activities. A farming system considers not only the different types of crops and livestock but also the interactions and synergies between them. Farming systems take into account factors like available resources, labor, technology, and market opportunities. It is a broader concept that goes beyond just crop cultivation and encompasses various elements of agricultural production.

In addition to crop cultivation and livestock rearing, a farming system also incorporates elements such as agroforestry, which involves integrating trees and shrubs into farming landscapes. Agroforestry practices not only provide additional sources of income through timber, fruits, and nuts but also contribute to soil health, biodiversity, and carbon sequestration. This multilayered approach not only enhances the resilience of the farm ecosystem but also offers multiple benefits to farmers, communities, and the environment.

Moreover, a farming system may include fisheries as part of its diversification strategy. Integrated fish farming in ponds or tanks can complement traditional crop-livestock systems by providing an additional source of protein and income. Fish waste can also serve as a nutrient-rich fertilizer for crops, creating a symbiotic relationship between aquaculture and agriculture. By incorporating fisheries into the farming system, farmers can maximize the use of available resources and optimize the productivity of their land.

In essence, a farming system is a holistic approach that integrates various agricultural components to create a more resilient and sustainable system. By diversifying beyond traditional crop and livestock production, farmers can enhance productivity, income, and environmental stewardship. This comprehensive view of farming acknowledges the interconnectedness of different activities on the farm and seeks to optimize their interactions for the benefit of both farmers and the ecosystem.

5.2 Cropping system

A cropping system refers to the way crops are planned, managed, and rotated on a piece of land over time. It involves the selection of crops, their planting, maintenance, and harvesting, as well as the sequences and timing of these activities. Cropping systems can be designed based on factors such as climate, soil type, water availability, and market demands. Examples of cropping systems include monoculture (cultivation of a single crop), intercropping (growing two or more crops together in the same field), and crop rotation (systematic alternation of crops over time).

In addition to monoculture, intercropping, and crop rotation, there are other innovative cropping systems that farmers are adopting to improve sustainability and productivity. One such system is polyculture, which involves cultivating multiple crops simultaneously in the same field. Unlike intercropping, where crops are grown in distinct rows or patterns, polyculture integrates crops more closely, mimicking natural ecosystems. This approach enhances biodiversity, improves pest and disease resistance, and optimizes resource use efficiency.

Another emerging cropping system is precision agriculture, which utilizes advanced technologies such as GPS, sensors, and drones to optimize crop management. Precision agriculture enables farmers to tailor their practices to specific areas of the field based on soil conditions, moisture levels, and crop health. By applying inputs like water, fertilizers, and pesticides more precisely, farmers can reduce waste, increase yields, and minimize environmental impacts.

Furthermore, agroecological cropping systems emphasize the integration of ecological principles into agricultural practices. These systems focus on enhancing ecosystem services, such as natural pest control, soil fertility, and pollination, through diverse cropping patterns and ecological management practices. Agroecological approaches promote resilience to climate change, reduce the need for external inputs, and improve overall sustainability.

Additionally, conservation agriculture is gaining popularity as a cropping system that prioritizes soil health and biodiversity. This system involves minimal soil disturbance through practices like no-till or reduced tillage, maintaining permanent soil cover with crop residues or cover crops, and diverse crop rotations. Conservation agriculture helps to prevent soil erosion, retain moisture, and improve soil structure, thereby enhancing long-term productivity and sustainability.

In conclusion, the diversity of cropping systems reflects the dynamic nature of modern agriculture, where farmers continually innovate to address challenges and maximize efficiency. These systems highlight the importance of considering ecological, technological, and socioeconomic factors in designing sustainable agricultural practices. By adopting appropriate cropping systems, farmers can optimize resource use, improve resilience to climate change, and contribute to the long-term sustainability of food production.

5.3 Integrated farming system (IFS)

An IFS is a more advanced and holistic approach that combines various agricultural activities in a coordinated and synergistic manner to optimize resource use, increase productivity, and achieve sustainability. IFS integrates multiple components such as crop cultivation, livestock farming, aquaculture, and agroforestry. The goal of IFS is to create a balanced and interconnected system where the outputs from one component become inputs for another, reducing waste and promoting efficiency. It aims to maximize farm productivity while minimizing negative environmental impacts through recycling of nutrients, effective waste management, and reduced dependence on external inputs like chemical fertilizers and pesticides.

While cropping systems focus on the arrangement and rotation of crops, farming systems encompass the entirety of agricultural activities on a farm, and IFS takes a comprehensive and sustainable approach by integrating different agricultural components to create a harmonious and efficient farming system.

An **IFS** is a holistic and sustainable approach to agriculture that involves combining different agricultural activities and practices on a single farm to maximize productivity, optimize resource utilization, and promote environmental sustainability. The aim of IFS is to achieve higher yields, better economic returns, and improved livelihoods for farmers while also minimizing negative impacts on the environment.

The concept of IFS represents a transformative approach to agriculture that transcends traditional paradigms by recognizing the intricate interplay between various components of farming. IFS embodies the principles of sustainability, efficiency, and holistic management, acknowledging that the success of any agricultural system hinges on the harmonious coexistence of crops, livestock, natural resources, and socioeconomic factors. As global challenges such as climate change, resource depletion, and population growth continue to impact food production and rural livelihoods, the adoption of IFS becomes increasingly imperative.

IFS provides a roadmap for creating resilient and productive farming systems that are not only economically viable but also environmentally sound. By leveraging synergies between crops, livestock, aquaculture, agroforestry, and more, IFS offers a way to optimize resource utilization, enhance soil fertility, and mitigate the risks associated with monocropping and excessive external inputs. Moreover, IFS aligns with the principles of agroecology and sustainable agriculture, emphasizing the importance of diversity, natural processes, and community engagement.

The success of IFS relies on a collaborative approach involving farmers, researchers, policymakers, and local communities. As demonstrated by the diverse models and initiatives around the world, IFS is a flexible concept that can be adapted to suit the unique needs of different regions and ecosystems. By fostering innovation, knowledge sharing, and the integration of traditional wisdom with modern technology, IFS holds the potential to pave the way toward a more food-secure, resilient, and environmentally conscious future.

In essence, IFS symbolizes a shift in the agricultural paradigm, from linear and segmented practices to circular and interconnected systems. As we navigate the complex challenges of the twenty-first century, embracing the ethos of IFS offers a promising avenue to create thriving agricultural landscapes that sustain not only the current generation but also future generations, ensuring that food security, rural prosperity, and ecological harmony remain steadfast ideals guiding our collective agricultural journey.

5.4 Key components of an integrated farming system

The key components of an IFS encompass a diverse range of agricultural practices designed to optimize resource utilization, enhance productivity, and promote sustainability. These components typically include crop cultivation, livestock management, agroforestry, aquaculture, and beekeeping (Figure 5.1). By integrating multiple enterprises within a single farming system, IFS aims to create synergies between different components, improve resource efficiency, and increase resilience to environmental challenges. This holistic approach emphasizes the interconnectedness of various agricultural activities and highlights the importance of diversification for achieving long-term viability and success in farming operations.

Figure 5.1: Components of integrated farming system (IFS).

5.4.1 Crop cultivation

IFS involves growing a combination of crops that complement each other in terms of nutrient requirements, growth habits, and market demands. This can include cereals, pulses, fruits, vegetables, and other crops.

5.4.2 Livestock farming

Integrating livestock into the farming system can provide multiple benefits. Livestock can provide a source of organic manure for crop fields, utilize crop residues as feed, and diversify income streams through the sale of meat, milk, eggs, and other products.

5.4.3 Aquaculture and fisheries

Incorporating fish farming (aquaculture) and fish harvesting from ponds or water bodies on the farm can add another dimension to the integrated system. Fish can contribute to protein intake for the household and generate additional income.

5.4.4 Agroforestry

Planting trees and shrubs on the farm provides multiple benefits, including soil conservation, improved microclimate, fruit and fodder production, and potential timber and fuelwood resources.

5.4.5 Crop residue management

Crop residues, such as straw and stalks left after harvest, can be used as fodder for livestock or can be incorporated back into the soil as organic matter to improve soil fertility and structure.

5.4.6 Biodynamic nutrient cycling

Nutrients from different sources are cycled within the farm. For example, livestock waste is used to fertilize crops, and crop residues are used as feed for animals. This minimizes the need for external inputs like chemical fertilizers.

5.4.7 Resource recycling

The system emphasizes recycling resources to reduce waste and improve efficiency. Animal waste can be converted into compost or biogas, while crop residues can be used as feed or as a source of organic matter for soil.

5.4.8 Integrated pest management

IFS promotes the use of natural methods to control pests and diseases, reducing reliance on chemical pesticides. For instance, introducing beneficial insects, crop rotation, and using resistant crop varieties.

5.4.9 Diversification of income

Integrating multiple activities on the farm helps diversify income sources, reducing the risks associated with relying solely on one crop or livestock species.

5.4.10 Sustainability

IFS aims to maintain the ecological balance of the farm, promote soil health, conserve water resources, and preserve biodiversity.

5.4.11 Technology adoption

While IFS is rooted in traditional farming practices, modern technologies and innovations can be integrated, such as precision agriculture techniques, improved crop varieties, and efficient livestock management practices.

By combining these various components in a synergistic manner, IFSs aim to create a self-sustaining, resilient, and environmentally friendly agricultural system that benefits both farmers and the ecosystem.

5.5 Some examples of integrated farming systems

IFSs involve the combination of various agricultural activities to maximize resource use, increase productivity, and promote sustainability. Some examples of IFSs incorporating crops, livestock, fisheries, honeybees, trees, ducks, poultry, and other components are discussed further.

5.5.1 Agroforestry systems

5.5.1.1 Alley cropping

This means growing crops between rows of trees or shrubs; for example, planting of rows of fruit trees with crops like maize or beans grown in the alleyways.

5.5.1.2 Silvopasture

Integrating trees with pastureland for livestock grazing is called silvopasture. Livestock graze on pasture while trees provide shade, fodder, and timber.

5.5.1.3 Taungya system

This is a system of cultivating crops alongside young trees in forest plantations. Farmers grow food crops between rows of newly planted trees, providing additional income while the trees mature.

5.5.2 Aquaponics

Aquaponics is the combination of fish farming with hydroponic plant cultivation. Fish waste provides nutrients for plants, while plants filter and purify water for fish. Honeybees can be introduced to pollinate flowering plants in the system. Aquaponics is a sustainable farming system that combines aquaculture (fish farming) with hydroponics (soilless plant cultivation). In this integrated system, fish and plants work together in a symbiotic relationship.

Here is an example: Tilapia fish are raised in tanks, and their waste, which is rich in nutrients like nitrogen and phosphorus, is pumped into hydroponic beds. In these beds, vegetables such as lettuce, herbs, or tomatoes are grown without soil, using the nutrient-rich water from the fish tanks.

As the plants grow, they take up these nutrients, effectively purifying the water for the fish. The clean water is then recirculated back to the fish tanks, completing the cycle. This closed-loop system minimizes water usage, as water is continually recycled between the fish and plants.

Additionally, aquaponic systems can be further enhanced by introducing honeybees. Bees play a crucial role in pollinating flowering plants in the system, increasing the yield and quality of fruits and vegetables. Their presence enhances the overall productivity and biodiversity of the aquaponic system.

Overall, aquaponics demonstrates an innovative and sustainable approach to food production. It maximizes the efficient use of resources, reduces waste, and produces both fish and vegetables in a controlled environment. This system is suitable for urban settings, small-scale farms, and areas with limited access to arable land, offering a solution that integrates aquaculture and hydroponics into a harmonious and productive ecosystem.

5.5.3 Integrated crop-livestock systems

5.5.3.1 Crop rotation with livestock integration

Rotating crops such as corn and soybeans with pastures grazed by cattle or sheep helps to improve soil health, reduce pest and disease. Livestock manure fertilizes the soil for subsequent crop rotations, enriching it with nutrients and improving overall soil fertility and structure.

5.5.3.2 Dairy farming with forage production

Dairy farms grow forage crops like alfalfa or clover to feed their cattle while utilizing cattle manure to fertilize the forage crops.

5.5.4 Integrated rice-duck farming

Ducks are allowed to forage in rice paddies during the fallow period or after rice harvest. Ducks control pests like weeds, insects, and snails while also providing manure to fertilize the rice fields.

Integrated rice-duck farming is a traditional farming practice that combines the cultivation of rice with the rearing of ducks in the same field. This integrated system offers multiple benefits for both rice production and duck rearing.

Here is an example: After the rice harvest or during the fallow period, ducks are introduced into the rice paddies. The ducks are allowed to forage freely in the paddies, where they feed on weeds, insects, snails, and other pests that may harm the rice crops. This natural pest control method helps to reduce the need for chemical pesticides, promoting a more environmentally friendly and sustainable approach to pest management.

In return, the ducks provide additional benefits to the rice fields. Their droppings act as a natural fertilizer, enriching the soil with nutrients such as nitrogen and phosphorus. This natural fertilization improves soil fertility and enhances the growth of the next rice crop, leading to higher yields.

Moreover, the presence of ducks in the rice paddies helps to improve soil structure. As they move around and forage, the ducks aerate the soil and reduce soil compaction, which can benefit the root development of rice plants.

This integrated system not only increases rice productivity but also provides an additional source of income through duck meat and eggs. Farmers benefit from diversifying their income streams while simultaneously reducing input costs on pesticides and fertilizers.

Overall, integrated rice-duck farming exemplifies a sustainable and eco-friendly farming approach. It harnesses the natural behaviors of ducks to control pests and fertilize fields, promoting a balanced ecosystem within the rice paddies. This traditional practice is not only environmentally beneficial but also economically viable, offering a model of integrated farming that harmonizes crop cultivation with animal husbandry.

5.5.5 Integrated poultry systems

5.5.5.1 Free-range poultry with orchard

Poultry roam freely in orchards, feeding on insects and fallen fruits while also fertilizing the soil with their manure.

5.5.5.2 Poultry with vermicomposting

Poultry housing is combined with vermicomposting beds where earthworms break down poultry manure into nutrient-rich compost for crops.

5.5.6 Multistory farming systems

Multistory agroforestry means growing different crops at multiple canopy layers, including tall trees, medium-height shrubs, and ground-level crops. For example, coffee or cocoa trees grown beneath shade trees alongside crops like bananas or vegetables.

5.5.7 Beekeeping in agroecosystems

Beehives are integrated into agricultural landscapes to pollinate crops and produce honey. In addition to pollinating crops, honeybees also benefit from forage crops and flowering plants grown for their nectar and pollen.

These examples demonstrate the diverse ways in which different components of IFS can be combined to optimize resource use, enhance productivity, and promote ecological sustainability.

5.6 IFS in drylands and irrigated lands

IFS can be adapted to both dryland and irrigated lands, but the specific components and practices may vary based on the environmental conditions, water availability, and the needs of the local communities. Some examples of IFS for both dryland and irrigated areas are listed further.

5.6.1 Integrated farming system in dryland areas

In dryland areas where water availability is limited and rainfall is uncertain, the focus of IFS is often on maximizing the efficient use of available resources and conserving soil moisture. Some components of an IFS in dryland areas are listed below (Figure 5.2).

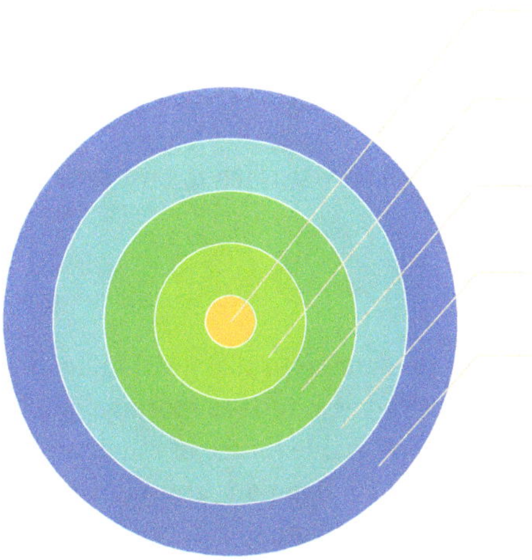

Figure 5.2: Components of integrated farming system in dryland areas.

5.6.1.1 Rainfed crops

Select drought-resistant crop varieties that can thrive with limited water availability. Crops like millets, sorghum, and legumes (pulses) are well-suited for such conditions.

5.6.1.2 Agroforestry

Plant drought-resistant trees and shrubs that can improve soil structure, provide shade, and help in moisture retention. These trees can also yield fruits, fodder, and other products.

5.6.1.3 Livestock rearing

Raise hardy livestock breeds that can graze on available forage and crop residues. Livestock can help recycle nutrients, provide manure, and generate additional income through milk, meat, and other products.

5.6.1.4 Microirrigation

Utilize drip or sprinkler irrigation methods for targeted water application, minimizing water wastage. This can be especially effective for cultivating high-value crops or gardens.

5.6.1.5 Soil conservation techniques

Implement soil conservation practices like contour bunding, mulching, and cover cropping to reduce soil erosion and moisture loss.

5.6.1.6 Rainwater harvesting

Collect and store rainwater in tanks or ponds during rainy seasons to ensure water availability during dry periods.

5.6.1.7 Example

In a dryland IFS, a farmer might cultivate drought-tolerant crops like pearl millet, intercropped with legumes like pigeon peas. Agroforestry trees, such as moringa, might be planted for their leaves and pods, which serve as a valuable source of nutrition for both humans and livestock. Livestock, such as goats or indigenous chicken breeds, can be reared for meat and eggs. Rainwater harvesting structures can provide additional water for household use and irrigation during the dry season.

5.6.2 Integrated farming system in irrigated areas

In irrigated areas, where water is more readily available, the focus of IFS might be on diversification and intensification of agricultural activities. Some components of an IFS in irrigated areas are discussed further (Figure 5.3).

5.6.2.1 High-value crops

Grow high-value crops that require consistent water supply such as vegetables, fruits, and cash crops like cotton or tobacco.

5.6.2.2 Livestock integration

Utilize available water resources to provide better-quality forage for livestock. Integrate dairy farming with crop production, using manure to enrich the soil and increase crop yields.

Figure 5.3: Components of an integrated farming system (IFS) in irrigated areas.

5.6.2.3 Fish farming

Introduce aquaculture practices in ponds or tanks to produce fish alongside crops. Fish can contribute to protein intake and provide an additional income source.

5.6.2.4 Agroforestry

Plant fruit trees and multipurpose trees that can provide additional income, improve microclimate, and help control soil erosion.

5.6.2.5 Intensive cropping

Utilize modern irrigation methods and efficient water management practices to intensively cultivate crops with high water requirements.

5.6.2.6 Example

In an irrigated IFS, a farmer might cultivate high-value vegetables like tomatoes and peppers, along with fruit trees like mangoes and citrus. Livestock such as dairy cows could be integrated, with their manure being used to fertilize crops. Fish ponds could provide a source of fish for consumption and sale. Agroforestry trees like guava or papaya might be planted along field borders to provide shade and additional income.

Remember that the specifics of an IFS will depend on factors like local climate, soil conditions, available resources, and the preferences and needs of the farmers in the region.

5.7 Integrated farming systems in winter and summer seasons

IFS can be adapted to different seasonal conditions, such as winter (cool temperature) and summer (high temperature). The components and strategies within an IFS can be tailored to optimize production and resource utilization based on the specific climatic conditions of each season. Here are examples of IFS for both winter and summer seasons.

5.7.1 Integrated farming system in winter season

In regions with cooler temperatures during the winter season, an IFS may focus on optimizing resource use, extending the growing season, and managing cold stress.

5.7.1.1 Greenhouse or high tunnel production

Using greenhouses or high tunnels can help create a controlled environment where crops can be grown throughout the winter. This extends the growing season and protects plants from cold temperatures and frost.

5.7.1.2 Cover crops

Planting cover crops during the winter can protect and enrich the soil. Leguminous cover crops can fix nitrogen, improving soil fertility for the next growing season.

5.7.1.3 Livestock integration

Animals can contribute to heat production in the farm system. Housing animals like poultry or small ruminants can provide warmth, and their waste can be composted to generate heat for greenhouse production.

5.7.1.4 Cold-resistant crops

Selecting cold-tolerant crops like certain varieties of brassicas (e.g., kale and cabbage) and root vegetables (e.g., carrots and turnips) can ensure harvest during the winter months.

5.7.1.5 Agroforestry

Planting windbreaks or shelterbelts using cold-resistant trees can protect crops and livestock from harsh winter winds.

5.7.2 Integrated farming system in summer season

In areas with hot summer temperatures, an IFS may focus on managing heat stress, water scarcity, and optimizing water use.

5.7.2.1 Drought-tolerant crops

Cultivating heat-tolerant and drought-resistant crops like certain varieties of maize, sorghum, and sunflowers can withstand high temperatures and limited water availability.

5.7.2.2 Mulching and soil conservation

Mulching can help retain soil moisture and regulate soil temperature. Implementing soil conservation practices like contour bunding or terracing can prevent soil erosion and conserve water.

5.7.2.3 Irrigation management

Efficient irrigation techniques such as drip or sprinkler irrigation can ensure that crops receive adequate water without excessive wastage.

5.7.2.4 Agroforestry for shade

Planting shade-providing trees or establishing agroforestry systems can help create microclimates that reduce heat stress on crops and livestock.

5.7.2.5 Livestock water management

Providing shaded areas and access to cool water sources for livestock can help mitigate heat stress in animals.

5.7.2.6 Crop rotation and timing

Planning crop rotations and planting schedules to avoid the hottest periods of the day can reduce heat stress on crops.

In both winter and summer seasons, the key is to design an IFS that takes advantage of the unique opportunities and challenges presented by the climate, and to integrate various components in a way that maximizes productivity, resource-use efficiency, and sustainability.

5.8 IFS in temperate, subtropical, and tropical areas

IFS can indeed vary based on the climatic conditions of different regions, such as temperate, subtropical, and tropical areas. The specific components and strategies within an IFS are influenced by factors like temperature, rainfall patterns, and the types of crops and livestock that thrive in each climate. Here are details and examples of how IFS can differ in these three types of regions.

5.8.1 Integrated farming system in temperate regions

Temperate regions experience distinct seasons, including cold winters and warm summers. Here is how IFS might be adapted to such conditions.

5.8.1.1 Seasonal crop rotation

Farmers may implement crop rotations that align with the changing seasons. For instance, they might grow cool-season crops like wheat and barley in the winter and transition to warm-season crops like corn and tomatoes in the summer.

5.8.1.2 Perennial crops

In temperate regions, some fruit trees (e.g., apples and cherries) and berries (e.g., strawberries and blueberries) are well-suited. Integrating these perennial crops into the system provides long-term benefits.

5.8.1.3 Livestock shelter

Providing adequate shelter for livestock during cold winters is essential. Livestock may be housed in barns or other structures to protect them from harsh weather.

5.8.1.4 Cold storage

If there is surplus production during the growing season, establishing cold storage facilities can help preserve crops for consumption during the colder months.

5.8.1.5 Cover crops

In winter, cover crops like clover and rye can protect and enrich the soil. Legumes can also fix nitrogen, enhancing soil fertility.

5.8.2 Integrated farming system in subtropical regions

Subtropical regions have relatively milder winters and warm to hot summers. Here is how IFS might be adapted.

5.8.2.1 Diversified crop cultivation

A wider range of crops can be cultivated due to the milder climate. For example, citrus fruits, avocados, and certain types of vegetables can thrive.

5.8.2.2 Rainwater harvesting

Subtropical regions often experience seasonal rainfall patterns. Rainwater harvesting techniques can help capture and store water for irrigation during drier periods.

5.8.2.3 Intercropping and mixed farming

Intercropping can be used to optimize space and resource use. Mixed farming systems that integrate crops and livestock can be more prevalent.

5.8.2.4 Agroforestry

Utilizing shade-providing trees in agroforestry systems can help regulate temperatures and reduce the heat stress on crops and livestock.

5.8.2.5 Livestock management

Ensuring access to water and shaded areas for livestock is essential during hot periods. Choosing heat-tolerant livestock breeds is important.

5.8.3 Integrated farming system in tropical regions

Tropical regions experience high temperatures year-round and may have distinct wet and dry seasons. Here is how IFS might be adapted.

5.8.3.1 Multiple crop cycles

Due to the consistent warm temperatures, multiple crop cycles can be managed in a year, allowing for continuous cultivation.

5.8.3.2 Drought-tolerant crops

Heat-tolerant and drought-resistant crops like millets, cassava, and certain types of beans are commonly cultivated.

5.8.3.3 Agroforestry for shade

Shade-providing trees in agroforestry systems can mitigate the intense tropical heat and offer shelter to crops and livestock.

5.8.3.4 Integrated aquaculture

Incorporating fish farming in ponds or water bodies can provide protein and enhance overall system resilience.

5.8.3.5 Monsoon farming

In regions with wet and dry seasons, water management is crucial. Rainwater harvesting and efficient irrigation methods are important strategies.

5.8.3.6 Pest and disease management

High humidity in tropical regions can lead to increased pest and disease pressure. Integrated pest management strategies are vital.

Remember, these examples provide a general idea of how IFS might differ in different climate zones. The specifics of each IFS will depend on local conditions, available resources, cultural practices, and economic considerations.

5.9 IFS in urban or city environments and rural areas

There are significant differences between IFS in urban or city environments compared to rural areas. These differences stem from the unique challenges and opportunities presented by urban settings, such as limited space, higher population density, and proximity to markets.

Some distinctions and examples of IFS in both urban and rural contexts are listed below.

5.9.1 Integrated farming systems in urban areas

Urban IFS must address the limitations of space, pollution, and the demand for fresh produce. They often focus on maximizing resource efficiency, utilizing vertical space, and producing food in close proximity to consumers.

5.9.1.1 Vertical farming

In cities where horizontal space is limited, vertical farming systems like hydroponics and aeroponics are used to grow crops in stacked layers. This conserves space and makes use of underutilized areas such as rooftops and indoor spaces.

5.9.1.2 Rooftop gardening

Urban agriculture can involve growing crops on rooftops, which not only optimizes space but also helps regulate building temperatures and reduce energy consumption.

5.9.1.3 Container gardening

Utilizing containers and pots, even on small balconies or patios, allows individuals to grow herbs, vegetables, and even small fruits.

5.9.1.4 Aquaponics

Integrating fish farming with hydroponics, known as aquaponics, creates a closed-loop system, where fish waste provides nutrients for plant growth.

5.9.1.5 Waste recycling

Urban IFS often incorporates composting of organic waste, turning kitchen scraps into nutrient-rich soil for plant cultivation.

5.9.1.6 Beekeeping and pollination

Urban beekeeping can provide honey and improve pollination for nearby plants, increasing overall urban biodiversity.

5.9.2 Integrated farming systems in rural areas

In rural areas, IFS tends to be more expansive, taking advantage of available land for diversified agriculture and livestock integration.

5.9.2.1 Crop-livestock integration

Combining crop cultivation with livestock farming can create a mutually beneficial relationship, where crop residues serve as livestock feed and animal waste enriches the soil.

5.9.2.2 Agroforestry

Rural IFS often incorporates agroforestry systems, planting trees that provide multiple benefits like fruit production, timber, and soil improvement.

5.9.2.3 Traditional agriculture

In some rural areas, traditional IFS practices are deeply rooted, where different crops and livestock are already integrated due to historical agricultural practices.

5.9.2.4 Soil and water management

Rural IFS may include soil conservation practices like terracing and contour bunding to prevent erosion, as well as water harvesting techniques to enhance irrigation.

5.9.2.5 Mixed cropping and crop rotation

Diversified cropping systems, such as intercropping and crop rotation, are more common in rural settings to optimize yield and resource utilization.

5.9.2.6 Livelihood diversification

Rural IFS can incorporate a variety of income-generating activities, such as dairy farming, poultry keeping, and handicraft production.

5.10 Integrated farming systems in lowland and upland areas

While both urban and rural IFS share the core principle of integrating various components for improved sustainability and efficiency, the specific strategies and techniques used are influenced by the unique characteristics of each setting. Urban IFS is often designed to fit within space and resource constraints, while rural IFS has more land to work with and may draw from local traditional practices.

There are differences between IFS in lowland areas and upland areas due to variations in climate, topography, soil types, and available resources. These differences influence the types of crops and livestock that can be raised, as well as the strategies employed for sustainable agricultural practices.

Some distinctions and examples of IFS in lowland and upland areas are discussed further.

5.10.1 Integrated farming systems in lowland areas

Lowland areas often have flatter terrain and may experience waterlogging or flooding during rainy seasons. IFS in lowlands focuses on managing water levels and nutrient-rich soil.

5.10.1.1 Paddy-rice farming

In areas with regular flooding, paddy-rice farming is common. Fish are sometimes integrated into rice paddies, benefiting from the water environment.

5.10.1.2 Aquaculture and wetland agriculture

Lowland IFS can include fish farming in ponds and wetland areas. Fish provide protein while their waste fertilizes the water and nearby crops.

5.10.1.3 Floodplain agriculture

Floodplain areas can be used for cultivating crops that tolerate waterlogging, such as water chestnuts and lotus. Livestock that graze on the floodplain vegetation can also be integrated.

5.10.1.4 Integrated livestock

Integrating ducks, poultry, or pigs into the system can help manage pests and utilize crop residues for feed.

5.10.1.5 Water management

Implementing proper water management techniques like bunds, dykes, and canal systems is crucial to prevent excessive flooding and control water flow.

5.10.2 Integrated farming systems in upland areas

Upland areas have steeper slopes and tend to experience soil erosion and water runoff. IFS in uplands aim to conserve soil, manage water, and prevent degradation.

5.10.2.1 Agroforestry

Upland IFS often involves planting trees that stabilize soil, protect against erosion, and provide multiple benefits like fruit and timber.

5.10.2.2 Terracing and slope farming

Due to steep slopes, terracing is common to prevent soil erosion. Crop cultivation on terraces helps retain moisture and nutrients.

5.10.2.3 Mixed cropping

Growing a mix of crops with varying root depths can help prevent soil erosion and optimize resource use.

5.10.2.4 Livestock integration

Grazing livestock can be managed in a way that prevents overgrazing and soil compaction. Livestock manure can improve soil fertility.

5.10.2.5 Soil conservation

Implementing conservation techniques like contour bunding, mulching, and cover cropping helps protect against soil erosion and water runoff.

5.10.2.6 Horticulture and cash crops

In uplands, horticultural crops like vegetables and cash crops such as coffee, tea, and spices can be integrated into the system.

It is important to note that while these examples provide general insights, the specifics of IFS in lowland and upland areas will depend on local conditions, traditional practices, and available resources. Sustainable agricultural practices aim to maximize productivity while minimizing negative impacts on the environment, regardless of the specific landscape.

5.11 Models of integrated farming systems

There are several models of IFS that have been developed and adopted around the world. These models provide frameworks and guidelines for implementing integrated and sustainable agricultural practices.

A few notable IFS models are described further.

5.11.1 M.S. Swaminathan Research Foundation (MSSRF) model

Developed by the M.S. Swaminathan Research Foundation in India, this model promotes the integration of various components, including crop cultivation, livestock, aquaculture, agroforestry, and renewable energy. It emphasizes ecological balance, resource conservation, and livelihood improvement for smallholder farmers.

5.11.2 Integrated farming and agroforestry model

This model combines agroforestry practices with traditional farming systems. It promotes the cultivation of trees along with crops and livestock, allowing for increased biodiversity, improved soil health, and diversified income sources.

5.11.3 Agroecological farming model

Agroecology is a holistic approach to farming that integrates ecological principles with agricultural practices. It emphasizes biodiversity, natural resource management, and the reduction of external inputs like synthetic pesticides and fertilizers.

5.11.4 Sustainable livelihoods approach

This approach focuses on improving the livelihoods of rural communities by integrating various income-generating activities, including crop production, livestock rearing, agroprocessing, and off-farm enterprises. It aims to enhance resilience and reduce poverty.

5.11.5 Permaculture design system

Permaculture is a design philosophy that seeks to create self-sustaining, productive ecosystems by mimicking natural patterns and processes. It integrates plants, animals, structures, and technology to create efficient and harmonious systems.

5.11.6 Zero-budget natural farming

Zero-budget natural farming, pioneered by Subhash Palekar in India, is a model that focuses on eliminating external inputs like synthetic fertilizers and pesticides. It utilizes natural processes to enhance soil fertility, manage pests, and promote sustainable farming practices.

5.11.7 System of rice intensification

While primarily focused on rice cultivation, system of rice intensification is an innovative system that emphasizes reduced plant density, early transplanting, and soil and water management techniques. It improves yields with fewer inputs, making it a form of IFS in rice-growing regions.

5.11.8 Conservation agriculture model

Conservation agriculture emphasizes minimal soil disturbance, permanent soil cover, and diversified crop rotations. It aims to improve soil health, conserve water, and reduce erosion.

5.11.9 Multistrata agroforestry model

This model involves growing trees of different heights alongside crops and livestock. It provides multiple benefits such as improved microclimate, soil fertility, and diversified income sources.

5.11.10 Urban agriculture and vertical farming models

These models address the challenges of limited space in urban areas by integrating various agricultural practices like rooftop gardening, hydroponics, and aquaponics.

These models demonstrate the diversity of approaches within the field of IFS, each tailored to the specific needs, resources, and conditions of different regions and communities.

5.12 Integrated farming system research (IFSR)

Integrated farming system research (IFSR) is an interdisciplinary approach to agricultural research that focuses on understanding and improving the overall farming systems rather than studying individual components in isolation. IFSR takes into account various components of agriculture, including crops, livestock, natural resources, socioeconomic factors, and local cultural practices. The goal of IFSR is to develop integrated and sustainable solutions that enhance agricultural productivity, rural livelihoods, and environmental sustainability.

In IFSR, researchers work closely with farmers and local communities to identify challenges, assess existing practices, and codesign interventions that are contextually relevant. The approach aims to create a holistic understanding of the entire farming system, considering factors like soil fertility, water availability, cropping patterns, livestock management, market access, and socioeconomic conditions.

For example, in Pakistan, farming research, including IFSR, is conducted by various agricultural research institutions, universities, and government agencies. The country's diverse agroclimatic zones and agricultural practices make research particularly important.

Some aspects of IFSR (Figure 5.4) are described further.

5.12.1 Agricultural research institutes

Organizations like the Pakistan Agricultural Research Council (PARC) and provincial agricultural research departments are responsible for conducting research on various aspects of farming, including crop breeding, pest management, and IFS.

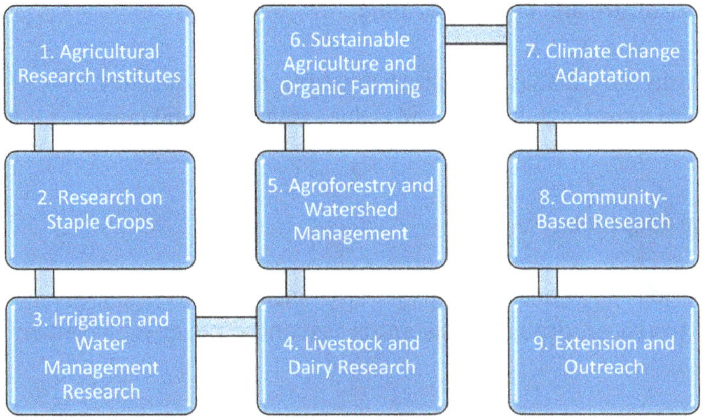

Figure 5.4: Aspects of integrated farming system research.

5.12.2 Research on staple crops

Given the importance of staple crops like wheat, rice, and cotton in Pakistan, research institutions conduct breeding programs to develop high-yielding and disease-resistant varieties. Efforts are made to improve crop management practices for enhanced productivity.

5.12.3 Irrigation and water management research

Water scarcity is a significant challenge in Pakistan. Research focuses on efficient irrigation techniques, water conservation, and sustainable water management practices.

5.12.4 Livestock and dairy research

Pakistan has a substantial livestock sector. Research in this area aims to improve livestock breeds, health management, and value addition to dairy products.

5.12.5 Agroforestry and watershed management

Research on agroforestry models and watershed management techniques is carried out to address soil erosion, degradation, and water scarcity.

5.12.6 Sustainable agriculture and organic farming

Researchers are exploring sustainable agriculture practices, organic farming, and reduced chemical inputs to improve soil health and reduce environmental impact.

5.12.7 Climate change adaptation

Given Pakistan's vulnerability to climate change, research includes developing climate-resilient agricultural practices, drought-resistant crops, and disaster management strategies.

5.12.8 Community-based research

Participatory approaches, including FSR, involve working closely with farming communities to understand local needs, challenges, and opportunities for sustainable agricultural development.

5.12.9 Extension and outreach

Research findings are often disseminated through extension services, training programs, and farmer field schools (FFS) to ensure that innovative practices are adopted on the ground. Extension and outreach activities are crucial in Pakistan to bridge the gap between research findings and their practical implementation in agriculture. Here is a discussion on the importance and ways of conducting extension and outreach in Pakistan.

5.12.9.1 Importance of extension and outreach

1. **Knowledge transfer**: Extension services play a vital role in transferring valuable research findings, technologies, and best practices from research institutions to farmers. This helps in improving agricultural productivity and sustainability.
2. **Capacity building**: Outreach programs provide farmers, agricultural workers, and rural communities with the necessary knowledge and skills to adopt modern and sustainable agricultural practices.
3. **Adoption of innovations**: By disseminating research findings, extension services facilitate the adoption of innovative technologies and practices, leading to improved crop yields, better pest management, and enhanced resilience to climate change.

4. **Problem-solving**: Extension workers act as intermediaries, assisting farmers in addressing specific challenges they face on the ground, such as pest infestations, soil degradation, or crop diseases.
5. **Empowerment**: Extension programs empower farmers by providing them with information on market trends, pricing, and opportunities for diversification, thereby enabling them to make informed decisions.

5.12.9.2 Strategies for extension and outreach in Pakistan

1. **Extension services**: Government agencies, agricultural universities, and NGOs can establish extension services that directly engage with farmers. Extension workers can visit villages, hold workshops, and provide one-on-one consultations to disseminate information.
2. **Training programs**: Conducting training programs for farmers on topics such as modern farming techniques, water management, integrated pest management, and postharvest handling can significantly enhance agricultural practices.
3. **Demonstration plots**: Setting up demonstration plots in villages allows farmers to observe firsthand the benefits of adopting new techniques or varieties. These plots serve as practical learning centers and encourage adoption.
4. **Farmer field schools**: FFS bring together groups of farmers to learn through practical, hands-on activities. FFS promote peer-to-peer learning, problem-solving, and the adoption of best practices.
5. **Use of** information and communication technologies **(ICT)**: In a tech-savvy era, ICTs such as mobile apps, SMS services, and online platforms can be utilized to disseminate agricultural information, market prices, and weather forecasts to farmers.
6. **Public-private partnerships**: Collaboration between government agencies, private companies, research institutions, and NGOs can strengthen extension services. This partnership can leverage resources, expertise, and outreach capabilities.
7. **Tailored approaches**: Considering the diversity of agriculture in Pakistan, extension programs should be tailored to specific regions, crops, and socioeconomic conditions. What works in one area may not be suitable for another.
8. **Incentives**: Providing incentives such as subsidies, seed distribution, or access to credit for adopting recommended practices can motivate farmers to participate in extension programs.
9. **Monitoring and evaluation**: Regular monitoring and evaluation of extension activities are essential to assess their effectiveness and make necessary improvements. Feedback from farmers should be actively sought and incorporated into program planning.

By prioritizing extension and outreach efforts in Pakistan's agricultural sector, the country can accelerate the adoption of sustainable and innovative practices. This, in turn, can lead to improved livelihoods for farmers, increased food security, and enhanced resilience to the challenges posed by climate change.

Overall, IFSR encompasses a wide range of topics to address the country's agricultural challenges and promote sustainable development in rural areas.

The global significance of IFSR lies in its potential to address multiple challenges facing agriculture, such as climate change adaptation, resource efficiency, and sustainable food production. By promoting integrated approaches that combine crop cultivation, livestock rearing, agroforestry, and other practices, IFSR aims to enhance agricultural resilience, productivity, and environmental sustainability on a global scale. Furthermore, IFSR facilitates knowledge sharing and capacity building among farmers, researchers, and policymakers, contributing to the adoption of innovative solutions and the advancement of agricultural systems worldwide.

This chapter delves into the orchestration of diverse elements, where poultry, fish, and crops intermingle, recycling nutrients and energies. As we witness the fusion of traditional wisdom and modern innovation, we uncover a blueprint for sustainable intensification – an inspiring path forward in the quest for food security and prosperity.

From the canvas of IFS, we embark on a voyage through the meticulous art of **Chapter 6**. As we peer into the intricate world of crop cultivation, we uncover a multidimensional tapestry of strategies. This chapter's discourse weaves the threads of pest management, nutrient optimization, and precision techniques, unveiling a holistic perspective that underscores the significance of synergy in achieving optimal yields and ecological balance.

References

Amanullah. (2024). Harvesting Success: A Comprehensive Guide to Agronomy Courses. Punjab Books, Lahore, PK.

Amanullah., Ondrasek, G., Al –Tawaha, A. R. M. S., eds. (2023). Integrated nutrients management: An approach for sustainable crop production and food security in changing climates. Lausanne: Frontiers Media SA. doi: 10.3389/978-2-8325-3169-3

Amanullah, & Khalid, S. (2020). Agronomy – Climate Change & Food Security. IntechOpen, London, United Kingdom.

Amanullah, & Fahad, S. (2018). Nitrogen in Agriculture-Updates. InTechOpen, London, UK.

Amanullah, & Fahad. (2018). Corn – Production and Human Health in Changing Climate. InTech, London, UK.

Devendra, C., & Thomas, D. (2002). Crop–animal systems in Asia: Importance of livestock and characterization of agro-ecological zones. Agricultural Systems, 71(1–2), 5–15.

FAO. (2011). Integrated farming systems for improved productivity, sustainability, and rural livelihoods. Retrieved from http://www.fao.org/3/i1897e/i1897e00.pdf

Kassam, A., Friedrich, T., & Derpsch, R. (2019). Global achievements in soil conservation with minimum tillage and no-till. In: Conservation Agriculture (pp. 265–285). Springer, Heidelberg, Germany.

R. & B. Stewart (2013). Principles of sustainable soil management in agroecosystems. CRC Press, Boca Raton, US.

Kumar, V., & Ladha, J. K. (2011). Direct seeding of rice: Recent developments and future research needs. Advances in Agronomy, 111, 297–413.

Lal, R. (2005). World crop residues production and implications of its use as a biofuel. Environment International, 31(4), 575–584.

Pretty, J. N., & Bharucha, Z. P. (2014). Integrated pest management for sustainable intensification of agriculture in Asia and Africa. Insect Science, 21(1), 1–7.

Pretty, J. N., & Hine, R. E. (2001). Reducing food poverty by increasing agricultural sustainability in developing countries. Agriculture, Ecosystems & Environment, 86(3), 257–274.

Uphoff, N. (2002). Agroecological Innovations: Increasing Food Production with Participatory Development. Earthscan. London, UK.

Chapter 6
Integrated crop management for global sustainability

Introduction

In the quest for achieving global sustainability in agriculture, integrated crop management (ICM) emerges as a pivotal approach that encompasses a spectrum of practices aimed at optimizing crop production while minimizing environmental impacts and ensuring long-term viability. As the world population continues to grow, and pressure on natural resources intensifies, the need for sustainable and resilient crop management strategies becomes increasingly imperative.

ICM represents a paradigm shift from conventional crop management approaches by advocating for holistic and integrated solutions that consider the complex interactions between crops, soil, water, pests, and other environmental factors. By adopting a multifaceted approach that combines agronomic practices, ecological principles, and technological innovations, ICM seeks to address the challenges of modern agriculture while promoting sustainability, resilience, and food security.

This chapter delves into the principles, practices, and benefits of ICMICM within the context of global sustainability. By examining case studies and success stories from diverse agroecological regions, we aim to showcase the transformative potential of ICM in enhancing agricultural productivity, conserving natural resources, and mitigating environmental impacts.

From precision agriculture and agroecological principles to integrated pest management and soil conservation techniques, ICM encompasses a wide array of practices that can be tailored to suit the specific needs and conditions of different farming systems. By emphasizing the importance of ecological balance, biodiversity conservation, and resilience to climate variability, ICM offers a roadmap toward a more sustainable and equitable agricultural future.

As we confront the challenges of feeding a growing global population while safeguarding the planet's finite resources, ICM stands as a beacon of hope and innovation, offering practical solutions that prioritize sustainability, resilience, and social equity. Through collaboration, knowledge sharing, and collective action, we can harness the transformative power of ICM to build a more sustainable and resilient food system for generations to come.

Crop management in integrated agriculture is a harmonious symphony where the intricate melodies of cultivation, sustenance, and ecological balance converge. This holistic approach goes beyond mere cultivation techniques; it seeks to optimize crop production while acknowledging the intricate interplay between crops, livestock, and the delicate environment they coexist within. ICM isn't a solitary endeavor; it's a collabora-

https://doi.org/10.1515/9783111448077-006

tive dance of science, observation, and adaptation that orchestrates bountiful harvests without compromising the health of the land and the livelihoods it supports.

6.1 Key aspects of integrated crop management

The key aspects of crop management in integrated agriculture are given further (Figure 6.1).

Figure 6.1: Key aspects of integrated crop management.

6.1.1 Crop rotation

Crop rotation is a fundamental aspect of integrated agriculture, playing a crucial role in maintaining soil health, optimizing nutrient cycling, managing pests and diseases, and promoting sustainable agricultural systems. It involves systematically changing the type of crops grown in a specific field over time. Crop rotation is a time-tested practice that dates back centuries and is a key component of integrated farming systems that prioritize diversified and holistic approaches to farming.

The key aspects and benefits of crop rotation within integrated agriculture are given below.

6.1.1.1 Soil health and fertility

Different crops have varying nutrient requirements and effects on soil health. Crop rotation helps prevent the depletion of specific nutrients from the soil, as different crops extract different nutrients. Leguminous crops, for example, fix nitrogen from the atmosphere and enrich the soil with this essential nutrient. This leads to balanced soil fertility and reduces the need for excessive chemical fertilizers.

6.1.1.2 Pest and disease management

Crop rotation disrupts the life cycle of pests and pathogens that are specific to certain crops. By rotating crops, the buildup of pests and diseases is minimized, reducing the need for chemical pesticides. Some crops can also act as natural repellents or attractants for pests, further enhancing pest management.

6.1.1.3 Weed control

Crop rotation can also help control weeds, as different crops may shade out certain weed species or disrupt their growth patterns. This reduces the reliance on herbicides and promotes more sustainable weed management practices.

6.1.1.4 Biodiversity and ecosystem services

Diversifying the crops grown through rotation enhances biodiversity on the farm. This, in turn, supports beneficial insects, soil microbes, and other organisms that contribute to ecosystem services like pollination and natural pest control.

6.1.1.5 Soil structure and erosion prevention

Different crops have varying root structures that impact soil structure. Deep-rooted crops can help break up compacted soil layers, improve water infiltration, and reduce soil erosion. Crop rotation helps maintain soil structure and prevents erosion, which is especially important in sloping or vulnerable landscapes.

6.1.1.6 Risk management

In regions prone to climate variability or extreme weather events, crop rotation spreads the risk associated with crop failure. If one crop is affected by adverse conditions, others in the rotation may still thrive, providing a safety net for farmers.

6.1.1.7 Income diversification

By growing different crops in rotation, farmers can diversify their income sources. This helps mitigate the financial risks associated with relying solely on one crop.

6.1.1.8 Long-term sustainability

Crop rotation is a ey strategy for enhancing the long-term sustainability of agricultural systems. It reduces the negative impacts of continuous monoculture and promotes more balanced and regenerative farming practices.

In integrated agriculture, crop rotation is often integrated with other practices such as cover cropping, green manure incorporation, and livestock integration. The specific crop rotation sequences will depend on factors like climate, soil type, market demand, and local conditions. Overall, crop rotation exemplifies the synergy achieved in integrated agriculture, where different components work harmoniously to achieve sustainable and resilient farming systems.

6.1.2 Multiple cropping, intercropping, and mixed cropping

Multiple cropping, intercropping, and mixed cropping are all strategies within the realm of integrated agriculture that involve cultivating multiple crops together in a single field. These approaches promote resource efficiency, biodiversity, and increased overall productivity.

6.1.2.1 Multiple cropping

Multiple cropping involves growing two or more crops in the same field within a single year. It takes advantage of the different growth periods and resource requirements of various crops to maximize land use and crop output. Multiple cropping can include sequential cropping (planting a second crop after the first is harvested) or relay cropping (overlapping the growth of two crops).

Benefits of multiple cropping

- Maximizes land-use efficiency by utilizing available growing seasons effectively.
- Diversifies income sources by allowing farmers to harvest multiple crops.
- Reduces risk by spreading the impact of adverse weather conditions across different crops.
- Optimizes resource use by making efficient use of water, nutrients, and sunlight.

6.1.2.2 Intercropping

Intercropping involves planting different crops simultaneously in the same field, typically in distinct rows or patterns. Intercropping allows plants to grow closely together, and each crop may have a different growth habit, root structure, and nutrient requirement.

Benefits of intercropping

- Optimizes resource utilization by reducing competition for nutrients and water between crops with differing requirements.
- Enhances biodiversity, which can attract beneficial insects and suppress pests.
- Improves soil health and structure by diversifying root systems and organic matter inputs.
- Increases overall productivity by utilizing niches in space and time.

6.1.2.3 Mixed cropping

Mixed cropping is a practice where two or more crops are sown together in the same field without a distinct row or pattern. In mixed cropping, the crops are not spatially separated but are mixed together throughout the field.

Benefits of mixed cropping

- Provides additional protection against crop failure as different crops may have varying resistance to pests, diseases, and adverse conditions.
- Enhances soil fertility through nutrient cycling between different crops.
- Reduces risk by spreading economic dependence on a single crop.
- Enhances landscape diversity and ecosystem services by promoting a varied plant community.

In integrated agriculture, the choice of multiple cropping, intercropping, or mixed cropping depends on factors such as climate, soil type, available resources, and the specific goals of the farming system. The main objective of these strategies is to optimize the use of resources, minimize risk, enhance biodiversity, and achieve sustainable agricultural productivity.

Successful implementation of these approaches often requires careful planning, crop selection, and management practices. The selection of compatible crops, consideration of growth habits, nutrient requirements, and potential interactions between crops are essential for maximizing the benefits of these integrated cropping systems.

6.1.3 Cover crops in integrated crop management

ICM is a comprehensive approach that involves the coordinated use of various practices to optimize crop production while minimizing negative impacts on the environment. Cover cropping is an essential component of ICM that contributes significantly to sustainable and holistic agricultural systems.

Here's a closer look at how cover crops fit within the framework of integrated crop management.

6.1.3.1 Soil health improvement

Cover crops play a crucial role in maintaining and enhancing soil health. They protect the soil from erosion caused by wind and water, especially during fallow periods when fields are left bare. The plant roots hold the soil in place, preventing valuable topsoil from being washed away. This erosion control is vital for maintaining soil structure and fertility.

6.1.3.2 Nutrient cycling and soil fertility

Certain cover crops, particularly legumes like clover and vetch, have the ability to fix atmospheric nitrogen into the soil through symbiotic relationships with nitrogen-fixing bacteria. This contributes to improved soil fertility and reduces the need for synthetic nitrogen fertilizers in subsequent crops. The cover crop biomass also adds organic matter to the soil as it decomposes, further enhancing nutrient availability.

6.1.3.3 Weed suppression

Cover crops, when managed properly, can act as natural weed suppressants. They compete with weeds for sunlight, nutrients, and water, reducing weed growth and the need for herbicides. This helps maintain a cleaner field and reduces the weed seed bank for future seasons.

6.1.3.4 Pest and disease management

Some cover crops, such as certain brassicas, can serve as "trap crops" that attract pests away from cash crops, reducing pest pressure. Additionally, cover crops can create habitats for beneficial insects that help control pest populations, contributing to natural pest management.

6.1.3.5 Water management

Cover crops improve water infiltration and retention in the soil, reducing runoff and enhancing water-use efficiency. This is particularly valuable in regions prone to drought or heavy rainfall.

6.1.3.6 Biodiversity enhancement

Cover crops contribute to biodiversity by diversifying the plant species within a cropping system. This can lead to a more balanced ecosystem, providing habitat for beneficial insects, pollinators, and other organisms that support ecosystem services.

6.1.3.7 Disease suppression

Certain cover crops release natural compounds that can suppress soil-borne diseases, reducing the need for chemical treatments and enhancing overall crop health.

In the context of ICM, cover crops are strategically selected based on the specific needs of the cropping system. Factors such as the main cash crops, local climate, soil type, and management goals determine which cover crops are most appropriate. The timing of planting, termination methods (such as mechanical or chemical), and the integration of cover crops into the crop rotation are all essential considerations.

Overall, cover cropping exemplifies the integration of ecological principles with crop management practices, contributing to improved sustainability, reduced environmental impact, and enhanced agricultural productivity within integrated crop management systems.

6.1.4 Mulching in integrated crop management

Mulching is a key component of ICM that involves covering the soil surface with a layer of organic or synthetic material. This practice offers numerous benefits for soil health, crop growth, weed control, water conservation, and overall sustainable agriculture.

Here's an in-depth look at the various aspects of mulching within ICM.

6.1.4.1 Soil health enhancement

Mulching helps improve soil structure by preventing compaction and reducing soil erosion caused by wind and water. The organic material in mulch contributes to soil organic matter, enhancing microbial activity, nutrient cycling, and overall soil fertility.

6.1.4.2 Weed suppression

Mulch acts as a physical barrier that inhibits weed germination and growth by blocking sunlight. This reduces competition for water, nutrients, and space between crops and weeds, resulting in reduced weed pressure.

6.1.4.3 Moisture conservation

One of the primary benefits of mulching is its ability to conserve soil moisture. The layer of mulch helps reduce evaporation by shading the soil surface, which is especially important in arid or water-stressed regions.

6.1.4.4 Temperature regulation

Mulching helps regulate soil temperature by providing insulation. It can keep the soil cooler in hot weather and warmer in cold weather, providing a more favorable environment for root growth and microbial activity.

6.1.4.5 Erosion prevention

Mulching prevents soil erosion by acting as a buffer against raindrops, reducing the impact of water on the soil surface. This is particularly important on slopes or in areas prone to heavy rainfall.

6.1.4.6 Disease and pest management

Certain mulch materials, such as straw or wood chips, can create a barrier between the soil and crop, reducing the spread of soil-borne diseases and deterring some pests from reaching the plants.

6.1.4.7 Improved crop quality

Mulching can help prevent soil splashing onto plants during rainfall, reducing the risk of soil-borne pathogens splashing onto leaves and fruits, which can lead to disease.

6.1.4.8 Organic matter contribution

Organic mulch materials, as they break down over time, contribute to soil organic matter content. This enhances nutrient availability, soil structure, and overall soil health.

6.1.4.9 Reduced erosion and runoff

Mulch minimizes the impact of heavy rain on the soil, reducing surface runoff and erosion. This contributes to better water quality in nearby water bodies by preventing sediment and nutrient runoff.

6.1.4.10 Customization and compatibility

Mulching materials can be selected based on crop type, local climate, and availability. Organic mulches like straw, leaves, or compost are commonly used, but synthetic mulches like plastic films may also be suitable depending on the context.

6.1.4.11 Weed management in organic systems

Mulching is a critical component of organic farming systems, where synthetic herbicides are not used. It provides effective weed control without relying on chemical interventions.

6.1.4.12 Application considerations

Proper mulch application involves evenly spreading the material over the soil surface, avoiding direct contact with plant stems, and replenishing mulch as needed over time.

By integrating mulching practices into ICM, farmers can enhance soil health, improve water-use efficiency, minimize weed pressure, and reduce the need for chemical inputs. It is important to choose appropriate mulch materials and application techniques based on specific cropping systems, local conditions, and the overall goals of sustainable and resilient agriculture.

6.1.5 Nutrient management

Nutrient management is a critical aspect of ICM that focuses on optimizing the use of fertilizers and other nutrient sources to enhance crop productivity while minimizing environmental degradation and resource wastage. This approach aims to provide crops with the necessary nutrients for healthy growth, maintain soil fertility, and mitigate the negative impacts of nutrient runoff.

The key aspects of nutrient management within integrated crop management are given as follows.

6.1.5.1 Balanced nutrition

ICM emphasizes supplying crops with a balanced combination of essential nutrients, including macronutrients (nitrogen, phosphorus, and potassium) and micronutrients (iron, zinc, copper, etc.). This balanced approach ensures optimal growth, development, and yield of crops.

6.1.5.2 Organic and inorganic sources

Nutrient management in ICM involves utilizing both organic and inorganic nutrient sources. Organic sources, such as compost, manure, and cover crops, contribute organic matter and nutrients to the soil. Inorganic sources, including mineral fertilizers, provide precise and concentrated nutrient supplementation based on crop requirements.

6.1.5.3 Precision application

Soil testing and nutrient analysis are integral to ICM. By understanding the nutrient content of the soil, farmers can make informed decisions about nutrient application.

Precision application techniques, such as variable rate fertilization, enable the targeted delivery of nutrients to specific areas of the field based on soil variability.

6.1.5.4 Timing and placement

ICM emphasizes applying nutrients at the right time and in the right place to match the crop's growth stages and nutrient demands. Proper timing minimizes nutrient losses and maximizes uptake efficiency by crops.

6.1.5.5 Avoiding nutrient imbalance

Applying excessive amounts of certain nutrients can lead to imbalances in the soil, negatively impacting crop health and environmental quality. ICM encourages the use of balanced nutrient ratios and the avoidance of nutrient excesses.

6.1.5.6 Environmental stewardship

Nutrient management in ICM aims to prevent nutrient runoff, which can lead to water pollution and eutrophication of water bodies. By applying nutrients judiciously and in accordance with crop needs, ICM helps safeguard water quality and protect ecosystems.

6.1.5.7 Integrated approach

ICM recognizes that nutrient management is closely linked to other aspects of farming, such as pest and disease management, soil health, and water management. An integrated approach considers how these factors interact and influence nutrient availability and uptake.

6.1.5.8 Farmer education

ICM emphasizes the importance of farmer education and training on proper nutrient management practices. Educated farmers are more likely to adopt sustainable nutrient management techniques and make informed decisions.

Overall, nutrient management within ICM underscores the importance of achieving a balance between productivity and environmental stewardship. By adopting precision application methods, incorporating organic matter, and considering the diverse factors that influence nutrient availability, farmers can optimize crop yields while contributing to long-term soil health, water quality, and sustainable agriculture.

6.1.5.9 Sustainable nutrient management practices

Sustainable nutrient management practices stand as guardians of soil health and custodians of agricultural productivity. With a mission to cultivate prosperity while safeguarding the environment, these practices embark on a journey to balance the scales of nutrient availability and ecological stewardship.

At their core, these practices seek to enhance the nutrient content of soils, ensuring that crops receive the sustenance they require for optimal growth and yield. This entails judiciously applying fertilizers, organic amendments, and other nutrient sources in quantities that match crop demands, thus preventing nutrient imbalances that can lead to runoff and water pollution.

The ethos of sustainability runs deep within these practices, extending their influence beyond the boundaries of the farm. By minimizing the leaching of excess nutrients into water bodies and curbing the release of greenhouse gases, they mitigate the environmental footprints of agriculture. This synergy between productivity and ecological resilience is the hallmark of sustainable nutrient management.

Through precision and innovation, these practices unlock a realm of possibilities. Technologies like soil testing guide farmers in making informed decisions about nutrient applications, preventing over-fertilization and its associated repercussions. Cover cropping and crop rotation, integral parts of sustainable nutrient management, enrich the soil with organic matter, fostering nutrient cycling and bolstering long-term fertility.

In embracing sustainable nutrient management practices, agriculture dons the mantle of a responsible steward of resources. It becomes a testament to the fact that abundant harvests and environmental preservation need not be at odds. With every application, every rotation, and every harvest, these practices carve a path toward a future where the sustenance of the land sustains us all.

Here are some key practices:

Organic matter management: Adding organic matter, such as compost, crop residues, and animal manure, improves soil structure, moisture retention, and nutrient cycling. Organic matter enhances soil fertility, promotes beneficial microbial activity, and reduces the need for synthetic fertilizers.

Soil conservation techniques: Implementing soil conservation practices, such as contour plowing, terracing, and bunding, helps prevent soil erosion and nutrient loss. These techniques stabilize the soil, maintain its structure, and protect it from the impact of heavy rainfall and wind.

Precision nutrient management: Precision nutrient management involves applying fertilizers based on the specific nutrient requirements of crops. Soil testing, plant tissue analysis, and nutrient management software help optimize fertilizer application, reducing excess use and minimizing nutrient runoff into water bodies.

Nutrient recycling: Recycling nutrients through practices like crop residue management and the use of organic fertilizers helps close nutrient cycles and reduce reliance on synthetic fertilizers. Recycling nutrients improves soil fertility, reduces nutrient losses, and contributes to sustainable nutrient management.

Integrated nutrient management: Integrated nutrient management (INM) emerges as a sophisticated symphony of nutrient harmonization, orchestrating a delicate balance between crop sustenance and soil vitality (Figure 6.2). This holistic approach transcends the confines of traditional fertilization practices, recognizing that the nutritional needs of crops are best met through a fusion of diverse nutrient sources.

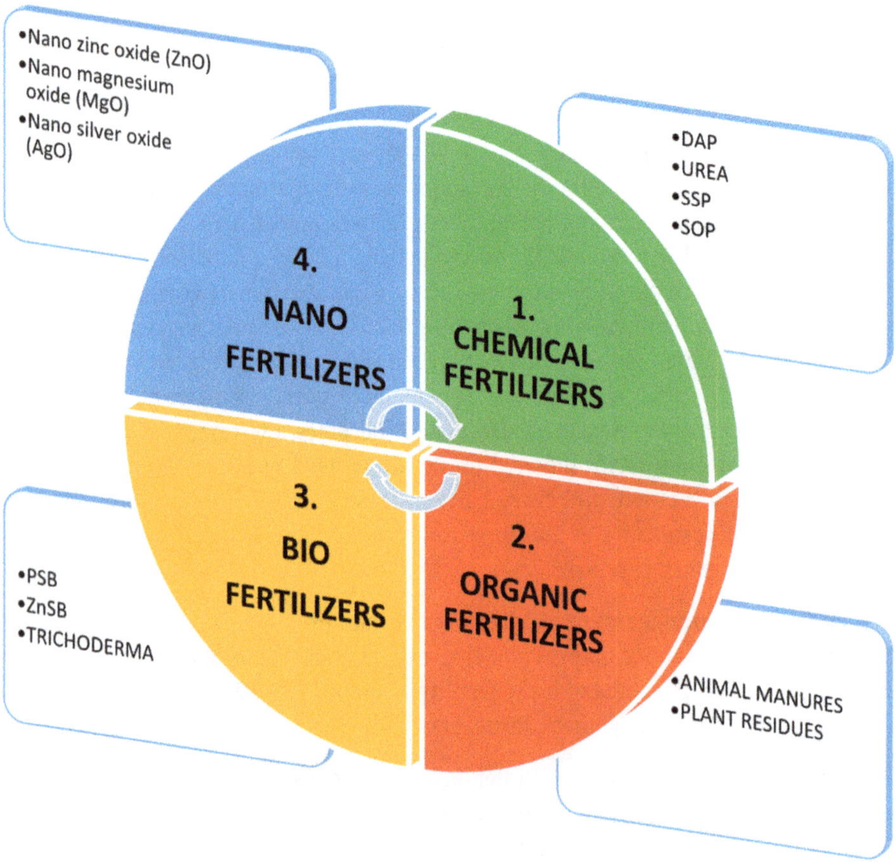

Figure 6.2: Integrated nutrient management.

At its core, INM seamlessly integrates the benefits of organic and inorganic nutrient inputs, aligning them with the specific demands of crops. Organic fertilizers, hailing from natural sources like manure, compost, and crop residues, enrich the soil with essential elements while enhancing its structure and water-holding capacity. Simultaneously, mineral fertilizers provide a targeted and efficient boost of specific nutrients, ensuring that crops receive what they require without unnecessary excess.

The artistry of INM lies in its ability to mitigate the drawbacks of each nutrient source. Organic materials, while invaluable for soil health, might release nutrients slowly. Mineral fertilizers, on the other hand, provide an immediate nutrient influx but could risk imbalances and environmental runoff. INM deftly harmonizes these contrasts, delivering a synergistic impact that not only enhances crop yields but also safeguards the environment.

Biofertilizers add another layer to this intricate composition. These living microorganisms, harnessed from nature's laboratories, contribute to nutrient mobilization, solubilization, and fixation. They transform the soil into a thriving ecosystem where nutrients are efficiently cycled, ensuring that crops have sustained access to the nourishment they need.

Beyond bolstering yields, INM nourishes the very soul of the land. It cultivates soil fertility, curtails the overuse of synthetic fertilizers, and minimizes the risk of nutrient leaching and pollution. In doing so, INM doesn't just secure the present; it invests in the future, ensuring that generations to come inherit a land teeming with life and abundance.

As a testament to the power of harmony, INM stands as a beacon of sustainable agriculture, demonstrating that by blending the best of nature's offerings with scientific precision, we can cultivate both prosperity and environmental resilience.

6.1.6 Pest and disease management

Integrated pest and disease management (IPM) practices aim to minimize reliance on chemical pesticides while effectively managing pests and diseases. IPM strategies include biological control, cultural practices, pest monitoring, and the use of resistant crop varieties. Timely interventions and integrated approaches help maintain pest and disease populations below economic thresholds.

IPM practices represent a holistic and sustainable approach to managing pests and diseases in agricultural systems. With the increasing concern over the environmental and health impacts of chemical pesticides, IPM strategies aim to minimize reliance on these inputs while effectively controlling pest and disease populations. By integrating various control methods and practices, IPM offers a comprehensive and environmentally friendly approach to pest and disease management.

6.1.6.1 Biological control

Biological control is a key component of IPM that involves using natural enemies, such as predators, parasitoids, and pathogens, to suppress pest populations. By harnessing the ecological relationships between pests and their natural enemies, biological control can effectively reduce pest populations while minimizing the need for chemical pesticides. Strategies such as releasing beneficial insects, conserving natural enemy habitats, and using microbial biopesticides enhance biological control effectiveness and promote ecological balance in agricultural ecosystems.

6.1.6.2 Cultural practices

Cultural practices play a crucial role in IPM by creating unfavorable conditions for pests and diseases while promoting crop health and resilience. Practices such as crop rotation, intercropping, and crop diversification disrupt pest cycles, reduce pest buildup, and enhance soil fertility, thereby reducing the likelihood of pest and disease outbreaks. Additionally, practices such as sanitation, weed management, and crop residue management help eliminate pest habitats and reduce disease inoculum, further contributing to effective pest and disease management.

6.1.6.3 Pest monitoring

Pest monitoring is essential for timely and targeted interventions in IPM programs. By regularly monitoring pest populations and disease incidence, farmers can detect potential outbreaks early and implement appropriate control measures before pest populations reach damaging levels. Monitoring techniques such as visual inspection, trapping, and pheromone-based monitoring systems help farmers assess pest and disease pressures accurately, enabling them to make informed management decisions and minimize crop losses.

6.1.6.4 Use of resistant crop varieties

Utilizing resistant crop varieties is a proactive approach to pest and disease management in IPM. Plant breeding efforts have led to the development of crop varieties with inherent resistance or tolerance to specific pests and diseases, reducing the need for chemical control measures. By selecting and planting resistant varieties, farmers can effectively reduce pest damage and disease incidence, enhance crop resilience, and minimize the risk of yield losses due to pest and disease pressures.

6.1.6.5 Integrated approaches

Integrated pest and disease management emphasize the synergistic integration of multiple control methods and practices to achieve effective and sustainable pest and disease control. By combining biological control, cultural practices, pest monitoring, and resistant crop varieties, farmers can adopt a proactive and holistic approach to pest and disease management that minimizes reliance on chemical pesticides while maintaining pest and disease populations below economic thresholds. Integrated approaches not only promote environmental sustainability but also contribute to long-term crop health, productivity, and profitability.

IPM practices offer a multifaceted and sustainable approach to managing pests and diseases in agricultural systems. By harnessing biological control, implementing cultural practices, conducting pest monitoring, utilizing resistant crop varieties, and integrating various control methods, farmers can effectively control pest and disease populations while minimizing environmental impacts and promoting agricultural sustainability. Through proactive and integrated approaches, IPM represents a cornerstone of modern agriculture that prioritizes ecological balance, crop health, and sustainable production systems.

6.1.6.6 Integrated pest management and disease control strategies

IPM is an approach that focuses on minimizing the use of chemical pesticides while effectively managing pests and diseases. It combines various pest control methods to maintain pest populations below economically damaging levels.

Here are some key components of IPM and disease control strategies:

Biological control: Biological control involves the use of natural enemies, such as predators, parasitoids, and beneficial microorganisms, to control pests and diseases. By promoting the presence and activity of these natural enemies, farmers can achieve pest control without relying heavily on chemical pesticides.

Crop rotation: Crop rotation is an important practice in IPM. It involves alternating different crops in a sequence to disrupt pest life cycles and reduce pest and disease pressures. Crop rotation helps break the continuous cropping cycle, mitigates the buildup of pests and diseases, and promotes a healthy agroecosystem.

Cultural practices: Cultural practices include techniques such as proper crop spacing, timely sowing, and pruning. These practices create unfavorable conditions for pests and diseases, reduce the spread of pathogens, and enhance plant vigor and resilience.

Monitoring and scouting: Regular monitoring and scouting of crops are essential for IPM. This involves inspecting plants for pest and disease presence, assessing population levels, and identifying the need for control measures. Monitoring helps in timely decision-making and targeted interventions.

Threshold-based control: IPM employs threshold-based control, where control measures are implemented only when pest populations reach a certain level that may cause economic damage. This approach ensures that pesticide applications are minimized and used judiciously, reducing environmental impacts.

Chemical control as a last resort: Chemical pesticides are used as a last resort in IPM. If pest populations exceed economic thresholds, selective and less toxic pesticides are applied, considering their efficacy, target specificity, and environmental impact. Integrated approaches minimize the risks associated with pesticide use.

6.1.7 Integrated weed management (IWM)

Integrated weed management (IWM) is a comprehensive and sustainable approach to managing weeds in agricultural systems. It involves the combination of various strategies and techniques to effectively control weed populations while minimizing the negative impacts on crops, the environment, and human health. IWM recognizes that weeds are a complex challenge that requires a multifaceted approach, rather than relying solely on herbicides.

The key components of IWM are given as folows.

6.1.7.1 Diverse weed control strategies

IWM integrates various methods of weed control to prevent the development of herbicide-resistant weeds and reduce reliance on chemical herbicides. These strategies may include cultural, mechanical, biological, and chemical methods.

6.1.7.2 Crop rotation and diversification

Planting different crops in a sequence disrupts weed life cycles and reduces the buildup of specific weed species. Certain crops may also suppress weeds through shading or allelopathic effects.

6.1.7.3 Cover crops

Planting cover crops between main crops helps smother weeds, prevent soil erosion, and improve soil structure. Cover crops compete with weeds for resources and provide a protective layer that limits weed growth.

6.1.7.4 Mulching

Applying organic or synthetic mulches on the soil surface inhibits weed germination and growth by blocking sunlight and reducing moisture availability.

6.1.7.5 Mechanical methods

Mechanical weed control techniques include plowing, tilling, mowing, and cultivating. These practices disrupt weed growth and bury weed seeds, reducing their viability.

6.1.7.6 Biological control

Introducing natural enemies, such as insects, pathogens, or animals that feed on weeds, can help manage weed populations. Biological control reduces the need for chemical interventions and promotes natural balance in ecosystems.

6.1.7.7 Herbicide use

While IWM aims to reduce reliance on herbicides, they can still be a component of the strategy. However, IWM promotes judicious herbicide use by rotating herbicide modes of action, using appropriate application methods, and targeting specific weed species.

6.1.7.8 Timing

IWM emphasizes the importance of timing weed control measures to coincide with critical stages of weed growth. Early intervention can prevent weeds from reaching maturity and producing more seeds.

6.1.7.9 Weed seed bank management

IWM aims to prevent the buildup of weed seeds in the soil by reducing seed production through effective control measures and encouraging seed decay.

6.1.7.10 Monitoring and assessment

Regular monitoring of weed populations helps determine the effectiveness of weed management strategies. Adjustments can be made based on observations and assessments.

6.1.7.11 Education and outreach

Educating farmers about the principles and benefits of IWM is crucial for its successful adoption. Training programs and information dissemination help farmers make informed decisions.

IWM recognizes that weed control is a dynamic process influenced by various factors. By adopting a holistic approach that combines multiple strategies, IWM aims to achieve sustainable weed management that preserves crop yields, reduces environmental impacts, and maintains the long-term efficacy of weed control methods.

6.1.8 Water management

Water management is a cornerstone of ICM that focuses on optimizing water use to ensure crop productivity while conserving this precious resource. In an era of growing water scarcity and changing climate patterns, efficient water management practices are vital for sustainable agriculture. Here's an in-depth look at the role of water management within ICM.

6.1.8.1 Irrigation scheduling

ICM emphasizes the importance of proper irrigation scheduling, which involves providing crops with the right amount of water at the right time. Monitoring soil moisture levels and crop water requirements helps prevent overwatering or underwatering, which can both negatively impact crop growth and yield.

6.1.8.2 Efficient irrigation methods

Utilizing efficient irrigation methods, such as drip irrigation, micro-sprinklers, and sub-surface irrigation, is a core principle of ICM. These methods deliver water directly to the root zone, minimizing evaporation, runoff, and water wastage.

6.1.8.3 Water-saving techniques

ICM encourages the adoption of water-saving techniques such as mulching, which reduces soil evaporation and maintains consistent soil moisture. Mulch also helps suppress weed growth, reducing competition for water.

6.1.8.4 Crop water requirements

Understanding the water needs of different crops at various growth stages is essential for effective water management. ICM promotes tailoring irrigation to match the specific water requirements of each crop, thus optimizing water use.

6.1.8.5 Rainwater harvesting

Incorporating rainwater harvesting systems is a sustainable practice encouraged by ICM. Collecting and storing rainwater for irrigation purposes can reduce reliance on external water sources and enhance resilience during dry periods.

6.1.8.6 Soil health and water retention

Maintaining soil health through practices like cover cropping and organic matter incorporation enhances soil structure and water retention. Healthy soils can hold more water and make it available to plants over a longer period.

6.1.8.7 Crop selection and drought resistance

ICM encourages farmers to choose crop varieties that are well-suited to the local climate and exhibit drought resistance. Proper crop selection can reduce water demand while maintaining yield potential.

6.1.8.8 Monitoring and technology

Utilizing advanced technologies, such as soil moisture sensors and weather forecasts, can assist farmers in making informed decisions about irrigation timing and amounts. Real-time data helps optimize water use.

6.1.8.9 Integrated approach

Water management is intertwined with other components of ICM, such as nutrient management and pest control. An integrated approach considers how water interacts with these aspects and seeks to balance resource use.

6.1.8.10 Farmer training and education

ICM emphasizes the importance of educating farmers about efficient water management practices. Knowledgeable farmers are more likely to adopt water-saving techniques and implement sustainable irrigation practices.

Efficient water management in ICM is not only about conserving water but also about enhancing agricultural productivity and minimizing environmental impacts. By using water judiciously, preventing water wastage, and incorporating practices that improve soil health and water retention, ICM contributes to long-term sustainability, climate resilience, and the continued ability of agriculture to meet the needs of a growing population.

6.1.9 Climate resilience

Climate resilience is a fundamental aspect of ICM that addresses the challenges posed by changing climate patterns, extreme weather events, and the need to ensure food security in the face of uncertain conditions. ICM strategies are designed to enhance the ability of agricultural systems to withstand and adapt to climate-related stresses.

Here's a detailed exploration of climate resilience within ICM.

6.1.9.1 Crop varietal selection

Selecting crop varieties that are well-adapted to local climate conditions is a key strategy in ICM. Resilient crop varieties exhibit traits such as drought tolerance, disease resistance, and the ability to withstand temperature fluctuations. These varieties are better equipped to cope with the challenges posed by changing climates.

6.1.9.2 Optimal planting dates

ICM emphasizes the importance of managing planting dates to align with optimal weather conditions. Timely planting can help crops avoid extreme heat, cold, or periods of high humidity, ensuring healthier establishment and growth.

6.1.9.3 Crop diversification

Diversifying crops within a farming system is a powerful strategy to mitigate climate risks. Growing a variety of crops with different growth habits and requirements reduces vulnerability to the impacts of specific weather events. If one crop is negatively affected, others may still thrive, helping to ensure overall productivity.

6.1.9.4 Resilient rotations

Implementing crop rotations that include both annual and perennial crops can enhance climate resilience. Perennial crops, such as trees or shrubs, provide stability to the ecosystem and help prevent soil erosion, while annual crops provide short-term yield and income.

6.1.9.5 Conservation tillage

Conservation tillage practices, such as no-till or reduced tillage, promote climate resilience by preserving soil structure, reducing erosion, and increasing organic matter content. These practices help maintain soil moisture and fertility, contributing to better crop performance during weather extremes.

6.1.9.6 Water management in climate extremes

In regions facing water scarcity or increased flooding due to climate change, ICM promotes efficient water management practices. This includes selecting drought-tolerant crops, implementing water-saving irrigation methods, and developing systems to manage excess water during heavy rainfall.

6.1.9.7 Integrated pest and disease management

Climate resilience involves reducing the vulnerability of crops to pests and diseases, which can proliferate under changing climatic conditions. ICM encourages integrated pest management practices that rely on a combination of cultural, biological, and chemical control measures.

6.1.9.8 Farmer training and knowledge transfer

Educating farmers about climate-resilient practices is a core component of ICM. Providing training on climate-smart techniques empowers farmers to adapt to changing conditions and make informed decisions.

6.1.9.9 Monitoring and early warning systems

ICM utilizes climate data and early warning systems to anticipate extreme weather events. This enables farmers to take proactive measures to protect crops, manage risks, and minimize losses.

6.1.9.10 Research and innovation

Ongoing research and innovation are essential for developing new strategies and technologies that enhance climate resilience. This includes breeding climate-resilient crop varieties and developing sustainable agricultural practices.

Incorporating climate resilience into ICM ensures that farming systems can continue to produce food in the face of climate challenges. By adapting practices, embracing innovation, and working in harmony with natural processes, ICM contributes to the long-term sustainability and adaptability of agriculture in a changing climate. By implementing integrated crop management practices, farmers can optimize crop production, conserve resources, enhance sustainability, and improve the overall resilience of their farming systems. Integrated approaches recognize the importance of crop management in the broader context of the agricultural ecosystem, leading to improved agricultural outcomes.

6.1.10 Conservation agriculture and agroforestry approaches

Conservation agriculture (CA) and agroforestry are sustainable farming approaches that aim to enhance agricultural productivity, conserve natural resources, and promote ecological balance.

Here are the key aspects of these approaches.

6.1.10.1 Conservation agriculture

CA emphasizes the principles of minimal soil disturbance, permanent soil cover, and crop diversification. It involves reduced tillage, mulching, and the use of cover crops to protect the soil, improve water retention, and enhance nutrient cycling. CA practices promote soil health, reduce erosion, and contribute to long-term sustainability.

6.1.10.2 Agroforestry

Agroforestry involves the integration of trees, crops, and livestock in a farming system. Agroforestry systems provide multiple benefits such as improved soil fertility, diversified income sources, increased biodiversity, and climate resilience. Examples include alley cropping, where crops are grown between rows of trees, and silvopastoral systems, where trees are integrated with livestock grazing.

6.1.10.3 Soil conservation and erosion control

CA and agroforestry contribute to soil conservation by minimizing soil erosion. Reduced tillage and permanent soil cover practices protect the soil from wind and water erosion, maintaining soil structure and fertility. The integration of trees in agroforestry systems also helps stabilize the soil and prevent erosion.

6.1.10.4 Biodiversity and ecological balance

CA and agroforestry systems promote biodiversity by creating habitats for beneficial insects, birds, and other organisms. The presence of trees in agroforestry systems enhances biodiversity, improves pollination, and provides ecosystem services such as carbon sequestration and water regulation.

6.1.10.5 Climate resilience

CA and agroforestry contribute to climate resilience by enhancing the adaptive capacity of farming systems. These approaches improve water retention, reduce the impact of extreme weather events, and help mitigate climate change through carbon sequestration in soils and trees.

6.1.11 Precision agriculture and use of advanced technologies in crop management

Precision agriculture involves the use of advanced technologies and data-driven approaches to optimize crop production, reduce input use, and minimize environmental impact (Figure 6.3).

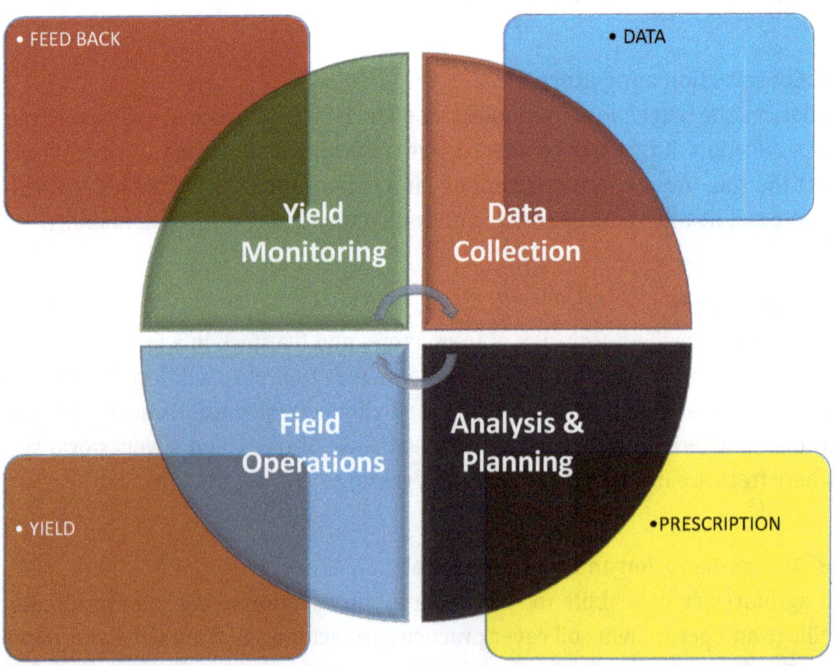

Figure 6.3: Key aspects of precision agriculture.

Here are the key aspects of precision agriculture and advanced technologies.

6.1.11.1 Remote sensing and GIS

Remote sensing technologies, such as satellite imagery and aerial drones, provide valuable information about crop health, growth patterns, and variability within fields. Geographic information systems (GIS) help in mapping and analyzing spatial data for better decision-making.

6.1.11.2 Global positioning system (GPS)

GPS technology enables precise location tracking and mapping of field boundaries, facilitating accurate data collection and spatial information management. GPS is essential for precision equipment guidance and targeted application of inputs.

6.1.11.3 Variable rate technology (VRT)

VRT allows the precise application of inputs, such as fertilizers, pesticides, and irrigation water, based on site-specific requirements. VRT adjusts input rates according to variability within fields, optimizing resource use and minimizing waste.

6.1.11.4 Sensor technology

Sensor-based technologies, such as soil moisture sensors, weather stations, and crop health sensors, provide real-time data on soil conditions, weather parameters, and plant health. Sensor data helps in decision-making related to irrigation scheduling, disease monitoring, and nutrient management.

6.1.11.5 Data analytics and decision support systems

Advanced data analytics and decision support systems process large amounts of data, including sensor data, historical records, and weather forecasts. These systems provide insights, generate recommendations, and assist farmers in making informed decisions about crop management practices.

6.1.11.6 Automated farm machinery

Automation technologies, such as robotic systems, autonomous tractors, and smart implements, enable precise and efficient operations in the field. Automated machinery performs tasks like planting, harvesting, and spraying with high accuracy, reducing labor requirements and improving efficiency.

Precision agriculture and advanced technologies enable farmers to make data-driven decisions, optimize resource utilization, improve crop health, and enhance overall farm efficiency. These technologies contribute to sustainable and environmentally friendly crop management practices.

At the crossroads of innovation and cultivation lies the landscape of integrated crop management – a sanctuary where science and stewardship intersect. The symphony of prudent pest control, judicious nutrient application, and strategic planning harmonizes to maximize yields while safeguarding ecosystems. This chapter's narrative unearths the potential for resilient agriculture, where knowledge, technology, and environmental consciousness converge to guide the journey toward sustainability.

Leaving behind the fields of crops, we enter the realm of **Chapter** 7 "integrated livestock management: sustainable practices," where the intricate dance between animals, humans, and landscapes takes center stage. As the curtain rises on integrated livestock management, the interdependence between livestock, fodder, and soil health unfolds – a dynamic display that underscores the importance of holistic approaches in enhancing livelihoods and ecological balance.

References

Altieri, M. A. (1999). The ecological role of biodiversity in agroecosystems. Agriculture, Ecosystems & Environment, 74(1–3), 19–31.

Amanullah. (2024). Harvesting Success: A Comprehensive Guide to Agronomy Courses. Punjab Books, Lahore, PK.

Amanullah., Ondrasek, G., Al –Tawaha, A. R. M. S., eds. (2023). Integrated nutrients management: An approach for sustainable crop production and food security in changing climates. Lausanne: Frontiers Media SA. doi: 10.3389/978-2-8325-3169-3

Amanullah, & Khan, U. (2023). Advancing sustainable agriculture with beneficial microbes: Enhancing crop growth and yield for food security and human health. Advances in Modern Agriculture, 4(2), 2426.

Amanullah, Khalid, S., Muhammad, A., Yar, M., et al. (2021). Integrated use of biofertlizers with organic and inorganic phosphorus sources improve dry matter partitioning and yield of hybrid maize. Communications in Soil Science and Plant Analysis, 52(21), 2732–2747.

Amanullah, Ullah, H., Soliman Elshikh, M., Alwahibi, M. S., Alkahtani, J., Muhammad, A., & Khalid, S. (2020). Nitrogen contents in soil, grains, and straw of hybrid rice differ when applied with different organic nitrogen sources. Agriculture, 10, 386.

Amanullah, Hidayatullah, Jan, A., Shah, Z., Khan, M. J., Parmar, B., & Fahad, S. (2019). Organic carbon sources and nitrogen management improve biomass of hybrid rice (*Oryza sativa* L.) under nitrogen deficient condition. In: M. Hasanuzzaman, M. Fujita, K. Nahar & J. K. Biswas (Eds.). Advances in Rice Research for Abiotic Stress Tolerance (pp. 447–468). Woodhead Publishing, Chennai.

Amanullah, & Fahad, S. (2018). Nitrogen in Agriculture-Updates. InTechOpen, London, UK.

Amanullah, & Fahad, S. (2018). Integrated nutrient management in corn production: Symbiosis for food security and grower's income in arid and semiarid climates. In: Amanullah & S. Fahad (Eds.). Corn – Production and Human Health in Changing Climate (pp. 03–12). InTech, London, UK.

Amanullah, Khan, A., Khalid, S., et al. (2019). Integrated management of phosphorus, organic sources, and beneficial microbes improve dry matter partitioning of maize. Communications in Soil Science and Plant Analysis, 50(20), 2544–2569.

Amanullah, & Khalid, S. (2016). Integrated use of phosphorus, animal manures and biofertilizers improve maize productivity under semiarid condition. In: L. L. Marcelo & S. Soloneski (Eds.). Organic Fertilizers – From Basic Concepts to Applied Outcomes (pp. 137–155). InTech, Rijeka, Croatia.

Bàrberi, P., & Lo Cascio, B. (2011). Long-term tillage and crop rotation effects on weed seedbank size and composition. Weed Research, 51(5), 573–581.

Gurr, G. M., Wratten, S. D., & Landis, D. A. (2017). Habitat management to suppress pest populations: Progress and prospects. Annual Review of Entomology, 62, 91–109.

Krasilnikov, P., Miguel A. Taboada, & Amanullah. (2021). Fertilizer Use, Soil Health and Agricultural Sustainability. Agriculture (SI). Multidisciplinary Digital Publishing Institute (MDPI), Basel, Switzerland.

Kromp, B. (1999). Carabid beetles in sustainable agriculture: A review on pest control efficacy, cultivation impacts and enhancement. Agriculture, Ecosystems & Environment, 74(1–3), 187–228.

Leakey, R. R. (2019). Integrated crop and resource management: What can it do for sustainable intensification of tropical agriculture? Global Food Security, 20, 98–106.

Mangan, F. X., & Adler, L. S. (2017). Management-intensive grazing increases native forb richness and diversity and decreases weed abundance. Agriculture, Ecosystems & Environment, 240, 1–7.

Norsworthy, J. K., Oliveira, M. J., Jha, P., Malik, M. S., & Buckelew, J. K. (2017). Integrating herbicide programs with harvest weed seed control and impact on *Amaranthus palmeri* seedbank depletion over six years of sorghum cropping. Pest Management Science, 73(10), 2160–2166.

Oerke, E. C. (2006). Crop losses to pests. The Journal of Agricultural Science, 144(1), 31–43.

Pickett, J. A., & Khan, Z. R. (2016). Plant volatile-mediated signalling and its application in agriculture: Successes and challenges. New Phytologist, 212(1), 856–870.

Power, A. G. (2010). Ecosystem services and agriculture: Tradeoffs and synergies. Philosophical Transactions of the Royal Society B: Biological Sciences, 365(1554), 2959–2971.

Pretty, J. (2008). Agricultural sustainability: Concepts, principles and evidence. Philosophical Transactions of the Royal Society B: Biological Sciences, 363(1491), 447–465.

Pretty, J. (2018). Agroecology: What it is and what it has to offer. In: Agroecology (pp. 1–28). Springer. Heidelberg, Germany.

Reynolds, H. L., Packer, A., Bever, J. D., & Clay, K. (2003). Grassroots ecology: Plant-microbe-soil interactions as drivers of plant community structure and dynamics. Ecology, 84(9), 2281–2291.

Soto, G., & Caswell, M. (2019). Is integrated pest management sustainable? An economic analysis. Journal of Environmental Management, 233, 11–21.

Tilman, D., Cassman, K. G., Matson, P. A., Naylor, R., & Polasky, S. (2002). Agricultural sustainability and intensive production practices. Nature, 418(6898), 671–677.

Tuck, S. L., Winqvist, C., Mota, F., Ahnström, J., Turnbull, L. A., & Bengtsson, J. (2014). Land-use intensity and the effects of organic farming on biodiversity: A hierarchical meta-analysis. Journal of Applied Ecology, 51(3), 746–755.

Uphoff, N. (2002). Agroecological Innovations: Increasing Food Production with Participatory Development. Earthscan. London, UK.

Vandermeer, J. H., & Perfecto, I. (2007). The agricultural matrix and a future paradigm for conservation. Conservation Biology, 21(6), 1389–1391.

Wezel, A., Bellon, S., Doré, T., & Francis, C. (2009). Agroecology as a science, a movement and a practice. A review. Agronomy for Sustainable Development, 29(4), 503–515.

Zou, Y., & Qiang, S. (2005). The balance of nutrient and energy of the integrated agricultural system. Chinese Journal of Ecology, 24(1), 81–86.

Chapter 7
Integrated livestock management: sustainable practices

Introduction

Livestock farming plays a pivotal role in global food production, livelihoods, and rural economies. However, the intensive and unsustainable practices associated with conventional livestock management pose significant challenges to environmental sustainability, animal welfare, and public health. In response, integrated livestock management (ILM) emerges as a holistic and forward-thinking approach that seeks to harmonize livestock production with environmental conservation, social equity, and economic viability.

This chapter delves into the principles, practices, and benefits of ILM within the context of sustainable agriculture. By examining case studies, best practices, and innovative solutions from diverse agroecological contexts, this chapter aims to showcase the transformative potential of ILM in promoting sustainable and resilient livestock production systems.

At its core, ILM emphasizes the integration of diverse components within livestock farming systems to optimize resource use, minimize environmental impacts, and enhance animal welfare. By adopting a multifaceted approach that combines livestock production, pasture management, waste recycling, and agroecological principles, ILM seeks to address the complex challenges facing the livestock sector while promoting sustainability, resilience, and social equity.

Through a comprehensive exploration of ILM, this chapter aims to inspire and empower farmers, policymakers, and stakeholders to embrace a more holistic and integrated approach to livestock farming. By prioritizing environmental sustainability, animal welfare, and social equity, ILM offers a pathway toward a more sustainable and equitable livestock production system that benefits both people and the planet.

Livestock management is an essential component of integrated agriculture, combining crop and livestock production to optimize resource utilization and promote sustainability.

7.1 Integrated crop-livestock systems

Integrated crop-livestock systems involve the integration of crops and livestock within a farming system. Animal waste can be used as organic fertilizer for crop production, while crop residues can be used as feed for livestock. This integration enhances nutrient cycling, reduces waste, and maximizes resource efficiency.

https://doi.org/10.1515/9783111448077-007

The following are the key aspects of integrated crop-livestock system in integrated agriculture (Figure 7.1).

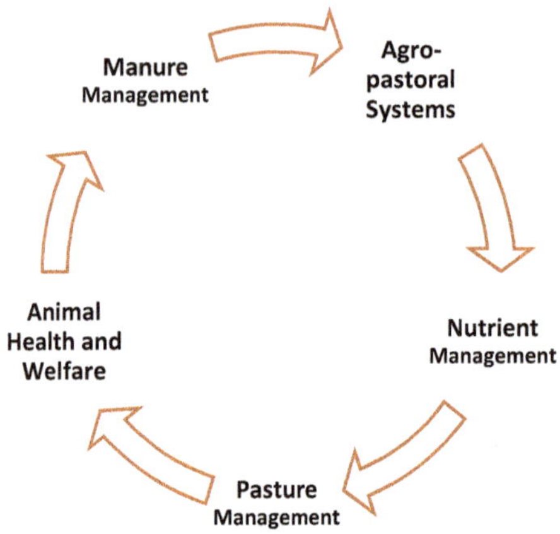

Figure 7.1: Key aspects of integrated crop-livestock system.

7.1.1 Agropastoral systems

Agropastoral systems involve combining livestock grazing with crop production. Livestock graze on crop residues or in fallow fields, providing natural weed and pest control, and nutrient recycling through their manure. Agropastoral systems benefit from the symbiotic relationship between crops and livestock.

7.1.2 Nutrient management

Proper nutrient management is crucial in livestock systems to ensure optimal animal health and productivity. Balanced rations, appropriate supplementation, and efficient feed conversion help maximize nutrient utilization and minimize environmental impacts. Nutrient-rich animal waste can be composted or used as organic fertilizer for crops.

7.1.3 Pasture management

Proper pasture management is important for sustainable livestock production. Rotational grazing, paddock systems, and improved forage varieties can optimize grazing efficiency, prevent overgrazing, and improve forage quality. Well-managed pastures enhance livestock nutrition, reduce soil erosion, and promote biodiversity.

7.1.4 Animal health and welfare

Integrated agriculture emphasizes the health and welfare of livestock. Proper vaccination, disease prevention, and veterinary care are essential to maintain animal health and productivity. Good animal husbandry practices, such as providing proper housing, access to clean water, and balanced diets, contribute to animal welfare.

7.1.5 Manure management

Efficient manure management is critical to prevent environmental pollution and optimize nutrient recycling. Proper collection, storage, and treatment of animal waste help minimize nutrient runoff, reduce odor emissions, and produce high-quality organic fertilizers for crop production.

7.1.6 Examples in different parts of the globe

Integrated crop-livestock systems, where crops and livestock are combined in a mutually beneficial way, are practiced in various countries across continents.

Here are examples of countries with notable integrated crop-livestock systems:

1. **The United States (North America):** In the United States, particularly in the Midwestern Corn Belt states such as Iowa and Illinois, integrated crop-livestock systems are common. Farmers grow corn and soybeans for feed and grain, which are then used to support intensive livestock production, especially for pork and poultry. Manure from livestock is often used as fertilizer for crops, creating a closed nutrient cycle.
2. **Brazil (South America):** Brazil is a global leader in integrated crop-livestock systems, especially in regions like Mato Grosso. Farmers here grow soybeans, corn, and other grains in rotation with pasture for beef cattle. The pasture phase allows forage production for cattle grazing, while cattle provide manure that improves soil fertility for subsequent crops.
3. **Australia (Oceania):** In Australia, particularly in the Northern Territory and Queensland, integrated crop-livestock systems are practiced. Farmers here combine cattle grazing with cropping systems such as sorghum, maize, and legumes. Cattle graze on crop residues after harvest, providing natural fertilizer for the next crop cycle.
4. **China (Asia):** China has regions, such as Sichuan Province, where integrated crop-livestock systems are prevalent. Farmers here grow crops like rice, wheat, and vegetables, while also raising pigs and poultry. Livestock manure is used to fertilize crops, and crop residues are fed to animals, creating a symbiotic relationship between crops and livestock.

5. **France (Europe):** In France, the Limousin region is known for its integrated crop-livestock systems. Farmers grow crops like maize and cereals alongside beef cattle production. Cattle graze on pasture and crop residues, which helps maintain soil fertility and reduces the need for external inputs.
6. **South Africa (Africa):** South Africa has integrated crop-livestock systems in regions like the Free State Province. Here, farmers grow maize and sunflowers alongside livestock such as beef and dairy cattle. Livestock manure is used to fertilize crops, and crop residues provide feed for animals during dry seasons.
7. **Argentina (South America):** Argentina's Pampas region is known for its integrated crop-livestock systems. Farmers here grow crops like soybeans, wheat, and corn in rotation with beef cattle production. Livestock graze on crop residues and cover crops, which helps improve soil health and fertility.
8. **India (Asia):** In India, states like Punjab and Haryana have integrated crop-livestock systems. Farmers grow crops such as wheat, rice, and pulses alongside dairy and buffalo farming. Crop residues are used as feed for animals, and animal manure is applied to fields as organic fertilizer.
9. **Russia (Europe/Asia):** Russia's Krasnodar Krai region practices integrated crop-livestock systems. Farmers grow crops like wheat, barley, and corn alongside cattle and sheep farming. Livestock provide manure for crop fertilization, and crop residues are used as feed for animals.
10. **Kenya (Africa):** In Kenya, regions like Nandi County practice integrated crop-livestock systems. Farmers here grow maize, beans, and vegetables alongside dairy cattle and poultry farming. Cattle graze on crop residues, and animal manure is used as organic fertilizer for crops.

These examples highlight how integrated crop-livestock systems can be adapted to different climatic conditions, farming practices, and cultural contexts across continents. Farmers in these countries benefit from improved soil fertility, reduced reliance on external inputs, and diversified income streams from both crops and livestock. Livestock management in integrated agriculture ensures the integration of animal production with crop production, promoting resource efficiency, sustainability, and overall farm productivity. The symbiotic relationship between crops and livestock creates a balanced and resilient farming system.

7.2 Integrated livestock-crop systems and their benefits

Integrated livestock-crop systems represent a holistic approach to farming that combines the benefits of livestock production and crop cultivation within a single agricultural system. In these integrated systems, livestock and crops are managed together, with each component complementing and supporting the other. Livestock provide valuable inputs such as manure, which can be used as organic fertilizer to enrich soil fertility and enhance crop productivity. Additionally, crop residues and by-products from crop harvests can serve as feed for livestock, reducing the need for external feed sources and promoting resource efficiency. Furthermore, the integration of livestock and crops can enhance nutrient cycling and soil health, leading to improved overall sustainability of the farming system. By leveraging the synergies between livestock and crop production, integrated livestock-crop systems offer a promising pathway toward more resilient, efficient, and sustainable agricultural practices.

The following are some key benefits of integrated livestock-crop systems (Figure 7.2).

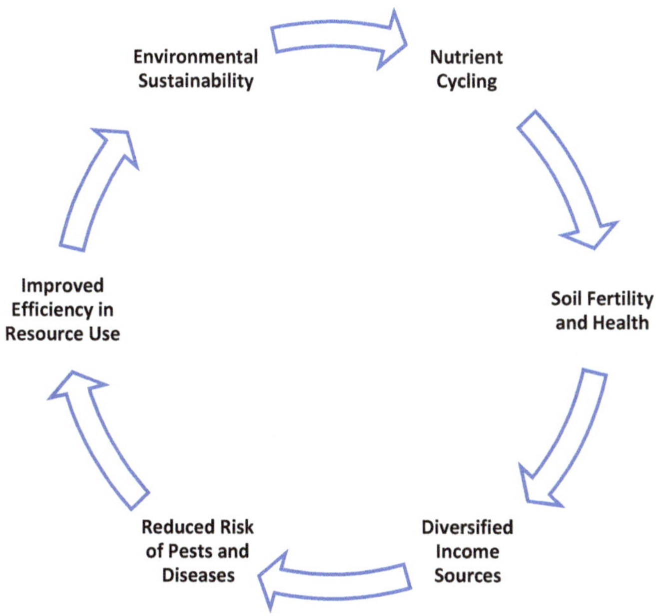

Figure 7.2: Key benefits of integrated livestock-crop system.

7.2.1 Nutrient cycling

Integrated systems facilitate the recycling of nutrients between crops and livestock. Animal manure, a valuable source of organic matter and nutrients, can be used as fertilizer for crop production, reducing the need for synthetic fertilizers. In turn, crop residues and by-products can be fed to livestock, closing nutrient cycles and optimizing resource utilization.

7.2.2 Soil fertility and health

Livestock manure, when properly managed, improves soil fertility and structure. It enhances soil organic matter content, nutrient availability, and microbial activity, leading to improved soil health. The integration of leguminous forages and cover crops in the system further contributes to nitrogen fixation and soil improvement.

7.2.3 Diversified income sources

Integrated systems provide multiple income streams. Farmers can generate income from crop sales, livestock sales, and value-added products. Diversified income sources contribute to economic stability and resilience, reducing the risks associated with relying solely on one enterprise.

7.2.4 Reduced risk of pests and diseases

Integrated systems can help manage pests and diseases through the diversification of crops and livestock. Crop rotations break pest and disease cycles, reducing the build-up of pathogens in the soil. Livestock grazing can help control weeds and reduce pest pressure in crops. The overall system resilience contributes to pest and disease management.

7.2.5 Improved efficiency in resource use

Integrated systems optimize resource utilization. Crop residues, by-products, and low-quality forages can be fed to livestock, reducing waste and improving feed efficiency. Livestock also provide draft power for agricultural operations, reducing the need for mechanization and fossil fuel consumption.

7.2.6 Environmental sustainability

Integrated livestock-crop systems contribute to environmental sustainability. They reduce the reliance on synthetic fertilizers, minimize nutrient runoff into water bodies, and promote soil conservation practices. The integration of trees and agroforestry components further enhances biodiversity, carbon sequestration, and climate resilience.

7.3 Sustainable livestock production practices

Sustainable livestock production practices strive to achieve a harmonious balance between economic viability, social responsibility, and environmental stewardship within the livestock farming sector. These practices recognize the interconnectedness of livestock production with broader ecological systems and human well-being. They prioritize efficient resource utilization, minimize environmental impacts such as pollution and habitat destruction, and promote animal welfare and social equity. Sustainable livestock production often involves the adoption of technologies and management practices that optimize feed efficiency, reduce greenhouse gas emissions, and enhance animal health and welfare.

Additionally, sustainable livestock farming embraces principles of responsible land management, water conservation, and biodiversity conservation to safeguard natural resources and ecosystems for future generations. By embracing sustainability principles, livestock producers can contribute to resilient food systems, mitigate environmental degradation, and ensure the long-term viability of the livestock industry.

The following are some key sustainable livestock production practices (Figure 7.3).

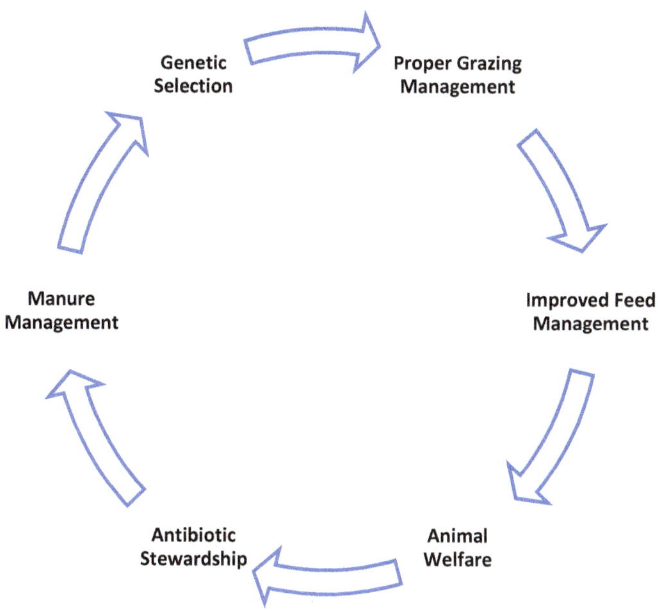

Figure 7.3: Key sustainable livestock production practices.

7.3.1 Proper grazing management

Implementing rotational grazing systems helps optimize pasture utilization, prevent overgrazing, and promote the regeneration of forage plants. Rotational grazing ensures that livestock have access to fresh forage while allowing sufficient rest and recovery periods for pastures.

7.3.2 Improved feed management

Efficient feed management practices reduce feed waste, improve feed conversion efficiency, and minimize environmental impacts. Balanced rations, proper feed storage, and feeding strategies based on livestock nutritional needs help optimize feed utilization and reduce nutrient losses.

7.3.3 Animal welfare

Ensuring good animal welfare is a fundamental aspect of sustainable livestock production. Providing appropriate housing, access to clean water, proper ventilation, and comfortable resting areas promotes animal health and well-being. Implementing proper handling practices and minimizing stress contribute to animal welfare.

7.3.4 Antibiotic stewardship

Responsible use of antibiotics in livestock production is crucial to prevent the emergence of antibiotic-resistant bacteria. Adhering to good animal husbandry practices, implementing biosecurity measures, and using antibiotics judiciously, as per veterinary guidance, helps maintain animal health while minimizing the risk of antimicrobial resistance.

7.3.5 Manure management

Proper management of livestock waste is essential to minimize environmental pollution and maximize nutrient recycling. Implementing practices such as manure storage, composting, and anaerobic digestion helps control odor, reduce greenhouse gas emissions, and produce valuable biofertilizers for crop production.

7.3.6 Genetic selection

Genetic selection plays a role in sustainable livestock production. Breeding programs focusing on traits such as disease resistance, feed efficiency, and adaptability to local conditions contribute to sustainable livestock systems. Genetic improvements can lead to more productive and resilient livestock breeds.

7.3.7 Examples of famous countries

Several countries are known for their advancements in sustainable livestock production practices, emphasizing animal welfare, environmental conservation, and efficient resource utilization.

Here are some countries that are famous for their sustainable livestock production practices:

1. **Denmark:** Denmark is a leader in sustainable livestock production, particularly in pig farming. The country has implemented strict environmental regulations, focusing on reducing emissions and improving waste management. Danish pig farms utilize biogas systems to convert manure into renewable energy. Additionally, Denmark emphasizes animal welfare standards, such as larger living spaces for animals.
2. **The Netherlands:** The Netherlands is recognized for its innovative and sustainable livestock farming practices. Dutch farmers are known for efficient use of resources, such as optimizing feed efficiency and reducing water usage. The country also emphasizes precision farming techniques to minimize environmental impact. The Netherlands has a strong focus on reducing ammonia emissions and has implemented measures to promote biodiversity on farms.
3. **New Zealand:** New Zealand is famous for its grass-fed livestock production systems, particularly for beef and lamb. The country's extensive grasslands provide natural grazing for animals, reducing the need for intensive feedlots. New Zealand has stringent animal welfare standards and organic farming practices. The country's agricultural sector also emphasizes water conservation and soil health.
4. **Sweden:** Sweden is known for its sustainable dairy and meat production practices. Swedish farmers prioritize animal welfare, with cows often allowed access to pasture. The country has implemented strict regulations on antibiotic use in livestock and focuses on organic farming methods. Sweden also promotes biodiversity on farms, with many farms incorporating wildlife habitats.
5. **Switzerland:** Switzerland is recognized for its sustainable and traditional farming practices. Swiss farmers often follow extensive grazing systems, allowing animals to graze on alpine pastures. The country has strong regulations on animal welfare and pesticide use. Swiss dairy farms are known for producing high-quality milk while maintaining environmental sustainability.

6. **Norway:** Norway is notable for its sustainable aquaculture practices, particularly in salmon farming. Norwegian aquaculture emphasizes responsible use of feed, reducing fish waste, and minimizing environmental impact on coastal waters. The country also has regulations in place to ensure the welfare of farmed fish. In addition to aquaculture, Norway also has sustainable dairy and meat production practices.
7. **Costa Rica:** Costa Rica is a leader in sustainable livestock production in the Latin American region. The country promotes agroforestry systems, where livestock graze alongside trees and other crops. This approach helps conserve biodiversity, improve soil health, and reduce greenhouse gas emissions. Costa Rica also emphasizes organic farming practices and has initiatives to promote pasture-based livestock systems.
8. **Uruguay:** Uruguay is known for its sustainable grass-fed beef production. The country's extensive grasslands and rotational grazing systems allow cattle to graze on natural pastures, resulting in high-quality beef with minimal environmental impact. Uruguay also has strict regulations on the use of growth hormones and antibiotics in livestock production.

These countries have made significant strides in adopting sustainable livestock production practices, showcasing the possibilities of environmentally friendly and socially responsible farming. Their approaches serve as models for promoting sustainability while meeting the growing global demand for animal products.

7.4 Health management and disease control in livestock

Health management and disease control are essential components of livestock production systems, as they directly impact animal welfare, productivity, and profitability. Livestock producers implement various measures to prevent, diagnose, and treat diseases, ensuring the overall health and well-being of their animals. This includes implementing biosecurity protocols to prevent the introduction and spread of diseases, vaccinating animals against common pathogens, and providing appropriate nutrition and housing conditions to support immune function. Regular monitoring and surveillance help identify potential health issues early, allowing for timely intervention and control measures. Additionally, collaboration with veterinarians and animal health professionals facilitates disease diagnosis, treatment, and management strategies. By prioritizing health management and disease control, livestock producers can minimize production losses, maintain animal welfare standards, and ensure the sustainability of their operations.

The following are key components of health management and disease control in livestock (Figure 7.4).

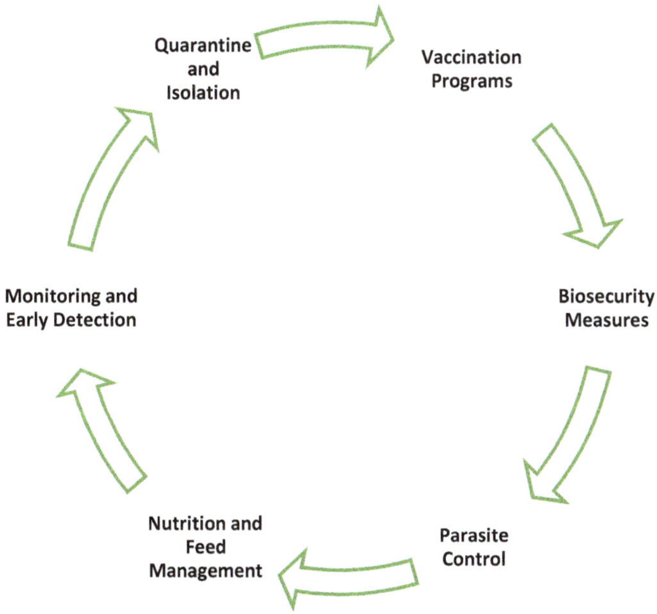

Figure 7.4: Key components of health management and disease control in livestock.

7.4.1 Vaccination programs

Vaccination is crucial for preventing infectious diseases in livestock. Implementing vaccination programs helps protect animals from common diseases and reduces the risk of disease outbreaks. Vaccination should be based on the specific disease risks and local epidemiology.

7.4.2 Biosecurity measures

Biosecurity practices help prevent the introduction and spread of diseases in livestock farms. Measures include controlling farm access, quarantining new animals, practicing proper sanitation and hygiene, and managing animal movements. Good biosecurity protocols reduce the risk of disease transmission.

7.4.3 Parasite control

Parasites, such as internal and external parasites, can negatively impact livestock health and productivity. Implementing parasite control measures, including strategic deworming, pasture management, and proper hygiene practices, helps minimize the risk of parasite infestations.

7.4.4 Nutrition and feed management

Proper nutrition plays a crucial role in maintaining animal health and immune function. Providing balanced diets, appropriate feed quality, and access to clean water helps enhance livestock immunity, reduce susceptibility to diseases, and promote overall health.

7.4.5 Monitoring and early detection

Regular monitoring of livestock health and early detection of disease signs are essential. Close observation of animals, including body condition, behavior changes, and feed intake, enables timely intervention and treatment. Prompt diagnosis and veterinary support help minimize the impact of diseases.

7.4.6 Quarantine and isolation

Quarantine and isolation practices are critical when introducing new animals to a farm. Isolating new animals for a specific period allows for health monitoring and prevents the potential spread of diseases to the existing livestock population.

7.4.7 Examples

Several countries excel in health management and disease control in livestock, implementing robust systems to monitor, prevent, and control diseases.

Here are some countries known for their effective approaches to livestock health management:

1. **The United States:** The United States has a sophisticated animal health management system led by the USDA (United States Department of Agriculture). The USDA's Animal and Plant Health Inspection Service oversees disease monitoring, prevention, and control programs. The country has a strong focus on disease surveillance, vaccination programs, and biosecurity measures. The National Animal Health Monitoring System conducts surveys to monitor the health and management practices of the US livestock.
2. **The United Kingdom:** The United Kingdom has stringent regulations and a comprehensive framework for livestock disease control. The Department for Environ-

ment, Food & Rural Affairs is responsible for overseeing livestock health. The Animal and Plant Health Agency conducts disease surveillance, implements control measures, and provides advice to farmers. The UK also has a robust veterinary network and research institutions dedicated to livestock health.

3. **Australia:** Australia is known for its effective biosecurity measures and disease control programs. The Australian Department of Agriculture, Water and the Environment oversees animal health policies. The country has a strong focus on biosecurity to prevent the introduction of diseases. Programs like the Australian Animal Health Laboratory and the National Livestock Identification System contribute to disease monitoring and traceability.
4. **The Netherlands:** The Netherlands has a well-developed system for livestock health management and disease control. The country emphasizes biosecurity measures, early detection, and rapid response to outbreaks. The Wageningen Bioveterinary Research institute conducts research and provides expertise on livestock diseases. The Dutch government also collaborates with farmers and industry stakeholders on disease prevention strategies.
5. **New Zealand:** New Zealand is recognized for its proactive approach to animal health management. The Ministry for Primary Industries oversees livestock health policies and programs. The country has a strong focus on biosecurity and has implemented the National Animal Identification and Tracing system for livestock traceability. New Zealand also has research institutions and veterinary services dedicated to livestock health.
6. **Denmark:** Denmark is known for its successful strategies in controlling diseases in livestock, particularly in pig farming. The country has strict biosecurity measures, vaccination programs, and surveillance systems. The Danish Veterinary and Food Administration oversees disease control and prevention. Denmark also has a strong focus on antibiotic reduction in livestock, contributing to better overall health management.
7. **Canada:** Canada has a comprehensive approach to livestock health management, led by the Canadian Food Inspection Agency (CFIA). The CFIA is responsible for disease surveillance, control programs, and regulations. Canada's National Farmed Animal Health and Welfare Council works on disease prevention strategies. The country also has research institutions and veterinary services dedicated to livestock health.

These countries have demonstrated success in managing livestock diseases through a combination of stringent regulations, robust surveillance systems, proactive biosecurity measures, and collaboration between government agencies, research institutions, and farmers. Their approaches serve as models for effective livestock health management worldwide.

7.5 Utilization of livestock waste for bioenergy and nutrient recycling

Livestock waste management is a critical aspect of sustainable agriculture, offering opportunities for both environmental stewardship and resource optimization. Livestock produce significant amounts of organic waste, including manure and bedding materials, which can be effectively utilized for bioenergy production and nutrient recycling. Anaerobic digestion is a common method used to convert livestock waste into biogas, a renewable energy source that can be used for heating, electricity generation, or vehicle fuel.

Additionally, composting livestock waste can produce nutrient-rich organic fertilizers, which can be used to improve soil health and fertility in agricultural fields. By harnessing the energy potential of livestock waste and recycling nutrients back into the soil, farmers can reduce greenhouse gas emissions, minimize reliance on fossil fuels, and enhance the sustainability of their agricultural operations. This integrated approach to waste management not only mitigates environmental impacts but also generates value from a previously underutilized resource, contributing to the circular economy and overall resilience of the agricultural system.

The following are the ways to utilize livestock waste (Figure 7.5).

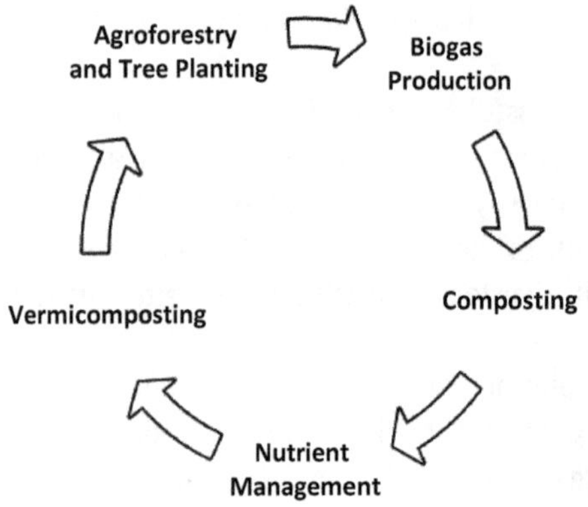

Figure 7.5: Different ways to utilize livestock waste.

7.5.1 Biogas production

Livestock waste, such as manure, can be used in anaerobic digesters to produce biogas. Biogas is a renewable energy source that can be used for heating, cooking, electricity generation, or as a transportation fuel. Biogas production helps reduce greenhouse gas emissions and provides an alternative energy source.

7.5.2 Composting

Composting livestock waste allows for the conversion of organic matter into nutrient-rich compost. Compost can be used as an organic fertilizer for crop production, improving soil fertility and structure. Composting reduces waste volume, controls odor, and provides a valuable soil amendment.

7.5.3 Nutrient management

Livestock waste contains valuable nutrients, such as nitrogen, phosphorus, and potassium. Proper management of livestock waste involves capturing and recycling these nutrients. Nutrient-rich animal manure can be applied as organic fertilizer to crop fields, reducing the need for synthetic fertilizers and closing nutrient cycles.

7.5.4 Vermicomposting

Vermicomposting is a process that uses earthworms to decompose organic waste, including livestock manure. The process produces nutrient-rich vermicompost, which enhances soil fertility and improves plant growth. Vermicomposting is a sustainable method of waste management and nutrient recycling.

7.5.5 Agroforestry and tree planting

Utilizing livestock waste as organic fertilizer can support agroforestry and tree planting initiatives. Nutrient-rich manure enhances tree growth, provides a sustainable nutrient source, and contributes to the establishment of agroforestry systems. Agroforestry combines tree cultivation with crop or livestock production, promoting ecological balance and diversifying farm income.

7.5.6 Examples

There are several good examples of utilizing livestock waste for bioenergy production and nutrient recycling around the world.

Here are a few examples:

1. **Biogas production from livestock manure:** Germany has a successful biogas industry, where livestock waste, particularly from dairy and swine farms, is used to produce biogas through anaerobic digestion. Biogas plants are widespread, and they not only produce renewable energy in the form of biogas but also generate

electricity and heat. The residual digestate from the process is rich in nutrients and can be used as organic fertilizer, closing the nutrient cycle.

2. **Anaerobic digestion projects:** Many farms in the United States, especially dairy farms, have adopted anaerobic digestion systems to convert manure into biogas. For example, the AgSTAR program by the Environmental Protection Agency promotes the use of anaerobic digesters in livestock operations. These digesters not only produce biogas for electricity and heat but also produce nutrient-rich digestate that can be used as fertilizer.
3. **Biogas for rural energy:** In India, particularly in rural areas, biogas plants are widely used to convert livestock waste, such as cow dung, into biogas for cooking and lighting. The National Biogas and Manure Management Program in India promotes the installation of biogas plants in households and communities, improving energy access while also managing livestock waste.
4. **Nutrient recycling:** The Netherlands has implemented innovative nutrient recovery technologies in livestock farming. For instance, technologies like struvite recovery from manure allow for the extraction of phosphorus, which is a valuable nutrient. This phosphorus can then be reused in fertilizers, reducing the reliance on mined phosphorus and closing the nutrient loop.
5. **Integrated farming systems:** Denmark has been successful in promoting integrated farming systems where livestock waste is used to fertilize crops and vice versa. Farms integrate crop production and animal husbandry, optimizing nutrient-use efficiency. This not only reduces the environmental impact of livestock waste but also improves overall farm sustainability.
6. **Sustainable livestock production:** In Kenya, there are initiatives promoting sustainable livestock production, including the utilization of livestock waste for biogas production. The Kenya Biogas Program, for example, supports the installation of biogas digesters in smallholder farms, improving energy access, reducing deforestation from wood fuel, and producing organic fertilizer for crops.

These examples demonstrate that the utilization of livestock waste for bioenergy production and nutrient recycling is not only feasible but also beneficial for both the environment and agricultural sustainability. Countries around the world are implementing various technologies and programs to harness the potential of livestock waste, turning it into a valuable resource for renewable energy and nutrient management.

In the pastoral embrace of ILM, animals and the environment embrace as partners, shaping a narrative of symbiosis and productivity. The chapters of animal husbandry, nutrition, and land stewardship intertwine, painting a portrait of harmonious coexistence. As this chapter unravels the potential for improved livestock welfare, enhanced food security, and empowered rural communities, it beckons us to explore the possibilities that lie at the crossroads of tradition and innovation.

From the pastures where livestock roam, we venture into the aquatic world in **Chapter 8** a realm where water and life intertwine in intricate harmony. Integrated fisheries and aquaculture management unveil the nuances of balancing aquatic biodiversity, sustainable harvesting, and responsible aquaculture practices. This chapter immerses us in the depths of aquatic ecosystems, revealing the intricate relationship between livelihoods, aquatic resources, and environmental stewardship.

References

Amanullah, Khan U. (2023). Advancing sustainable agriculture with beneficial microbes: Enhancing crop growth and yield for food security and human health. Advances in Modern Agriculture, 4(2), 2426.

Devendra, C., & Thomas, D. (2002). Crop–animal systems in Asia: Importance of livestock and characterization of agro-ecological zones. Agricultural Systems, 71(1–2), 5–15.

Herrero, M., Havlík, P., Valin, H., Notenbaert, A., Rufino, M. C., Thornton, P. K., . . . & Gerber, P. (2013). Biomass use, production, feed efficiencies, and greenhouse gas emissions from global livestock systems. Proceedings of the National Academy of Sciences, 110(52), 20888–20893.

Iqbal, Z., & Khan, M. S. (2016). Role of integrated livestock-crop farming systems in improving livelihoods of rural communities in developing countries. Sustainability, 8(11), 1154.

Knips, V., Cai, J., & Pikaar, I. (2015). Integrated approaches for sustainable livestock production and nutrient recycling. Animal, 9(05), 886–896.

Lal, R. (2015). Integrated crop-livestock systems in the United States Corn Belt. Journal of Soil and Water Conservation, 70(4), 78A–84A.

MacLeod, M., Viner, D., & Smith, P. (2010). Use of US croplands for biofuels increases greenhouse gases through emissions from land-use change. Science, 319(5867), 1238–1240.

McAllister, T. A., & Beauchemin, K. A. (2006). Consideration of greenhouse gases and nitrogen cycle in the life cycle assessment of cattle production. In: Greenhouse Gases and Animal Agriculture (pp. 101–123). Elsevier. Amsterdam, The Netherlands.

Place, F., & Hoffmann, I. (2011). Ten years of smallholder dairy development in Kenya: A holistic perspective. ILRI Discussion Paper, (20).

Steinfeld, H., Gerber, P., Wassenaar, T., Castel, V., Rosales, M., & De Haan, C. (2006). Livestock's long shadow: Environmental issues and options. Food & Agriculture Organization (FAO), Rome, Italy.

Thornton, P. K., & Herrero, M. (2010). Potential for reduced methane and carbon dioxide emissions from livestock and pasture management in the tropics. Proceedings of the National Academy of Sciences, 107(46), 19667–19672.

Chapter 8
Integrated fisheries and aquaculture management: ensuring food security

Introduction

Fisheries and aquaculture play a vital role in global food security, providing a significant source of protein, essential nutrients, and livelihoods for millions of people worldwide. However, the sustainability of fisheries and aquaculture systems is increasingly threatened by overexploitation, habitat degradation, pollution, and climate change. In response, integrated fisheries and aquaculture management (IFAM) emerges as a holistic and adaptive approach that seeks to balance the ecological, social, and economic dimensions of aquatic food production while ensuring the long-term sustainability of marine and freshwater resources.

This chapter delves into the principles, practices, and benefits of IFAM within the context of food security and sustainable development. By examining case studies, best practices, and innovative solutions from diverse aquatic ecosystems, this chapter aims to showcase the transformative potential of IFAM in promoting sustainable and resilient fisheries and aquaculture systems.

At its core, IFAM emphasizes the integration of diverse components within aquatic food production systems to optimize resource use, minimize environmental impacts, and enhance social and economic benefits. By adopting a multifaceted approach that combines fisheries management, aquaculture production, habitat restoration, and community engagement, IFAM seeks to address the complex challenges facing aquatic food production while promoting food security, resilience, and environmental sustainability.

Through a comprehensive exploration of (IFAM), this chapter aims to inspire and empower policymakers, fisheries managers, aquaculture practitioners, and stakeholders to embrace a more holistic and integrated approach to aquatic food production. By prioritizing ecosystem health, social equity, and economic viability, IFAM offers a pathway toward a more sustainable and equitable food system that benefits both present and future generations. Fisheries and aquaculture play a crucial role in integrated agriculture systems.

8.1 Key aspects of fisheries and aquaculture

Here are key aspects of fisheries and aquaculture in integrated agriculture.

https://doi.org/10.1515/9783111448077-008

8.1.1 Aquaponics

Aquaponics is a system that combines fish farming (aquaculture) with hydroponic plant cultivation. In this system, fish waste provides nutrients for plants, while plants help purify the water for the fish. Aquaponics optimizes resource utilization, enhances nutrient cycling, and enables the production of both fish and crops in a symbiotic system.

Aquaponics represents a sustainable and innovative approach to food production, particularly in areas where arable land and freshwater resources are limited. By integrating aquaculture and hydroponics, aquaponic systems maximize space utilization and resource efficiency, making them well-suited for urban and peri-urban environments. In addition to producing fish and vegetables, aquaponics also offers environmental benefits such as efficient water use and nutrient recycling. The closed-loop nature of aquaponic systems minimizes water consumption and eliminates the need for synthetic fertilizers, reducing the environmental footprint of food production. Moreover, aquaponics can contribute to food security and resilience by providing a reliable source of fresh produce and protein year-round, regardless of climate or seasonal variations. As such, aquaponics holds promise as a sustainable farming practice that aligns with the principles of integrated agriculture and offers a viable solution to the challenges of modern food production.

8.1.2 Pond aquaculture

Pond aquaculture involves the farming of fish in ponds. Fish are reared in water bodies, utilizing the natural productivity of the ecosystem. Integrated pond aquaculture systems can combine fish farming with other agricultural practices, such as rice cultivation or livestock grazing, creating an integrated farming system that maximizes resource use and productivity.

Pond aquaculture, also known as fish farming or pond culture, has been practiced for centuries and remains a vital component of global aquaculture production. In pond aquaculture, fish are raised in earthen ponds or artificial enclosures, where they feed on natural organisms present in the water or supplementary feed provided by farmers. This method of fish farming leverages the natural productivity of aquatic ecosystems, utilizing sunlight, water, and nutrients to support fish growth. Integrated pond aquaculture systems take this approach a step further by combining fish farming with other agricultural activities such as rice cultivation, vegetable farming, or livestock grazing. By integrating pond aquaculture with other farming practices, farmers can optimize land and water use, enhance nutrient cycling, and diversify their income streams. Furthermore, pond aquaculture systems can provide additional benefits such as flood mitigation, water filtration, and habitat creation, contributing to ecosystem health and resilience. Overall, pond aquaculture represents a versatile and sustainable approach to fish production that aligns with the principles of integrated agriculture and offers opportunities for livelihood improvement and food security in rural communities.

8.1.3 Nutrient recycling

Fish waste and uneaten feed from aquaculture operations can be used as nutrient-rich organic fertilizer for crop production. Proper collection and management of fish waste allow for its recycling as a valuable nutrient source. This nutrient recycling contributes to sustainable agriculture and reduces the need for synthetic fertilizers.

Nutrient recycling plays a crucial role in sustainable agriculture by closing the loop between aquaculture and crop production. In aquaculture systems, fish excrete waste and uneaten feed, which contain valuable nutrients such as nitrogen, phosphorus, and potassium. Rather than allowing these nutrients to accumulate and potentially cause water pollution, farmers can collect and utilize fish waste as organic fertilizer for crops. This practice not only reduces the environmental impact of aquaculture but also provides a cost-effective and sustainable source of nutrients for crop plants. By recycling fish waste back into the soil, farmers can improve soil fertility, enhance crop growth, and minimize reliance on chemical fertilizers. Additionally, nutrient recycling promotes resource efficiency and supports the principles of integrated agriculture by harnessing the symbiotic relationship between aquaculture and crop production. Overall, nutrient recycling from aquaculture to agriculture represents a valuable strategy for enhancing nutrient cycling, reducing environmental pollution, and promoting sustainable food production systems.

8.1.4 Biodiversity conservation

Integrated agriculture systems that include fisheries and aquaculture components contribute to biodiversity conservation. The presence of fish in water bodies promotes biodiversity and ecosystem health. Proper management of fisheries and aquaculture operations ensures the sustainable use of aquatic resources and minimizes negative impacts on aquatic ecosystems.

Biodiversity conservation is a key component of integrated agriculture systems that incorporate fisheries and aquaculture practices. The inclusion of fish within water bodies enhances biodiversity by fostering a balanced ecosystem and supporting the presence of diverse aquatic species. Sustainable management practices in fisheries and aquaculture operations are essential for preserving biodiversity and promoting ecosystem health. By implementing measures such as habitat protection, species conservation, and responsible harvesting practices, farmers can ensure the long-term sustainability of aquatic resources while minimizing negative impacts on aquatic ecosystems. Furthermore, integrated agriculture systems that incorporate fisheries and aquaculture components contribute to the conservation of aquatic biodiversity by maintaining the natural balance of ecosystems and supporting the health of aquatic habitats. Overall, biodiversity conservation is integral to the sustainability of integrated agriculture systems and plays a crucial role in safeguarding the ecological integrity of aquatic environments for present and future generations.

8.1.5 Integrated pest management

Integrated pest management (IPM) approaches can be applied in aquaculture to minimize the use of chemical pesticides. Techniques such as biological control, proper pond management, and water quality monitoring help control pests and diseases in aquaculture systems, promoting environmental sustainability.

IPM is a crucial component of sustainable aquaculture practices, aimed at reducing reliance on chemical pesticides and promoting environmental health. By employing a combination of biological, cultural, and mechanical control methods, aquaculturists can effectively manage pests and diseases while minimizing adverse impacts on aquatic ecosystems. Biological control involves the use of natural predators or parasites to regulate pest populations, thereby maintaining ecological balance within aquaculture systems.

Additionally, proper pond management practices such as regular water quality monitoring, habitat manipulation, and stocking density optimization contribute to pest and disease control by creating unfavorable conditions for their proliferation. Furthermore, IPM strategies prioritize preventive measures such as biosecurity protocols and quarantine procedures to minimize the introduction and spread of pests and diseases in aquaculture facilities. Overall, the adoption of IPM approaches in aquaculture enhances environmental sustainability by reducing chemical inputs, preserving ecosystem integrity, and ensuring the long-term viability of aquatic resources.

8.1.6 Livelihood diversification

Fisheries and aquaculture provide opportunities for livelihood diversification, especially in rural communities. Integrated agriculture systems that include fisheries components can enhance income generation, create employment opportunities, and contribute to rural development.

Livelihood diversification is a key benefit of integrating fisheries and aquaculture into agricultural systems, particularly in rural areas where traditional livelihood options may be limited. By incorporating fisheries components into integrated agriculture systems, communities can diversify their income sources, reduce reliance on single crops or activities, and build resilience against economic shocks. Fisheries and aquaculture activities offer various opportunities for income generation, including fish farming, fish processing, marketing, and value-added product development.

Moreover, the establishment of fish ponds or aquaculture facilities can create employment opportunities for local residents such as pond managers, fish feed suppliers, and fishery extension workers. These employment opportunities contribute to rural development by stimulating economic growth, improving household incomes, and reducing poverty. Additionally, the integration of fisheries and aquaculture into agricultural systems can enhance food security and nutrition by providing access to protein-rich fish and seafood products. Overall, livelihood diversification through fisheries and aquaculture integration fosters economic empowerment, social resilience, and sustainable development in rural communities.

The integration of fisheries and aquaculture in integrated agriculture systems offers multiple benefits including resource optimization, nutrient recycling, biodiversity conservation, and livelihood opportunities. These systems provide a holistic approach to sustainable food production and contribute to the overall resilience of integrated agriculture.

8.2 Sustainable fish farming practices

Sustainable fish farming practices are essential for meeting the growing demand for seafood while minimizing environmental impacts and promoting long-term viability. As global fisheries face increasing pressure from overfishing and habitat degradation, sustainable aquaculture offers a promising alternative to wild-caught seafood. This chapter explores the principles, techniques, and benefits of sustainable fish farming, highlighting practices that prioritize ecological stewardship, animal welfare, and social responsibility. By adopting sustainable fish farming practices, aquaculture producers can contribute to food security, economic development, and environmental conservation efforts worldwide. Sustainable fish farming practices aim to minimize environmental impacts, ensure fish health and welfare, and support long-term viability.

Here are key aspects of sustainable fish farming practices (Figure 8.1).

8.2.1 Water quality management

Maintaining good water quality is crucial for fish health and growth. Sustainable fish farming practices include monitoring and managing water parameters such as temperature, dissolved oxygen levels, pH, and ammonia concentrations. Proper filtration, aeration, and regular water testing help ensure optimal water quality.

Water quality management is a fundamental aspect of sustainable fish farming practices within integrated agriculture systems. The health and growth of fish are directly influenced by the quality of the water environment in which they are raised.

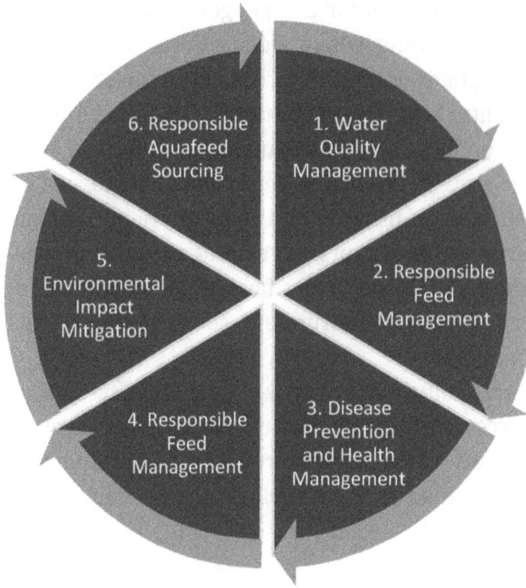

Figure 8.1: Key aspects of sustainable fish farming practices.

Therefore, it is essential to monitor and manage key water parameters to maintain optimal conditions for fish health and productivity. Parameters such as temperature, dissolved oxygen levels, pH, and ammonia concentrations play critical roles in determining water quality.

To ensure good water quality, sustainable aquaculture practices include the implementation of proper filtration systems to remove excess waste and debris as well as the use of aeration devices to oxygenate the water. Additionally, regular water testing is conducted to monitor parameter levels and detect any deviations from optimal conditions.

Effective water quality management not only promotes the health and welfare of fish but also contributes to the overall sustainability of aquaculture operations. By maintaining clean and healthy water environments, integrated agriculture systems can enhance fish growth, minimize disease risks, and support the long-term viability of aquaculture activities.

8.2.2 Responsible feed management

Feed management is important to minimize nutrient waste and prevent water pollution. Sustainable fish farming practices involve using feeds with high nutritional efficiency, proper feeding techniques to avoid overfeeding, and implementing feed conversion ratio (FCR) optimization strategies. This helps reduce the environmental impact of fish farming and improve feed utilization.

Responsible feed management is integral to sustainable fish farming practices within integrated agriculture systems. Efficient feed management strategies are employed to minimize nutrient waste and mitigate the risk of water pollution, thereby promoting environmental sustainability.

One key aspect of responsible feed management is the selection of feeds with high nutritional efficiency. Utilizing feeds that are formulated to meet the specific dietary requirements of fish while minimizing excess nutrients helps to optimize feed utilization and reduce the overall environmental footprint of fish farming operations.

Proper feeding techniques are also essential to prevent overfeeding, which can lead to nutrient runoff and water contamination. Fish farmers employ feeding schedules and portion control methods to ensure that fish receive the appropriate amount of feed without excess waste.

Furthermore, FCR optimization strategies are implemented to improve the efficiency of feed conversion into fish biomass. This involves monitoring and adjusting feeding practices to achieve the highest possible growth rates with the least amount of feed input.

By implementing responsible feed management practices, integrated agriculture systems can minimize nutrient waste, reduce environmental impacts, and improve the overall efficiency and sustainability of fish farming operations.

8.2.3 Disease prevention and health management

Sustainable fish farming practices prioritize disease prevention and health management. Implementing biosecurity measures, such as proper farm hygiene, quarantine protocols, and regular health monitoring, helps prevent the introduction and spread of diseases. Vaccination programs, proper nutrition, and prompt treatment of diseases contribute to fish health and welfare.

Disease prevention and health management are paramount in sustainable fish farming practices within integrated agriculture systems. Robust biosecurity measures are implemented to safeguard against the introduction and spread of diseases within fish farming operations. These measures include maintaining proper farm hygiene, implementing quarantine protocols for new stock, and conducting regular health monitoring to detect and address any signs of illness promptly.

Vaccination programs are commonly employed to bolster the immune systems of farmed fish and reduce their susceptibility to infectious diseases. Vaccines are developed specifically for prevalent fish diseases and administered according to recommended schedules to provide optimal protection.

Furthermore, ensuring proper nutrition is essential for maintaining fish health and resilience to diseases. Balanced and high-quality feeds are formulated to meet the nutritional requirements of different fish species, promoting optimal growth and immune function.

In the event of disease outbreaks, prompt diagnosis and treatment are crucial for mitigating their impact. Fish farmers work closely with veterinarians and aquatic health professionals to identify diseases early and implement appropriate treatment strategies. This may involve the use of therapeutics, antibiotics, or other medications under the guidance of qualified professionals.

By prioritizing disease prevention and health management, integrated agriculture systems can minimize the risk of disease outbreaks, enhance fish welfare, and contribute to the long-term sustainability of fish farming operations.

8.2.4 Environmental impact mitigation

Sustainable fish farming practices aim to minimize the environmental impact of fish farms. These include proper site selection, avoiding sensitive habitats, managing effluents, and implementing strategies to reduce the release of nutrients, chemicals, and sediments into water bodies. Utilizing appropriate technologies, such as recirculating aquaculture systems (RAS), can help reduce water usage and minimize waste discharge.

Environmental impact mitigation is a critical aspect of sustainable fish farming practices within integrated agriculture systems. To minimize the ecological footprint of fish farms, various measures are implemented to mitigate potential environmental impacts.

One fundamental strategy is proper site selection for fish farms, taking into account factors such as water quality, habitat integrity, and proximity to sensitive ecosystems. Avoiding areas with high ecological value helps prevent negative impacts on biodiversity and ecosystem functioning.

Effluent management is another key consideration in environmental impact mitigation. Fish farms produce wastewater containing nutrients, organic matter, and sometimes chemicals from feed and medications. Proper effluent management practices, such as using settling ponds, wetlands, or constructed wetlands, can help remove pollutants and improve water quality before discharge into surrounding water bodies.

Additionally, strategies are employed to reduce the release of nutrients, chemicals, and sediments into water bodies. This may involve optimizing feeding practices to minimize nutrient waste, using low-impact feed formulations, and implementing erosion control measures to prevent sediment runoff.

Innovative technologies, such as RAS, play a crucial role in environmental impact mitigation. RAS recirculate and treat water within closed-loop systems, reducing water usage and minimizing waste discharge. These systems help conserve water resources and minimize the risk of pollution, making them a sustainable option for intensive fish farming operations.

By implementing these environmental mitigation measures, integrated agriculture systems can minimize their ecological footprint, protect natural ecosystems, and promote the long-term sustainability of fish farming practices.

8.2.5 Responsible aquafeed sourcing

Sustainable fish farming practices consider the sourcing and composition of aquafeeds. Using feeds made from sustainable and responsibly sourced ingredients, such as fishmeal and fish oil derived from well-managed fisheries, reduces the pressure on wild fish stocks. The development of alternative feed sources, such as plant-based feeds or feeds derived from waste streams, also contributes to sustainability.

Responsible aquafeed sourcing is integral to sustainable fish farming practices within integrated agriculture systems. It involves careful consideration of the origin and composition of aquafeeds to minimize environmental impact and promote long-term sustainability.

One key aspect of responsible aquafeed sourcing is selecting feeds made from sustainably and responsibly sourced ingredients. This includes choosing fishmeal and fish oil derived from well-managed fisheries that adhere to sustainable harvesting practices. By supporting fisheries with robust management plans and certification schemes, fish farmers can help reduce the pressure on wild fish stocks and promote the health of marine ecosystems.

Furthermore, the development of alternative feed sources plays a crucial role in responsible aquafeed sourcing. Plant-based feeds, for example, can provide a sustainable alternative to traditional fish-based feeds, reducing reliance on wild fish as a feed ingredient. Additionally, feeds derived from waste streams, such as agricultural byproducts or algae, offer opportunities to minimize waste and utilize resources more efficiently.

By prioritizing responsible aquafeed sourcing, integrated agriculture systems can reduce their environmental footprint, support sustainable fisheries management, and contribute to the overall sustainability of fish farming practices. This approach aligns with the principles of responsible stewardship and conservation, ensuring the long-term health and viability of aquaculture operations.

8.2.6 Examples

Several countries have implemented sustainable fish farming practices to promote environmental conservation, efficient resource use, and the production of healthy seafood.

Here are a few examples from different regions:

1. **Norway:** Norway is a global leader in salmon aquaculture. The country has implemented sustainable practices such as open-ocean net pens with strong environmental regulations. Norwegian salmon farms focus on reducing disease outbreaks, using low-density stocking, and employing advanced technology for monitoring water quality. They also prioritize sustainable sourcing of fish feed, often using fish trimmings and plant-based ingredients.
2. **Chile:** Chile is another major player in salmon farming, particularly Atlantic salmon. Chilean salmon farms have adopted sustainable practices to minimize environmental impacts. They use wellboats for waste collection, implement strict biosecurity measures to prevent disease outbreaks, and employ sea lice management techniques. Some farms are also moving toward land-based RAS for more controlled and sustainable production.
3. **Vietnam:** Vietnam is known for its pangasius (catfish) farming, especially in the Mekong Delta. Sustainable practices in Vietnamese pangasius farming include reducing antibiotic use through better disease management, implementing certification programs (like ASC – Aquaculture Stewardship Council), and promoting responsible feed sourcing. Farmers are also encouraged to use efficient pond management techniques to minimize environmental impacts.
4. **Canada:** Canada has a strong focus on sustainable shellfish farming, particularly mussels and oysters. Shellfish farming is considered environmentally friendly as these organisms help filter water and improve water quality. Canadian shellfish farms adhere to strict regulations for water quality monitoring and use of biodegradable materials. Integrated multitrophic aquaculture, where different species are farmed together to utilize nutrients efficiently, is also practiced.
5. **Scotland:** Scotland is a major producer of farmed Atlantic salmon. Scottish salmon farms emphasize sustainability through practices such as cleaner fish usage (wrasse and lumpfish to control sea lice), fallowing sites to allow seabed recovery and optimizing feed to minimize waste. The industry also supports the Scottish Salmon Producers Organisation, which promotes responsible aquaculture practices.
6. **Australia:** Australia has sustainable barramundi farming operations, particularly in Queensland and Northern Territory. Sustainable practices include using RAS to minimize water usage and pollution, selective breeding for disease resistance, and sourcing sustainable fish feed. These farms often have certifications such as best aquaculture practices or ASC.
7. **Netherlands:** The Netherlands is known for its innovative approach to indoor fish farming, especially for species like tilapia and African catfish. Dutch fish farms use advanced RAS technology to recirculate water, remove waste, and control water quality parameters. This results in efficient resource use, minimal environmental impact, and year-round production. The farms also prioritize fish welfare and traceability.

8. **Japan:** Japan has sustainable yellowtail (hamachi) farming operations, particularly in Ehime and Kagoshima prefectures. Japanese farms emphasize good aquaculture practices to minimize environmental impacts, such as reducing fish escapees, using eco-friendly feed, and implementing strict biosecurity measures. Some farms also practice integrated aquaculture with seaweed or shellfish.

These examples demonstrate how countries around the world are implementing sustainable practices in fish farming to protect the environment, ensure food security, and produce high-quality seafood for consumers. Each region adapts its practices to local conditions, species, and market demands while striving for sustainability in aquaculture.

8.3 Integrated fish-plant systems and aquaponics

Integrated fish-plant systems and aquaponics are innovative approaches that combine fish farming with hydroponic plant cultivation. Integrated fish-plant systems and aquaponics represent innovative and sustainable approaches to food production that harness the symbiotic relationship between aquatic organisms and plants. In these systems, fish and plants are cultivated together in a mutually beneficial environment, where fish waste provides nutrients for plant growth, and plants help to filter and purify the water for the fish. This chapter explores the principles, design, and benefits of integrated fish-plant systems and aquaponics, offering insights into how these methods contribute to food security, resource efficiency, and environmental sustainability.

Here are key aspects of integrated fish-plant systems and aquaponics (Figure 8.2).

8.3.1 Nutrient cycling

Integrated fish-plant systems and aquaponics utilize the nutrient-rich waste produced by fish as a fertilizer for plant growth. Fish waste provides essential nutrients for plant uptake, while plants filter and purify the water for the fish. This symbiotic relationship between fish and plants enhances nutrient cycling, reduces waste, and maximizes resource utilization.

Nutrient cycling is a fundamental aspect of sustainable agriculture, particularly in integrated fish-plant systems and aquaponics. These innovative farming methods leverage the natural symbiosis between fish and plants to create a closed-loop system where nutrients are efficiently recycled.

In integrated fish-plant systems, fish waste serves as a valuable source of nutrients for plant growth. As fish consume feed and produce waste, organic matter and nutrients accumulate in the water. This nutrient-rich water is then circulated to plant beds where plants, such as vegetables or herbs, are grown hydroponically. The plants

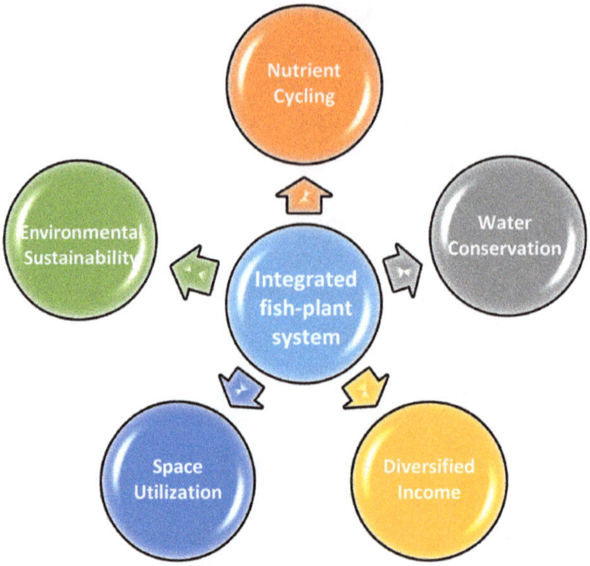

Figure 8.2: Key aspects of integrated fish-plant systems and aquaponics.

absorb nutrients, including nitrogen, phosphorus, and potassium, from the water, effectively filtering and purifying it in the process. The cleaned water is then recirculated back to the fish tanks, completing the nutrient cycle.

Similarly, aquaponics systems integrate fish farming with hydroponic plant cultivation to achieve nutrient cycling. In aquaponics, fish waste provides essential nutrients, primarily nitrogen in the form of ammonia, for plant uptake. Beneficial bacteria in the system convert ammonia into nitrate, a form of nitrogen that plants can readily absorb. As plants uptake nutrients, they help maintain water quality for the fish by removing excess nutrients and organic matter.

By harnessing nutrient cycling, integrated fish-plant systems and aquaponics minimize nutrient loss and waste, promoting resource efficiency and sustainability. These systems offer a holistic approach to agriculture that optimizes nutrient utilization, reduces environmental impact, and enhances productivity. As the demand for sustainable food production methods grows, nutrient cycling in integrated aquaculture systems plays a crucial role in shaping the future of agriculture.

8.3.2 Water conservation

Integrated fish-plant systems and aquaponics promote water conservation compared to traditional farming systems. The recirculating nature of these systems allows for the reuse of water, minimizing water consumption. The plants help to filter and clean the water, creating a closed-loop system that maximizes water efficiency.

Water conservation is a critical component of sustainable agriculture, and integrated fish-plant systems and aquaponics offer significant advantages in this regard. Unlike traditional farming methods that rely on continuous water replenishment, these integrated systems operate in a closed-loop fashion, where water is recycled and reused.

In integrated fish-plant systems, water circulates between the fish tanks and the plant beds, facilitating nutrient exchange and maintaining water quality. As the water passes through the plant beds, plants uptake nutrients and filter impurities, effectively cleansing the water for reuse in the fish tanks. This closed-loop system minimizes water loss through evaporation or runoff, leading to significant water savings compared to conventional farming practices.

Similarly, aquaponics systems employ a recirculating water system, where water from the fish tanks is continuously pumped to the plant beds and then returned to the fish tanks. This closed-loop circulation ensures that water is used efficiently, with minimal wastage. Additionally, the natural filtration provided by plants helps to maintain water quality, reducing the need for frequent water changes.

By conserving water and minimizing wastage, integrated fish-plant systems and aquaponics contribute to sustainable water management practices in agriculture. These systems offer a viable solution for regions facing water scarcity or limited access to freshwater resources, allowing farmers to produce food more efficiently while reducing their environmental footprint. As water conservation becomes increasingly important in the face of climate change and growing water demands, integrated aquaculture systems emerge as a promising approach to sustainable agriculture.

8.3.3 Diversified income

Integrated fish-plant systems and aquaponics offer multiple income streams. Farmers can generate revenue from both fish sales and the sale of hydroponic crops. This diversification of income sources contributes to economic stability and resilience.

Diversified income streams are a significant advantage of integrated fish-plant systems and aquaponics, providing farmers with multiple avenues for generating revenue. In these integrated systems, farmers can profit from both fish sales and the sale of hydroponically grown crops, maximizing their earning potential.

The cultivation of fish allows farmers to tap into the lucrative market for seafood products, catering to consumer demand for fresh and sustainably sourced fish. Whether it's tilapia, catfish, or trout, the fish produced in aquaponics systems are prized for their high quality and nutritional value. By selling fish directly to consumers, restaurants, or local markets, farmers can command premium prices and generate substantial income.

In addition to fish sales, aquaponic farmers can also profit from the sale of hydroponic crops grown in conjunction with the fish. Leafy greens, herbs, and vegetables thrive in the nutrient-rich water of aquaponic systems, offering farmers a steady supply of high-quality produce to sell. With growing consumer interest in locally grown and organic food, aquaponically grown crops are in high demand, presenting farmers with an opportunity to capture a niche market and fetch premium prices for their products.

By diversifying their income streams through fish and crop sales, farmers can mitigate risks associated with market fluctuations and seasonal variability. This economic stability enhances the resilience of farming operations, ensuring a steady source of income throughout the year. Moreover, the ability to generate revenue from multiple sources strengthens the financial viability of integrated fish-plant systems and aquaponics, making them an attractive option for farmers seeking sustainable and profitable agricultural practices.

8.3.4 Space utilization

Integrated fish-plant systems and aquaponics optimize space utilization by vertically stacking or incorporating the fish tanks and plant beds. This vertical integration allows for the production of both fish and plants in a compact area, making these systems suitable for urban and limited land availability.

Space utilization is a key advantage of integrated fish-plant systems and aquaponics, particularly in environments where land availability is limited. These systems optimize space by vertically stacking or incorporating the fish tanks and plant beds, allowing for the efficient production of both fish and plants in a compact area.

In traditional farming methods, separate space is often required for fish ponds or tanks and crop fields, leading to inefficiencies in land use. However, integrated systems capitalize on vertical space, with fish tanks positioned above or adjacent to the plant beds. This vertical integration maximizes the productive capacity of the available land, enabling farmers to cultivate fish and grow crops simultaneously without the need for extensive acreage.

The compact nature of integrated fish-plant systems and aquaponics makes them well-suited for urban and peri-urban environments, where space constraints are common. These systems can be implemented on rooftops, balconies, or in small backyard gardens, allowing urban dwellers to engage in food production in limited spaces. By utilizing vertical space effectively, aquaponic farmers can make efficient use of available land resources, enabling sustainable and productive agriculture even in densely populated areas.

8.3.5 Environmental sustainability

Integrated fish-plant systems and aquaponics contribute to environmental sustainability. They minimize the use of chemical fertilizers and pesticides, reduce water consumption, and promote efficient nutrient recycling. These systems have a smaller environmental footprint compared to separate fish and plant production systems.

Environmental sustainability is a cornerstone of integrated fish-plant systems and aquaponics. These systems promote sustainability by minimizing the use of chemical fertilizers and pesticides, thus reducing the environmental impact associated with conventional agriculture. By relying on natural processes and biological interactions, such as the nutrient cycling between fish and plants, aquaponics eliminates the need for synthetic inputs that can pollute waterways and harm ecosystems.

Furthermore, integrated systems are inherently water-efficient, as they utilize a recirculating system where water is continuously circulated between the fish tanks and plant beds. This minimizes water consumption compared to traditional irrigation methods, where water is often lost through runoff or evaporation. The efficient use of water in aquaponics helps conserve this precious resource, particularly in regions prone to water scarcity or drought.

Moreover, the nutrient-rich wastewater from the fish tanks serves as a natural fertilizer for the plants, eliminating the need for chemical fertilizers. This nutrient recycling not only reduces waste but also promotes soil health and fertility in plant beds. As a result, integrated fish-plant systems and aquaponics have a smaller environmental footprint compared to conventional agriculture, contributing to overall sustainability and ecosystem health.

8.4 Conservation of aquatic ecosystems

The conservation of aquatic ecosystems is vital for preserving biodiversity, supporting ecosystem services, and ensuring the sustainable management of water resources. Aquatic ecosystems, including rivers, lakes, wetlands, and oceans, are home to a diverse array of plant and animal species, playing a crucial role in global nutrient cycling and climate regulation. However, these ecosystems face numerous threats, including pollution, habitat destruction, overfishing, and climate change. This chapter explores the importance of conserving aquatic ecosystems, discussing strategies for protecting and restoring these valuable natural habitats. By prioritizing the conservation of aquatic ecosystems, we can safeguard biodiversity, enhance water quality, and promote the resilience of ecosystems in the face of environmental challenges. Conservation of aquatic ecosystems is crucial for the preservation of biodiversity, ecosystem services, and long-term sustainability.

Here are key aspects of conservation of aquatic ecosystems (Figure 8.3).

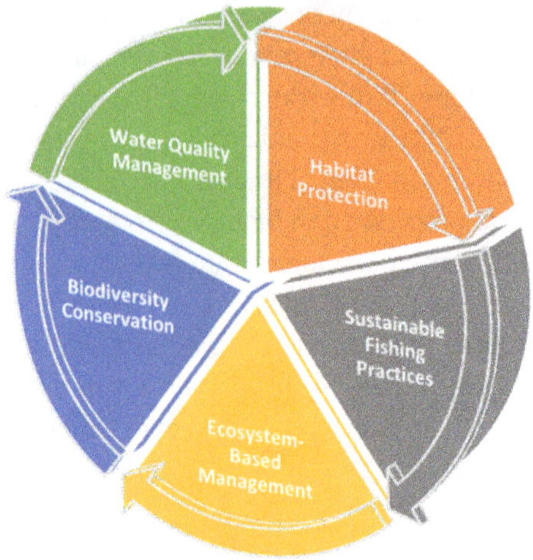

Figure 8.3: Key aspects of conservation of aquatic ecosystems.

8.4.1 Habitat protection

Protecting and preserving aquatic habitats, such as wetlands, rivers, lakes, and coral reefs, is essential for maintaining healthy ecosystems. Conservation efforts focus on preventing habitat destruction, reducing pollution, and restoring degraded habitats to ensure the survival of aquatic species and the overall functioning of ecosystems.

8.4.2 Sustainable fishing practices

Sustainable fishing practices aim to maintain fish populations at levels that support their reproduction and long-term viability. This involves implementing catch limits, size restrictions, and fishing gear regulations to prevent overfishing, minimize by-catch, and protect vulnerable species. Sustainable fishing practices contribute to the conservation of aquatic biodiversity.

8.4.3 Ecosystem-based management

Ecosystem-based management approaches consider the interactions and interdependencies within aquatic ecosystems. These approaches recognize the need to manage fisheries and other activities in a holistic manner, considering ecological relationships, food webs, and the impacts of human activities on the ecosystem as a whole.

8.4.4 Biodiversity conservation

Biodiversity conservation efforts focus on protecting and preserving the variety of species and ecosystems within aquatic environments. This includes the identification and protection of key habitats, conservation of endangered species, and the promotion of sustainable practices that minimize negative impacts on biodiversity.

8.4.5 Water quality management

Managing and maintaining good water quality is crucial for the conservation of aquatic ecosystems. This involves monitoring and controlling pollution sources, reducing nutrient runoff from agricultural and urban areas, and implementing water treatment and restoration measures. Maintaining healthy water quality supports the survival of aquatic organisms and promotes ecosystem resilience.

8.5 Value-added fish products and marketing strategies

In today's dynamic market landscape, the demand for value-added fish products continues to rise, driven by evolving consumer preferences, changing dietary habits, and a growing awareness of health and sustainability. Value-added fish products, which encompass a wide range of processed, ready-to-eat, and convenience seafood items, offer consumers convenience, variety, and nutritional benefits. This chapter explores the concept of value-added fish products and delves into innovative marketing strategies aimed at capturing market share and meeting consumer demands.

By understanding the nuances of value addition and implementing effective marketing strategies, fisheries and aquaculture businesses can unlock new opportunities for growth, profitability, and sustainability in the competitive seafood market. Creating value-added fish products and implementing effective marketing strategies contribute to the profitability and competitiveness of the fish farming industry.

Here are key aspects of value-added fish products and marketing strategies (Figure 8.4).

8.5.1 Product diversification

Developing value-added fish products involves diversifying the range of products beyond fresh or frozen whole fish. This can include processed products such as fillets, smoked fish, fish cakes, fish sausages, fish jerky, and ready-to-cook or ready-to-eat fish-based meals. Product diversification increases consumer choices and enhances market opportunities.

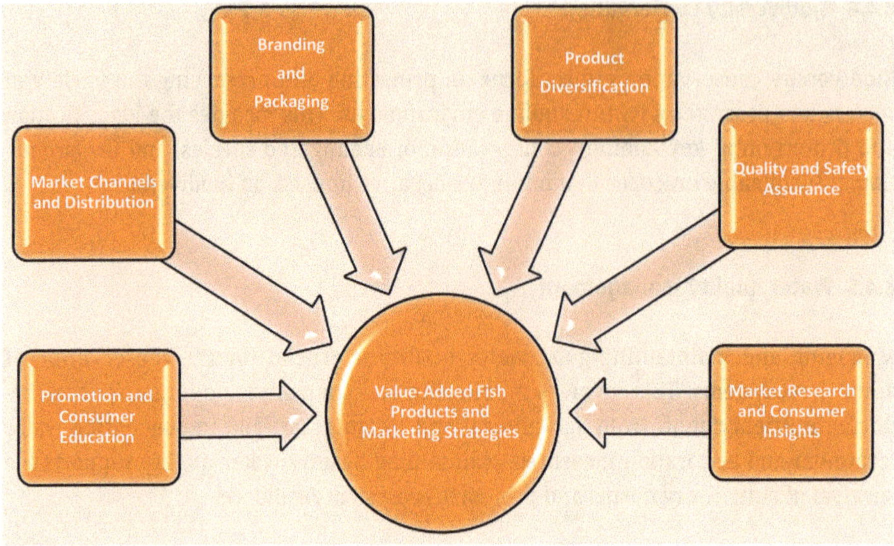

Figure 8.4: Key aspects of value-added fish products and marketing strategies.

8.5.2 Quality and safety assurance

Ensuring the quality and safety of value-added fish products is paramount. Implementing proper processing, handling, and storage practices, as well as adhering to quality control and food safety standards, helps maintain product integrity and consumer confidence. Certification schemes and traceability systems add value to the products.

8.5.3 Market research and consumer insights

Conducting market research and understanding consumer preferences is essential for developing successful marketing strategies. Identifying target markets, consumer trends, and preferences for fish products enables effective product positioning, branding, and communication strategies.

8.5.4 Branding and packaging

Developing a strong brand image and attractive packaging are key elements of successful marketing. Branding helps differentiate products in the market, builds consumer trust, and creates brand loyalty. Eye-catching and informative packaging helps attract consumers and communicates product attributes and benefits.

8.5.5 Market channels and distribution

Identifying appropriate market channels and distribution networks is crucial for reaching target consumers. This may include selling products through retail stores, restaurants, farmers' markets, online platforms, or direct sales. Establishing partnerships with distributors, retailers, and food service providers helps access wider markets.

8.5.6 Promotion and consumer education

Promoting value-added fish products involves targeted marketing campaigns, product demonstrations, and consumer education initiatives. Highlighting the unique features, nutritional benefits, and culinary versatility of the products helps raise awareness and generate consumer interest.

8.5.7 Examples

Value-added fish products and effective marketing strategies can add significant economic value to the seafood industry. Here are some good examples of countries globally that have successfully developed value-added fish products and implemented effective marketing strategies:

1. **Norway:** Norway is renowned for its high-quality smoked salmon, which is a value-added product made from fresh Atlantic salmon. Norwegian companies such as Marine Harvest and Lerøy Seafood Group are leaders in smoked salmon production. They market their products internationally, emphasizing the premium quality and sustainable production methods, targeting high-end restaurants and gourmet markets.
2. **Japan:** Japan is famous for its traditional sushi and sashimi, which are value-added products made from various fish species like tuna, salmon, and mackerel. Japanese companies, such as Tsukiji Fish Market vendors and sushi chains like Sushiro, focus on freshness, presentation, and unique flavors. Marketing strategies include showcasing the artistry of sushi chefs, emphasizing health benefits, and promoting cultural experiences.
3. **United States:** In the US, companies like Blue Apron and HelloFresh offer meal kits that include ready-to-cook seafood dishes, such as shrimp scampi or grilled salmon with vegetables. These value-added products cater to busy consumers

looking for convenient, healthy meal options. Marketing strategies involve promoting easy preparation, freshness, and sustainable sourcing.

4. **Thailand:** Thailand produces high-quality fish sauce, a fermented condiment made from anchovies and salt. Brands like Tiparos and Squid are well-known for their fish sauce products, which are used as a flavor enhancer in Asian cuisine worldwide. Thai fish sauce is marketed for its umami-rich flavor, natural fermentation process, and versatility in cooking.
5. **Spain:** Spain is a major producer of canned seafood products including tuna, sardines, and mussels. Brands like Ortiz and Conservas de Cambados are famous for their premium canned seafood, which is marketed as gourmet delicacies. These products often highlight traditional methods of preparation, sustainable fishing practices, and premium quality ingredients.
6. **South Korea:** South Korea is known for Ganjang Gejang, a dish made from raw crabs marinated in soy sauce. This value-added product has gained popularity internationally for its unique flavor and texture. Korean companies market Ganjang Gejang as a premium seafood delicacy, emphasizing the artisanal preparation and health benefits of seafood.
7. **Denmark:** Danish companies like Royal Greenland are leaders in developing innovative fish fillet products. They offer products such as marinated herring fillets, seafood salads, and breaded fish portions. These value-added products target convenience-focused consumers and are marketed for their taste, freshness, and ease of preparation.
8. **Peru:** Peru produces high-quality fish roe, particularly from species like herring and flying fish. Peruvian caviar is considered a gourmet product, known for its unique flavors and vibrant colors. Companies such as Caviar Perú and Sechura Caviar market their products internationally, highlighting the sustainable sourcing and artisanal production methods.
9. **Italy:** Italy produces Bottarga, a delicacy made from the salted, cured roe of grey mullet or tuna. Bottarga is used as a seasoning in Italian cuisine, adding a rich, savory flavor to dishes like pasta and salads. Italian companies market Bottarga as a premium product, emphasizing its artisanal production and unique taste profile.
10. **Vietnam:** Vietnam is known for its fish sauce and fish paste products, which are essential ingredients in Vietnamese cuisine. Brands like Three Crabs and Red Boat produce high-quality fish sauce, while fish paste is used in dishes like nem (spring rolls). Vietnamese products are marketed for their authentic taste, natural fermentation process, and versatility in cooking.

These examples demonstrate the diversity of value-added fish products and effective marketing strategies used globally. Each country has leveraged its unique seafood resources and culinary traditions to create premium products that cater to different consumer preferences and markets.

Amid the tranquil depths and shimmering waters, **Chapter 8** casts a spotlight on the nexus of aquatic life and human sustenance. Integrated fisheries and aquaculture management illuminate the art of nurturing aquatic ecosystems while fostering livelihoods. Through the lenses of responsible harvesting, conservation, and sustainable aquaculture, this chapter charts a course that harmonizes aquatic bounty with ecological guardianship – a journey that embodies the spirit of interconnectedness.

As the currents of aquatic life gently ebb, we find ourselves on the threshold of Chapter 9 a realm where trees, ecosystems, and livelihoods intertwine in the symphony of integrated forestry management. This chapter ushers us into the shaded groves and arboreal landscapes, where prudent management of forests and woodlands unfolds. Here, the chapters of conservation, livelihoods, and biodiversity interlace, revealing the profound role that well-managed forests play in nurturing our planet's vitality.

References

Alam, M. J., Asaduzzaman, M., & Thompson, P. M. (2016). An overview of integrated aquaculture-agriculture (IAA) systems in Bangladesh. International Journal of Agricultural and Biological Engineering, 9(2), 1–8.

Amanullah (2024). Harvesting Success: A Comprehensive Guide to Agronomy Courses. Punjab Books, Lahore, PK.

Amanullah., Ondrasek, G., Al –Tawaha, A. R. M. S., eds. (2023). Integrated nutrients management: An approach for sustainable crop production and food security in changing climates. Lausanne: Frontiers Media SA. doi: 10.3389/978-2-8325-3169-3

Amanullah, & Khalid, S. (2020). Agronomy – Climate Change & Food Security. IntechOpen, London, United Kingdom.

Amanullah, & Fahad, S. (2018). Nitrogen in Agriculture-Updates. InTechOpen, London, UK.

Amanullah, & Fahad. (2018). Corn – Production and Human Health in Changing Climate. InTech, London, UK.

Beveridge, M. C. M., Thilsted, S. H., Phillips, M. J., Metian, M., & Troell, M. (2013). Meeting the food and nutrition needs of the poor: The role of fish and the opportunities and challenges emerging from the rise of aquaculture. Journal of Fish Biology, 83(4), 1067–1084.

Costa-Pierce, B. A. (2002). Ecological Aquaculture: The Evolution of the Blue Revolution. Blackwell Science.

Halwart, M., & Gupta, M. V. (2004). Culture of Fish in Rice Fields. Food & Agriculture Org. Rome, Italy.

Kestemont, P., & Kirschbaum, F. (2007). Perspectives for integrated multi-trophic aquaculture. In: Handbook on Ingredients for Aquaculture Feeds (pp. 21–45). Springer. Heidelberg, Germany.

Little, D. C., Edwards, P., & De Silva, S. S. (2003). An Impact Assessment of Integrated Agriculture-aquaculture in Malawi. Food & Agriculture Org. Rome, Italy.

Love, D. C., Fry, J. P., Milli, M. C., Neff, R. A., & Pister, K. (2011). Feed conversion efficiency in aquaculture: Do we measure it correctly? Environmental Research Letters, 6(1), 014017.

Naylor, R. L., Goldburg, R. J., Primavera, J. H., Kautsky, N., Beveridge, M. C., Clay, J., . . . & Troell, M. (2000). Effect of aquaculture on world fish supplies. Nature, 405(6790), 1017–1024.

Troell, M., Naylor, R. L., Metian, M., Beveridge, M., Tyedmers, P. H., Folke, C., . . . & Kautsky, N. (2014). Does aquaculture add resilience to the global food system? Proceedings of the National Academy of Sciences, 111(37), 13257–13263.

FAO. 2019. Report of the Special Session on Advancing Integrated Agriculture Aquaculture through Agroecology, Montpellier, 25 August 2018. FAO Fisheries and Aquaculture Report No. 1286. Rome, Italy.

Chapter 9
Integrated forestry management: promoting biodiversity and conservation

Introduction

Forests are among the most biodiverse and ecologically valuable ecosystems on Earth, providing essential habitat for countless plant and animal species, regulating climate, purifying air and water, and supporting the livelihoods of millions of people worldwide. However, unsustainable forestry practices, deforestation, habitat fragmentation, and illegal logging pose significant threats to forest biodiversity and ecosystem integrity. In response, integrated forestry management (IFM) emerges as a holistic and adaptive approach that seeks to balance the ecological, social, and economic dimensions of forest management while promoting biodiversity conservation and sustainable use of forest resources.

IFM emphasizes the integration of diverse components within forest management systems to optimize biodiversity conservation, ecosystem services, and socioeconomic benefits. By adopting a multifaceted approach that combines forest conservation, sustainable harvesting, habitat restoration, and community engagement, IFM seeks to address the complex challenges facing forest ecosystems while promoting resilience, biodiversity conservation, and sustainable livelihoods.

Through a comprehensive exploration of IFM, this chapter aims to inspire and empower policymakers, forest managers, conservation practitioners, and stakeholders to embrace a more holistic and integrated approach to forest management. By prioritizing ecosystem health, biodiversity conservation, and community participation, IFM offers a pathway toward a more sustainable and resilient forest ecosystem that benefits both present and future generations.

9.1 Importance

Forests are crucial on a global scale for a multitude of reasons, playing vital roles in environmental, social, economic, and even cultural aspects. The ideal forest cover for a healthy society and ecosystems can vary based on factors such as regional climate, biodiversity, land use, and population density. However, a general consensus among experts is that maintaining at least 30% of land area under forest cover is crucial for ecological balance, biodiversity conservation, climate regulation, and sustainable resource management.

https://doi.org/10.1515/9783111448077-009

Here's a breakdown:

- **Biodiversity conservation:** Forests support diverse ecosystems, providing habitats for numerous plant and animal species. Maintaining sufficient forest cover is essential to preserve biodiversity.
- **Climate regulation:** Forests act as carbon sinks, sequestering carbon dioxide and helping mitigate climate change. They also play a role in regulating local and regional climates, influencing temperature and precipitation patterns.
- **Water cycle regulation:** Forests are crucial for regulating the water cycle, influencing rainfall patterns, preventing soil erosion, and maintaining healthy watersheds. They are essential for ensuring a sustainable supply of freshwater.
- **Soil health and fertility:** Forests help maintain soil health by preventing erosion, promoting nutrient cycling, and providing organic matter. They contribute to soil fertility and productivity.
- **Air and water quality:** Forests improve air quality by absorbing pollutants and releasing oxygen. They also filter and purify water, contributing to clean and healthy water sources.

Here are some examples highlighting the importance of forests on a global scale.

9.1.1 Biodiversity conservation

- **Example:** The Amazon Rainforest is one of the most biodiverse ecosystems on Earth, home to millions of species of plants, animals, and insects. It's estimated that one in 10 known species lives in the Amazon.
- **Importance:** Forests like the Amazon are essential for conserving biodiversity, providing habitats for countless species, many of which are endemic and found nowhere else.

9.1.2 Climate regulation

- **Example:** Boreal forests, found in countries like Canada, Russia, and Scandinavia, store vast amounts of carbon in their soils and vegetation.
- **Importance:** Forests act as carbon sinks, helping to mitigate climate change by absorbing carbon dioxide from the atmosphere. They play a crucial role in regulating global climate patterns.

9.1.3 Oxygen production

- **Example:** The world's tropical rainforests, such as those in the Congo Basin and Southeast Asia, are known as the "lungs of the Earth."
- **Importance:** Forests produce a significant amount of the world's oxygen through photosynthesis, providing the oxygen necessary for human and animal respiration.

9.1.4 Water cycle and watershed protection

- **Example:** The Andean forests in South America, including the cloud forests, are crucial for water supply to millions of people living downstream.
- **Importance:** Forests play a vital role in the water cycle, regulating rainfall patterns, reducing soil erosion, and maintaining healthy watersheds. They help supply fresh water to rivers and streams, which is essential for agriculture, drinking water, and ecosystems.

9.1.5 Economic importance

- **Example:** The timber industry in countries like the United States, Canada, and Sweden is a major contributor to their economies.
- **Importance:** Forests provide valuable timber and non-timber products, supporting industries such as logging, furniture making, and paper production. They also create jobs and income for millions of people worldwide.

9.1.6 Medicinal resources

- **Example:** The forests of Southeast Asia, such as those in Indonesia and Malaysia, are rich in medicinal plants like the Cinchona tree (source of quinine) and various species of orchids.
- **Importance:** Forests are a source of many pharmaceuticals and traditional medicines. Indigenous communities around the world rely on forests for their medicinal plants and traditional healing practices.

9.1.7 Cultural and spiritual significance

- **Example:** The boreal forests of Russia hold cultural significance for indigenous peoples like the Nenets and Evenks.

- **Importance:** Forests are often deeply ingrained in the cultures and traditions of indigenous communities. They provide spiritual and cultural value, serving as sites for rituals, ceremonies, and traditional knowledge transmission.

9.1.8 Tourism and recreation

- **Example:** The forests of Yosemite National Park in the United States attract millions of visitors each year for hiking, camping, and wildlife viewing.
- **Importance:** Forests offer recreational opportunities, ecotourism, and nature-based experiences that contribute to local economies. They are destinations for birdwatching, hiking, photography, and other outdoor activities.

9.1.9 Habitat for indigenous communities

- **Example:** The forests of the Amazon Basin are home to numerous indigenous tribes such as the Yanomami and Kayapo.
- **Importance:** Forests provide essential habitats for indigenous communities, supporting their traditional ways of life, cultural practices, and livelihoods. These communities often have deep spiritual and historical connections to the forests.

9.1.10 Erosion control and soil fertility

- **Example:** The forests of the Himalayas, such as those in Bhutan and Nepal, help stabilize steep slopes and prevent landslides.
- **Importance:** Forests play a crucial role in soil conservation, preventing erosion, and maintaining soil fertility. Their root systems help hold soil in place, reducing the risk of landslides and maintaining productive agricultural land downstream.

These examples illustrate the multifaceted importance of forests on a global scale. They are not only essential for environmental health and biodiversity but also for human well-being, economic development, cultural heritage, and the overall balance of ecosystems worldwide. Protecting and sustainably managing forests is crucial for the future of the planet and all its inhabitants.

9.2 Countries with sufficient and deficient forest cover

9.2.1 Countries with sufficient forest cover

1. **Russia:** Russia has the largest forest area in the world, covering about 45% of its land area. The country's vast boreal forests play a significant role in global carbon storage and biodiversity conservation.
2. **Canada:** Canada has extensive forest cover, with forests occupying around 38% of the country's land area. Canadian forests are diverse and provide habitat for numerous species as well as supporting the forestry industry.
3. **Brazil:** Despite facing deforestation challenges in the Amazon, Brazil still has significant forest cover, with about 56% of its land covered by forests. The Amazon rainforest is a crucial biodiversity hotspot.
4. **Australia:** Australia has about 17% of its land covered by forests. While not as high as some other countries, Australia's forests are ecologically diverse, with unique species and ecosystems.
5. **Sweden:** Sweden has approximately 69% of its land area covered by forests, mainly composed of coniferous and mixed forests. Swedish forests are sustainably managed and contribute to the country's bioeconomy.

9.2.2 Countries with deficient forest cover

1. **Haiti:** Haiti has one of the lowest forest cover percentages in the world, with only about 3% of its land area covered by forests. Deforestation due to agricultural expansion and charcoal production has led to severe environmental degradation.
2. **Nigeria:** Nigeria's forest cover has been significantly reduced over the years, with estimates of forest cover ranging from 5% to 10% of the land area. Deforestation is driven by agriculture, logging, and urban expansion.
3. **India:** India's forest cover is estimated to be around 24% of its land area, below the global average. The country faces challenges of deforestation, encroachment, and habitat loss for wildlife.
4. **China:** China has about 22% of its land covered by forests, but the quality of these forests varies. Deforestation, particularly in the past, has led to ecological challenges such as soil erosion and habitat loss.
5. **Indonesia:** Indonesia's forest cover is around 49% of its land area, but it has one of the highest deforestation rates globally. The expansion of palm oil plantations, logging, and wildfires has contributed to significant forest loss.
6. **Pakistan:** Forestry in Pakistan encompasses a diverse range of ecosystems, from the mangroves of the coastal areas to the alpine forests of the northern mountain ranges. Forests cover approximately 5.2% of the country's land area, contributing to vital ecological services and offering economic opportunities. The northern re-

gions, particularly in Khyber Pakhtunkhwa and Gilgit-Baltistan, are home to coniferous forests that not only provide timber for construction and furniture but also support the tourism industry through their breathtaking landscapes. Additionally, the mangrove forests along the coast, notably in the Indus Delta, play a crucial role in shoreline protection, providing a buffer against coastal erosion and storm surges.

Despite these valuable assets, Pakistan's forests face significant challenges, including deforestation, land degradation, and inadequate policy implementation. Addressing these challenges through sustainable forest management practices, community engagement, and effective policy enforcement is essential for safeguarding these diverse ecosystems, supporting local livelihoods, and ensuring the continued provision of vital ecosystem services.

Maintaining sufficient forest cover, ideally around 30% of land area, is crucial for the health of societies and ecosystems. Countries with high forest cover, such as Russia, Canada, and Brazil, benefit from the ecological services and biodiversity these forests provide. On the other hand, countries with deficient forest cover, like Haiti, Nigeria, and India, face challenges related to deforestation, loss of biodiversity, soil erosion, and water scarcity. Efforts to protect and restore forests are essential globally to ensure sustainable development, biodiversity conservation, and climate resilience.

9.3 Economic development

Forestry plays a multifaceted and indispensable role in economic development, contributing to sustainable growth, poverty alleviation, and environmental conservation. As a renewable natural resource, forests provide a wide array of products and services that underpin various sectors of the economy, including timber, non-timber forest products, ecotourism, and carbon sequestration. Moreover, forests support livelihoods for millions of people worldwide, particularly in rural and indigenous communities, by providing employment opportunities, income generation, and social benefits. Through innovative approaches and sustainable management practices, forestry can continue to drive economic development, enhance resilience to global challenges, and foster inclusive prosperity for generations to come.

Forests play a role in economic development through various channels (Figure 9.1).

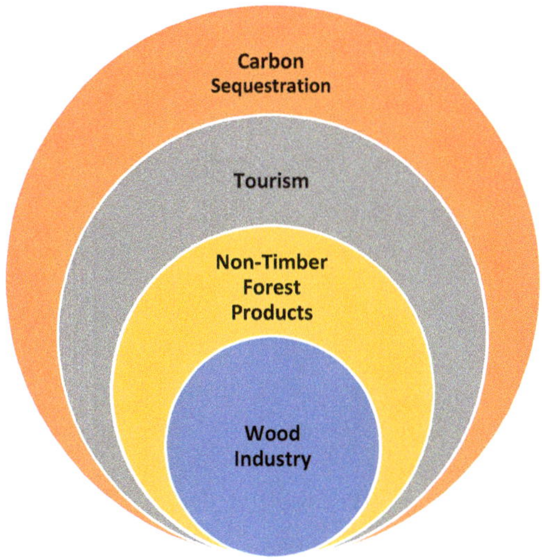

Figure 9.1: Role of forests in economic development through various channels.

9.3.1 Wood industry

The wood industry plays a significant role in the global economy, providing timber for construction, furniture, paper products, and various wood-based materials. Here are some points on a global scale:

- **Construction:** Timber is a primary material for construction worldwide. From traditional wooden houses to modern skyscrapers, wood is used in various forms. For example, Canada is a major exporter of softwood lumber to the United States and other countries for residential construction.
- **Furniture:** The furniture industry heavily relies on wood. Countries like Sweden and Italy are known for their furniture production, using high-quality wood to create designs exported globally.
- **Paper and packaging:** Wood pulp from trees is a key ingredient in paper and packaging production. The United States, China, and Brazil are among the top producers of wood pulp.

9.3.2 Non-timber forest products (NTFPs)

NTFPs are essential for many communities, providing sources of income and sustenance. Here's a global perspective:

- **Fruits and nuts:** Forests yield a variety of fruits and nuts such as berries, mushrooms, and chestnuts. In countries like Chile, the collection and sale of wild mushrooms contribute significantly to local economies.

- **Medicinal plants:** Many medicinal plants are sourced from forests. For instance, the Peruvian Amazon is rich in medicinal plants like cat's claw and dragon's blood, which are used in traditional medicine and have economic value.
- **Honey and bee products:** Forests support bee populations, leading to honey production. Countries like Ethiopia and New Zealand have vibrant honey industries, with forests providing diverse floral sources for bees.

9.3.3 Tourism

Forests are popular destinations for eco-tourists and nature enthusiasts, offering opportunities for hiking, birdwatching, and wildlife viewing. Here are some global examples:

- **National parks:** Countries like the United States (Yellowstone National Park) and Kenya (Maasai Mara National Reserve) have iconic forested national parks that attract millions of tourists annually, boosting local economies.
- **Ecotourism:** Costa Rica is a prime example of a country that has embraced ecotourism in its rainforests. Visitors come to experience the diverse flora and fauna, supporting local businesses and conservation efforts.

9.3.4 Carbon sequestration

Forests are crucial for absorbing and storing carbon dioxide from the atmosphere, aiding in climate change mitigation. Here's a global perspective:

- **Economic value:** The value of carbon sequestered by forests is significant. According to the World Bank, the global value of carbon stored in forests is estimated at over $100 trillion.
- **Carbon markets:** Initiatives like REDD+ (reducing emissions from deforestation and forest degradation) allow countries to receive payments for carbon sequestration efforts. For example, Indonesia has received funding through REDD+ for forest conservation.
- **International agreements**: Forests play a vital role in achieving international climate goals. The Paris Agreement recognizes the importance of forests in mitigating climate change, highlighting the economic benefits of preserving forests.

The wood industry, non-timber forest products, tourism, and carbon sequestration are all vital aspects of forests' global significance. They contribute to economies, support livelihoods, provide diverse products, and play a critical role in climate change mitigation. For sustainable forest management, it's essential to balance these economic benefits with conservation efforts to ensure forests' long-term health and resilience.

The current global forestry situation is characterized by a mix of challenges and opportunities, with efforts underway to address issues such as deforestation, biodiversity loss, climate change impacts, sustainable forest management, and the role of forests in supporting livelihoods and biodiversity. Here are some key points regarding the current global forestry situation.

9.4 Forestry challenges

Forestry is a critical component of environmental sustainability, providing essential ecosystem services, supporting biodiversity, and contributing to economic development. However, the forestry sector faces numerous challenges that threaten its sustainability and the health of forest ecosystems worldwide. From deforestation and forest degradation to climate change and invasive species, these challenges require urgent attention and concerted efforts to address. In this chapter, we delve into the multifaceted challenges confronting the forestry sector, exploring their causes, impacts, and potential solutions. By understanding and addressing these challenges, we can work toward ensuring the long-term health, resilience, and productivity of forest ecosystems for current and future generations.

9.4.1 Deforestation and forest degradation

Deforestation, driven by agricultural expansion, infrastructure development, and illegal logging, remains a significant threat to forests worldwide. Forest degradation, caused by unsustainable logging practices and wildfires, further exacerbates the loss of forest cover and biodiversity.

Deforestation continues to be a pressing issue globally, driven primarily by agricultural expansion, infrastructure development, and the demand for timber products. As forests are cleared to make way for agricultural land, urban areas, and roads, vast tracts of valuable forest ecosystems are lost, along with the biodiversity they support. Additionally, illegal logging operations contribute significantly to deforestation, particularly in regions with weak law enforcement and governance structures.

Forest degradation, often a consequence of unsustainable logging practices and wildfires, compounds the problem by reducing the resilience and productivity of remaining forest areas. Unsustainable logging methods, such as clear-cutting and selective logging, lead to the fragmentation of forest landscapes and the loss of key habitats for many plant and animal species. Wildfires, exacerbated by factors like climate change and land-use change, pose a severe threat to forests, especially in drought-prone regions. Addressing deforestation and forest degradation requires a multifaceted approach that combines effective land-use planning, law enforcement, community engagement, and sustainable forest management practices. It is essential to promote alternatives to deforestation, such as agroforestry and sustainable land management, while also restoring degraded forest ecosystems to their former health and productivity. Additionally, efforts to combat climate change must include measures to protect and restore forests, as they play a crucial role in carbon sequestration and climate regulation.

9.4.2 Land degradation

Soil erosion, degradation, and inappropriate land-use practices affect the health and productivity of forest ecosystems. Land degradation, encompassing soil erosion, degradation, and inappropriate land-use practices, poses significant threats to the health and productivity of forest ecosystems worldwide. Soil erosion, often accelerated by factors such as deforestation, unsustainable agricultural practices, and urbanization, results in the loss of fertile topsoil, reducing the land's ability to support vegetation and sustain biodiversity. Additionally, soil degradation, caused by factors like overgrazing, excessive tillage, and improper land management, leads to the depletion of soil nutrients, compaction, and loss of soil structure, further exacerbating the degradation process.

Inappropriate land-use practices, including deforestation for agriculture or infrastructure development without adequate reforestation efforts, disrupt ecological balances and diminish the capacity of forests to provide essential ecosystem services such as carbon sequestration, water regulation, and habitat provision. Addressing land degradation requires holistic approaches that promote sustainable land management practices, reforestation efforts, and the restoration of degraded landscapes to ensure the long-term health and resilience of forest ecosystems.

9.4.3 Biodiversity loss

Deforestation and habitat fragmentation pose significant threats to biodiversity, leading to the loss of numerous plant and animal species that rely on forests for their habitats. As forests are cleared for agriculture, urbanization, and infrastructure development, species lose their homes and struggle to survive in fragmented landscapes. This loss of biodiversity not only diminishes the natural heritage of our planet but also disrupts essential ecosystem functions. Biodiversity loss can lead to reduced ecosystem resilience, making ecosystems more vulnerable to disturbances such as climate change and invasive species.

Moreover, many ecosystems rely on a diverse array of species to provide ecosystem services such as pollination, water purification, and carbon sequestration. The decline in biodiversity undermines the capacity of ecosystems to provide these crucial services, posing risks to human well-being and ecological stability. Addressing biodiversity loss requires concerted efforts to conserve and restore forest habitats, protect endangered species, and promote sustainable land-use practices that prioritize biodiversity conservation.

9.4.4 Lack of awareness

Limited awareness about the importance of forests and sustainable forest management practices hinders conservation efforts. Many people may not fully grasp the crucial role that forests play in maintaining biodiversity, regulating climate, and providing ecosystem services essential for human well-being. This lack of awareness can lead to activities such as illegal logging, encroachment, and unsustainable land-use practices that degrade forest ecosystems.

Additionally, communities residing near forests may not have access to education and information about sustainable forest management, perpetuating harmful practices and undermining conservation efforts. Increasing awareness through education, outreach programs, and community engagement initiatives is essential to foster a greater understanding of the value of forests and the need for their protection.

9.4.5 Inadequate policy implementation

While there are policies in place for forest conservation and management, their effective implementation is often hindered by various challenges. In some cases, corruption and lack of political will may undermine efforts to enforce forest regulations and combat illegal activities such as logging and land encroachment. Additionally, limited financial resources and institutional capacity may constrain the ability of governments to implement forestry policies effectively.

Furthermore, conflicting interests among stakeholders, such as indigenous communities, conservation organizations, and commercial enterprises, can complicate policy implementation and enforcement. Addressing these challenges requires strengthening governance structures, enhancing law enforcement capabilities, and fostering collaboration among stakeholders to ensure that forest policies are effectively implemented and enforced.

9.4.6 Lack of institutional capacity

Weak governance, inadequate law enforcement, and limited institutional capacity for forest management and conservation hinder efforts to address deforestation, illegal logging, and other threats to forests. Strengthening institutional capacity, improving regulatory frameworks, and enhancing collaboration among stakeholders are essential for effective forest management and conservation.

9.4.7 Climate change

Changing climate patterns, including increased temperatures and altered rainfall patterns, can impact forest health and regeneration. Climate change poses a serious threat to forests, altering weather patterns, increasing the frequency and intensity of natural disasters such as wildfires and storms, and disrupting ecosystems. Rising temperatures and changing precipitation patterns can impact forest health, productivity, and species composition, leading to shifts in forest distribution and ecosystem dynamics.

9.4.8 Unsustainable logging practices

Unsustainable logging practices, including illegal logging, clear-cutting, and selective logging, can degrade forest ecosystems, impair soil and water quality, and diminish timber resources. Poorly regulated logging operations can also lead to habitat destruction, erosion, and loss of biodiversity.

The detrimental impact of unsustainable logging practices on forest ecosystems cannot be overstated. Illegal logging, which often occurs in violation of established regulations and without regard for sustainability, poses a significant threat to forests worldwide. This illicit activity not only undermines efforts to conserve forest resources but also fuels corruption and organized crime in affected regions. Clear-cutting, another prevalent logging method, involves the removal of entire stands of trees, leaving vast areas devoid of vegetation and disrupting the natural habitat of countless plant and animal species. Selective logging, while less destructive than clear-cutting, still poses risks to forest integrity by targeting specific tree species without considering their ecological importance or long-term viability.

These unsustainable practices not only degrade forest ecosystems but also compromise soil stability, water quality, and carbon sequestration capacities. Addressing unsustainable logging requires robust enforcement of forestry regulations, adoption of sustainable forest management practices, and support for alternative livelihoods that reduce dependence on timber extraction. Additionally, promoting certification schemes like the Forest Stewardship Council (FSC) can incentivize responsible logging practices and encourage the sustainable use of forest resources.

9.4.9 Land-use conflicts

Conflicts over land use, including competing demands for forest resources for agriculture, mining, infrastructure development, and conservation, can result in deforestation, forest degradation, and social unrest. Resolving land-use conflicts requires balancing the needs of different stakeholders and promoting sustainable land management practices.

In many regions, land-use conflicts arise from competing interests and demands placed on forest resources. Rapid urbanization, agricultural expansion, and industrial development often encroach upon forested areas, leading to conflicts over land tenure, resource allocation, and environmental conservation. Additionally, indigenous communities and local populations may assert their customary rights to forest lands, challenging government policies and private interests. These conflicts can escalate into social unrest, legal disputes, and environmental degradation, further exacerbating the pressures on forest ecosystems.

Resolving land-use conflicts necessitates a holistic approach that considers the socioeconomic, cultural, and ecological dimensions of land management. Stakeholder engagement, participatory decision-making processes, and conflict resolution mechanisms are essential for fostering consensus and promoting sustainable land-use practices. Moreover, integrating land-use planning with environmental conservation strategies can help mitigate conflicts and ensure the equitable and sustainable management of forest resources for present and future generations.

9.4.10 Invasive species and pest infestations

Invasive species and pest infestations pose a threat to forest ecosystems, causing damage to trees, altering ecosystem dynamics, and reducing biodiversity. Invasive pests such as bark beetles and pathogens can devastate forested areas, leading to tree mortality and ecosystem disruptions.

The introduction of invasive species and pest infestations poses a significant threat to forest ecosystems worldwide. Invasive species, including insects, pathogens, and plants, can disrupt native ecosystems, outcompete indigenous species, and alter the composition and structure of forest communities. For example, invasive insects like the emerald ash borer and the Asian longhorned beetle have caused widespread damage to forested areas in North America by attacking and killing native tree species. Similarly, fungal pathogens such as sudden oak death and chestnut blight have decimated oak and chestnut forests, respectively, in various regions. These invasive species and pest infestations can lead to the loss of forest biodiversity, decline in ecosystem services, and economic losses for forest-dependent communities.

Effective management strategies, including early detection and rapid response measures, biological control methods, and quarantine protocols, are essential for preventing the spread of invasive species and mitigating their impacts on forest ecosystems. Additionally, promoting native species diversity, enhancing forest resilience through silvicultural practices, and implementing integrated pest management approaches can help reduce the vulnerability of forests to invasive species and pest outbreaks.

9.4.11 Weak enforcement

Weak enforcement of regulations and lack of monitoring and surveillance contribute to illegal logging and encroachments. In many regions, weak enforcement of forest regulations and inadequate monitoring and surveillance mechanisms contribute to rampant illegal logging and encroachments on forest lands. Limited resources, corruption, and governance challenges often undermine the effective implementation of laws aimed at protecting forests. As a result, illegal logging activities go unchecked, leading to widespread deforestation, habitat destruction, and loss of biodiversity. Moreover, the absence of strict enforcement measures creates opportunities for land grabbers and illegal loggers to exploit forest resources for their economic gain, exacerbating environmental degradation and social conflicts.

9.4.12 Inadequate monitoring and surveillance

Insufficient monitoring and surveillance systems further exacerbate the problem of illegal logging and encroachments on forest lands. The lack of proper monitoring technologies, such as satellite imagery and remote sensing, hinders authorities' ability to detect and deter illegal logging activities effectively. In addition, limited human and financial resources allocated to forest monitoring agencies limit their capacity to patrol and protect vast forested areas adequately. As a result, illegal loggers and land grabbers operate with impunity, further depleting forest resources and undermining conservation efforts. Strengthening monitoring and surveillance capabilities through the use of advanced technologies, increased funding, and capacity building initiatives is crucial for addressing illegal logging and safeguarding forest ecosystems.

9.4.13 Lack of research and data

Insufficient research and data on forest ecosystems hinder informed decision-making and effective forest management strategies. Limited scientific studies and data collection efforts on forest biodiversity, ecosystem dynamics, and the impacts of human activities constrain our understanding of forest ecosystems' complexities. Without comprehensive data on forest resources, biodiversity, and ecological processes, policymakers, conservationists, and land managers struggle to develop evidence-based policies and conservation plans.

Moreover, the absence of up-to-date information on forest degradation, deforestation rates, and the status of endangered species impedes efforts to monitor changes in forest ecosystems and prioritize conservation interventions. Investing in research initiatives, biodiversity monitoring programs, and data collection efforts is essential for enhancing our understanding of forest ecosystems and informing sustainable forest management practices. By generating reliable data and scientific knowledge, researchers can support evidence-based decision-making and contribute to the conservation and sustainable use of forest resources.

Addressing these challenges requires concerted efforts at the local, national, and international levels, as well as innovative approaches that integrate forest conservation, sustainable management, and community engagement. By addressing the root causes of deforestation, promoting sustainable livelihoods for forest-dependent communities, and enhancing forest resilience to climate change, it is possible to overcome forestry challenges and ensure the long-term health and sustainability of forest ecosystems.

9.5 Opportunities and progress

1. **Sustainable forest management (SFM):** Many countries and organizations are promoting SFM practices to balance environmental, social, and economic objectives. SFM aims to conserve forests for future generations while providing livelihoods and ecosystem services.
2. **Forest certification:** Forest certification schemes like the FSC and Programme for the Endorsement of Forest Certification (PEFC) help promote responsible forest management practices, ensuring that forest products come from sustainably managed sources.
3. **Reforestation and restoration initiatives:** There is a growing global emphasis on reforestation, afforestation, and forest restoration projects to restore degraded lands, enhance biodiversity, and sequester carbon. Initiatives like the Bonn Challenge aim to restore 350 million hectares of degraded and deforested land by 2030.
4. **Climate change mitigation:** Forests play a crucial role in mitigating climate change by sequestering carbon dioxide. International agreements like the Paris Agreement recognize the importance of forests in climate change mitigation strategies.
5. **Community-based forest management:** In many regions, community-based forest management approaches empower local communities to manage forests sustainably, promoting both conservation and livelihoods.
6. **Global forest watch (GFW):** GFW, an online platform, provides real-time data and monitoring of forests worldwide. This tool helps governments, NGOs, and stakeholders track deforestation, forest fires, and other threats to forests.

Examples by region

- **Amazon rainforest:** The Amazon, one of the world's largest tropical forests, faces ongoing threats from deforestation, agriculture, and wildfires. Efforts are

underway to protect and conserve this vital ecosystem through initiatives like the Amazon Conservation Vision 2030.
- **Asia-Pacific region:** Countries like Indonesia and Malaysia are working to address deforestation driven by palm oil production. Both countries have pledged to reduce deforestation and promote sustainable palm oil practices.
- **Africa:** Many African countries are focusing on sustainable forest management and restoration. The African Forest Landscape Restoration Initiative (AFR100) aims to restore 100 million hectares of degraded land by 2030.
- **Europe:** European countries have made progress in sustainable forest management, with many forests being actively managed for multiple purposes including timber production, biodiversity conservation, and recreation.

The global forestry situation is complex, with a mix of challenges and opportunities. While deforestation, biodiversity loss, and climate change impacts remain significant concerns, there are also efforts and initiatives worldwide to promote sustainable forest management, conservation, and restoration. Collaboration among governments, NGOs, local communities, and stakeholders is crucial to address these challenges and ensure the long-term health and resilience of forests globally.

On a global scale, various initiatives are underway to tackle similar challenges faced by forests worldwide. Afforestation campaigns, community-based forestry projects, strengthened policy frameworks, and enhanced awareness of sustainable forest management (SFM) practices are being promoted in many countries. However, concerted and holistic efforts are essential to ensure that forests across the globe continue to deliver ecological, economic, and social benefits for present and future generations. By implementing robust conservation measures, fostering community engagement, and improving governance mechanisms, countries can safeguard their forests and mitigate the adverse impacts of deforestation, degradation, and biodiversity loss on a global scale.

9.6 Integrated forest management

Integrated forest management (IFM) is an approach to managing forests that aims to balance ecological, economic, and social objectives while ensuring the sustainable use of forest resources. IFM considers the multifaceted nature of forests and recognizes their roles in providing ecosystem services, supporting biodiversity, mitigating climate change, and contributing to local livelihoods.

9.6.1 Key principles of integrated forest management

Key principles of IFM are as follows (Figure 9.2).

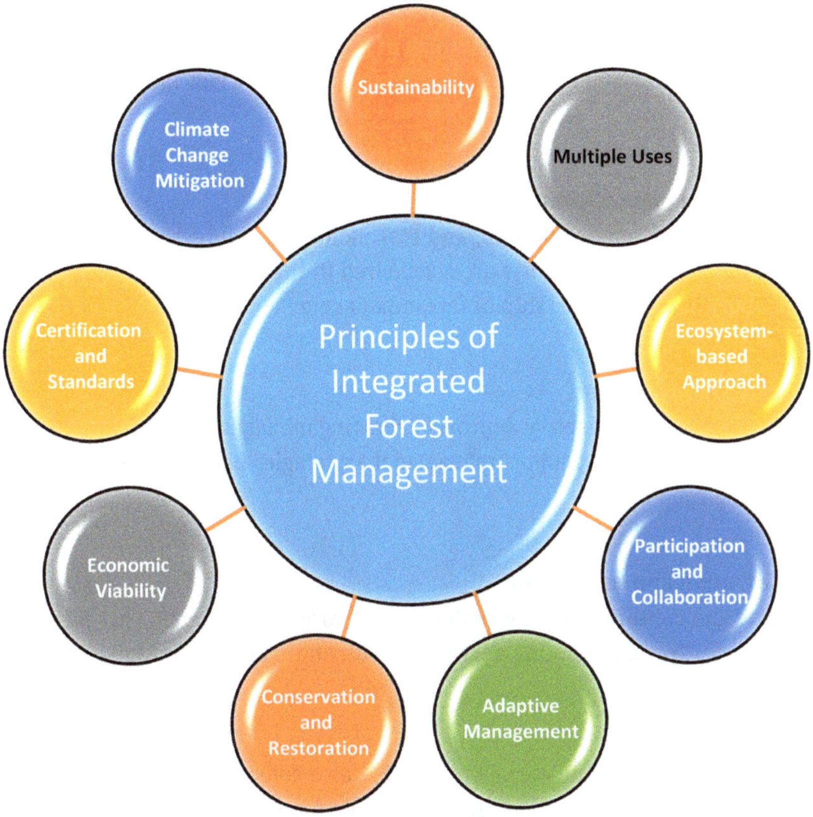

Figure 9.2: Key principles of integrated forest management.

9.6.1.1 Sustainability

IFM focuses on long-term sustainability, ensuring that the utilization of forest resources does not exceed their capacity for regeneration. It involves practices that allow forests to renew themselves naturally and maintain their ecological integrity.

9.6.1.2 Multiple uses

IFM recognizes that forests serve various purposes beyond timber production including non-timber forest products (such as fruits, mushrooms, and medicinal plants), water catchment, biodiversity conservation, recreation, and cultural values. Balancing these multiple uses is a fundamental aspect of IFM.

9.6.1.3 Ecosystem-based approach

IFM takes into account the ecosystem dynamics of forests. It considers factors like species composition, habitat diversity, and ecological processes when making management decisions.

9.6.1.4 Participation and collaboration

Local communities, indigenous peoples, and stakeholders are integral to IFM. Their knowledge, preferences, and interests are considered in the decision-making process, fostering collaboration and ownership of forest management efforts.

9.6.1.5 Adaptive management

IFM involves a flexible and adaptive approach, allowing managers to learn from experiences, monitor outcomes, and adjust management strategies based on new information and changing conditions.

9.6.1.6 Conservation and restoration

IFM emphasizes the conservation and restoration of forest ecosystems. This includes protecting biodiversity, rehabilitating degraded areas, and managing invasive species.

9.6.1.7 Economic viability

While prioritizing ecological and social values, IFM also considers economic sustainability. Responsible timber harvesting and sustainable forest enterprises can contribute to local economies without depleting forest resources.

9.6.1.8 Certification and standards

Forest certification schemes, such as the FSC, promote IFM practices by setting criteria for sustainable and responsible forest management. These standards help guide forest managers toward environmentally and socially responsible practices.

9.6.1.9 Climate change mitigation

IFM plays a role in mitigating climate change by promoting carbon sequestration through healthy forest ecosystems and sustainable timber management.

IFM recognizes that forests are complex systems with interconnected components. It seeks to harmonize the often competing demands of conservation, resource extraction, and human needs while safeguarding the long-term health and vitality of forest ecosystems.

IFM is a holistic approach to forest management that aims to balance ecological, social, and economic aspects for sustainable forest ecosystems. Here's a discussion with global examples illustrating the importance and benefits of IFM.

9.6.2 Importance of integrated forestry management

1. **Ecosystem health:** IFM recognizes the interconnectedness of forest ecosystems and aims to maintain their health and resilience. By considering the entire ecosystem including soils, water, flora, and fauna, IFM promotes sustainable forest health.
2. **Biodiversity conservation:** IFM emphasizes the conservation of biodiversity, recognizing forests as habitats for a wide range of species. This approach involves maintaining diverse forest ecosystems, protecting endangered species, and restoring degraded habitats.
3. **Community participation:** IFM involves local communities in decision-making processes, recognizing their traditional knowledge and rights. Engaging communities in forest management fosters sustainable practices and ensures the well-being of forest-dependent communities.

9.6.3 Examples of integrated forestry management

1. **Costa Rica:** Costa Rica's Payment for Ecosystem Services program is an example of successful IFM. The program incentivizes forest conservation and restoration by paying landowners for the ecosystem services their forests provide such as carbon sequestration and biodiversity conservation. This approach has led to increased forest cover and biodiversity protection while benefiting local communities economically.
2. **Canada:** The Canadian Model Forest Network is a collaborative initiative involving government, industry, Indigenous groups, and local communities. These model forests serve as testing grounds for sustainable forest management practices. An example is the Algonquin to Adirondacks Collaborative (A2A), where stakeholders work together to manage forests sustainably, protect wildlife corridors, and promote recreation and tourism.
3. **Germany:** Germany's SFM approach integrates ecological, economic, and social aspects. The country's mixed forests are managed with an emphasis on maintaining biodiversity, supporting local livelihoods, and providing ecosystem services. Forest certification schemes such as the FSC and PEFC ensure compliance with sustainable management practices.
4. **Brazil:** The Brazilian Amazon Sustainable Landscape Program (ASL) focuses on IFM in the Amazon rainforest. This program brings together government agencies, NGOs, and local communities to promote sustainable land-use practices, reduce deforestation, and improve livelihoods. Projects within the ASL program include agroforestry systems, community-managed conservation areas, and sustainable logging practices.

5. **Australia:** In Australia, the FSC certification is used to promote IFM practices. Forests certified under FSC adhere to rigorous standards that prioritize biodiversity conservation, protection of cultural values, and sustainable timber harvesting. An example is the Gondwana Link project, which aims to reconnect fragmented landscapes to support biodiversity conservation and sustainable land management.

IFM offers a holistic approach to forest management, addressing the complex interplay of ecological, social, and economic factors. The examples from Costa Rica, Canada, Germany, Brazil, and Australia demonstrate the diverse applications of IFM worldwide. By prioritizing ecosystem health, biodiversity conservation, and community participation, IFM provides a pathway toward sustainable and resilient forest ecosystems that benefit both present and future generations. Policymakers, forest managers, and stakeholders are encouraged to embrace IFM to ensure the long-term health and vitality of our forests.

9.6.4 Benefits of integrated forest management

IFM offers a range of benefits that contribute to the sustainable use, conservation, and enhancement of forest ecosystems.

Key benefits of IFM.

9.6.4.1 Biodiversity conservation

IFM promotes diverse forest habitats that support a wide array of plant and animal species. By considering ecosystem dynamics, IFM helps maintain biodiversity, protect endangered species, and enhance habitat quality.

9.6.4.2 Ecosystem services

Forests provide essential ecosystem services such as clean water, carbon sequestration, climate regulation, and soil conservation. IFM ensures that these services are maintained for the benefit of both local communities and the broader environment.

9.6.4.3 Sustainable resource use

IFM balances the extraction of forest resources, such as timber and non-timber forest products, with their capacity for regeneration. This prevents overexploitation and ensures a continuous supply of valuable resources.

9.6.4.4 Climate change mitigation

Well-managed forests play a significant role in mitigating climate change by capturing and storing carbon dioxide from the atmosphere. IFM supports healthy forest growth and sustainable timber harvesting, both of which contribute to carbon sequestration.

9.6.4.5 Resilience to disturbances

IFM fosters forest resilience to natural disturbances such as wildfires, pests, and diseases. Diverse ecosystems and appropriate management practices can enhance a forest's ability to recover from such events.

9.6.4.6 Local livelihoods

Many communities rely on forests for their livelihoods. IFM can create opportunities for sustainable livelihoods through activities like eco-tourism, non-timber forest product collection, and community-based forest enterprises.

9.6.4.7 Cultural and spiritual values

Forests hold cultural and spiritual significance for many communities. IFM acknowledges these values and incorporates traditional knowledge and practices into management decisions.

9.6.4.8 Reduced land degradation

IFM practices like reforestation, afforestation, and erosion control help combat land degradation, preventing soil erosion and promoting soil fertility.

9.6.4.9 Community engagement

IFM involves local communities in decision-making processes, fostering a sense of ownership and responsibility for forest resources. This participatory approach often leads to more effective and sustainable management.

9.6.4.10 Consistent timber supply

Sustainable timber management under IFM ensures a consistent supply of high-quality timber over the long term, supporting industries while safeguarding forest health.

9.6.4.11 Improved water management

Forests play a crucial role in regulating water cycles. IFM helps maintain water quality, reduce erosion, and enhance water availability downstream.

9.6.4.12 Enhanced landscape aesthetics

IFM promotes well-maintained and diverse forests that contribute to scenic landscapes, recreational opportunities, and ecotourism.

9.6.4.13 Long-term benefits

IFM's focus on sustainability and adaptive management leads to long-term benefits, preserving forests for future generations. IFM is not just about immediate gains but also about securing long-term benefits for future generations. By prioritizing sustainability and adaptive management, IFM practices ensure that forests remain healthy, productive, and resilient over time.

This long-term perspective offers several key benefits:

Biodiversity conservation: IFM strategies prioritize the preservation and enhancement of forest biodiversity, ensuring that diverse plant and animal species continue to thrive within forest ecosystems. By maintaining intact habitats, protecting critical wildlife corridors, and promoting habitat restoration, IFM safeguards biodiversity for future generations to enjoy and benefit from.

Ecosystem services: Forests provide a wide range of ecosystem services including carbon sequestration, water regulation, soil stabilization, and air purification. IFM practices such as reforestation, agroforestry, and watershed management help maintain these vital ecosystem services, ensuring the continued provision of clean air, water, and other resources for future generations.

Climate resilience: As climate change poses increasingly severe threats to forests, IFM plays a critical role in enhancing climate resilience. By promoting diverse forest ecosystems, reducing vulnerability to extreme weather events, and enhancing carbon storage, IFM helps forests adapt to changing climatic conditions, ensuring their survival and productivity for future generations.

Socioeconomic benefits: IFM practices support sustainable livelihoods and economic opportunities for local communities dependent on forest resources. By integrating community participation, equitable benefit-sharing, and traditional knowledge into forest management plans, IFM ensures that forests continue to provide for the needs and aspirations of present and future generations.

Cultural and spiritual values: Forests hold deep cultural and spiritual significance for many communities around the world. IFM recognizes the intrinsic value of forests as places of cultural heritage, spiritual renewal, and connection to nature. By preserving traditional land-use practices, sacred sites, and cultural landscapes, IFM honors these values and ensures their transmission to future generations.

IFM (IFM) offers a vision for sustainable forest management that extends far beyond the present moment. By focusing on long-term sustainability, IFM practices secure forests as invaluable resources for future generations, ensuring that they continue to provide essential ecosystem services, support biodiversity, and enrich the lives of people around the world for years to come.

IFM offers a holistic approach to forest management that recognizes the intricate interplay between ecological integrity, economic viability, and social well-being. By integrating principles of sustainable forestry, biodiversity conservation, and community engagement, this approach seeks to balance the needs of both present and future generations. Through careful planning and management, IFM aims to optimize the multiple benefits provided by forests including timber production, carbon sequestration, watershed protection, and recreational opportunities. Moreover, by fostering collaboration among stakeholders, including government agencies, local communities, and indigenous groups, IFM promotes inclusive decision-making processes that reflect diverse perspectives and values. Ultimately, this approach contributes to the resilience of forest ecosystems, the sustainability of livelihoods dependent on forests, and the overall health of the planet.

In the verdant embrace of integrated forestry management, trees stand as silent sentinels of sustenance and shelter. This chapter's discourse unfolds the significance of balancing wood production, carbon sequestration, and habitat preservation – a delicate choreography that resonates with the rhythm of ecological stewardship. As we delve into the heart of this chapter, we bear witness to the intricate relationship between forests, rural development, and the promise of a greener future.

With the whispers of rustling leaves still echoing, we find ourselves immersed in the landscape of Chapter 10. Here, the chapters of land, water, and minerals converge, fostering a holistic perspective that acknowledges the intricate tapestry of our planet's treasures. Through the exploration of sustainable land use, water conservation, and responsible resource extraction, this chapter offers a roadmap toward safeguarding the legacy of our natural endowments.

References

Aggarwal, R. K., & Pandey, A. K. (2004). Forest management and rural development: Experiences of Uttarakhand, India. Small-Scale Forest Economics, Management and Policy, 3(1), 99–118.

Amanullah. (2024). Harvesting Success: A Comprehensive Guide to Agronomy Courses Punjab Books, Lahore, PK.

Amanullah., Ondrasek, G., Al –Tawaha, A. R. M. S., eds. (2023). Integrated nutrients management: An approach for sustainable crop production and food security in changing climates. Lausanne: Frontiers Media SA. doi: 10.3389/978-2-8325-3169-3

Amanullah, & Khalid, S. (2020). Agronomy – Climate Change & Food Security IntechOpen, London, United Kingdom.

Amanullah, & Fahad, S. (2018). Nitrogen in Agriculture-Updates InTechOpen, London, UK.

Amanullah, & Fahad, S. (2018). Corn – Production and Human Health in Changing Climate InTech, London, UK.

Arnold, J. E., Perez, M. R., & Belcher, B. (2014). Toward understanding the roles of forests in rural livelihoods. World Development, 64, S32–S45.

Charnley, S., & Poe, M. R. (2007). Community forestry in theory and practice: Where are we now?. Annual Review of Anthropology, 36, 301–336.

Colfer, C. J. P., & Capistrano, D. (2005). Gender perspectives on access to forests in Indonesia: Implications for women's participation in sustainable forest management. Ecology and Society, 10(1), 21.

DeFries, R. S., Rudel, T., Uriarte, M., & Hansen, M. (2010). Deforestation driven by urban population growth and agricultural trade in the twenty-first century. Nature Geoscience, 3(3), 178–181.

FAO. (2018). Global Forest Resources Assessment 2020: Key Findings Food and Agriculture Organization of the United Nations, Rome, Italy.

Gurung, A. B. (2003). The Nepal community forestry program: A decade of success. Journal of Forest and Livelihood, 3(1), 10–12.

Katila, P., Galloway, G., de Jong, W., & Pacheco, P. (2011). Seeing REDD+ as a project of environmental governance. Environmental Science & Policy, 14(2), 100–110.

Myers, N., Mittermeier, R. A., Mittermeier, C. G., da Fonseca, G. A., & Kent, J. (2000). Biodiversity hotspots for conservation priorities. Nature, 403(6772), 853–858.

Chapter 10
Integrated natural resources management: a global perspective

Introduction

Integrated natural resources management (INRM) is a holistic approach to managing natural resources that recognizes the interconnectedness of various environmental components and seeks to balance ecological, economic, and social objectives. INRM integrates principles and practices from different disciplines, including ecology, economics, sociology, and governance, to promote sustainable resource use and conservation. INRM seeks to promote the sustainable management of natural resources by integrating ecological, economic, and social considerations and fostering collaboration among stakeholders. By adopting an integrated approach, INRM aims to address pressing environmental issues, such as biodiversity loss, land degradation, water scarcity, and climate change, while supporting livelihoods and enhancing resilience to global change.

This chapter delves into the multifaceted realm of INRM from a global standpoint. With the growing recognition of the interconnectedness between environmental sustainability, social equity, and economic development, INRM emerges as a holistic approach that seeks to optimize the use of natural resources while ensuring their long-term conservation and equitable distribution. By exploring diverse case studies, best practices, and innovative solutions from around the world, this chapter aims to shed light on the transformative potential of INRM in addressing complex environmental challenges and promoting sustainable development on a global scale.

10.1 Natural resources

Natural resources refer to the materials or substances that exist in the environment naturally and are used by humans for their benefit. These resources are essential for the survival and development of civilizations, providing raw materials for industries, energy for powering our lives, and ecosystems that support biodiversity. From water and air to minerals, forests, and fossil fuels, natural resources form the foundation of our economy and society. They are crucial for agriculture, manufacturing, transportation, and even recreation. The significance of natural resources cannot be overstated, as they are the backbone of our existence, directly impacting our quality of life, economic growth, and the sustainability of our planet. Natural resources are essential elements of the Earth's ecosystems that are used by humans for various purposes. As we recognize their finite nature, there is an increasing focus on sustainable management and conservation to ensure these vital resources are available for future generations.

https://doi.org/10.1515/9783111448077-010

Here are some major global natural resources along with examples.

10.1.1 Water resources

- **Freshwater:** Essential for drinking, agriculture, industry, and ecosystems.
  - Example: rivers (Nile, Amazon), lakes (Lake Baikal, Lake Victoria), groundwater.

10.1.2 Minerals

- **Metals:** Used in manufacturing, construction, and technology.
  - Example: Iron ore, copper, gold, aluminum.
- **Nonmetals:** Used in construction, agriculture, and industry.
  - Example: Coal, oil, natural gas, uranium.

10.1.3 Forests

- **Timber:** Used for construction, furniture, paper, and fuel.
  - Example: Boreal forests (Russia, Canada), Tropical rainforests (Amazon, Congo Basin).

10.1.4 Soil

- **Arable land:** Vital for agriculture and food production.
  - Example: The fertile plains of the Nile Delta, the Great Plains of the United States.

10.1.5 Biodiversity

- **Plant and animal species:** Crucial for ecosystem balance, medicine, and genetic diversity.
  - Example: Amazon Rainforest (Brazil), Coral Reefs (Great Barrier Reef, Australia).

10.1.6 Renewable energy sources

- **Solar:** Used for electricity generation.
  - Example: Solar farms in the Mojave Desert (USA), Atacama Desert (Chile).

- **Wind:** Used for electricity generation.
  - Example: Wind farms in North Sea (Europe), Gobi Desert (China).
- **Hydropower:** Used for electricity generation.
  - Example: Three Gorges Dam (China), Itaipu Dam (Brazil/Paraguay).

10.1.7 Fisheries

- **Marine life:** Source of food and livelihood for millions.
  - Example: North Sea (Europe), Bay of Bengal (Asia).

10.1.8 Geothermal energy

- **Geothermal Heat:** Used for electricity generation and heating.
  - Example: Geothermal power plants in Iceland, Geysers (Yellowstone National Park, USA).

10.1.9 Diamonds and gemstones

- **Precious stones:** Used in jewelry and industry.
  - Example: Diamonds (Mines in South Africa, Russia), emeralds (Colombia), rubies (Myanmar).

10.1.10 Oceans

- **Sea Resources:** Provide food, minerals, transportation, and recreation.
  - Example: Pacific Ocean (Tuna fisheries), Atlantic Ocean (oil and gas reserves).

10.1.11 Air

- **Atmosphere:** Provides oxygen for respiration, carbon dioxide for photosynthesis.
  - Example: Global air quality and composition.

10.1.12 Phosphorus

- **Phosphate rock:** Essential for agriculture (fertilizers).
  - Example: Major deposits in Morocco, China, and the United States.

10.1.13 Rare earth elements

- **Critical for electronics:** Used in smartphones, computers, and renewable energy technologies.
  - Example: China is a major producer of rare earth elements.

10.1.14 Freshwater aquifers

- **Groundwater reserves:** Important for drinking water and irrigation.
  - Example: Ogallala Aquifer (USA), Guarani Aquifer (South America).

10.1.15 Platinum group metals

- **Used in catalytic converters:** Important for reducing vehicle emissions.
  - Example: Major deposits in South Africa, Russia, and Zimbabwe.

10.1.16 Copper

- **Versatile metal:** Used in electronics, construction, and plumbing.
  - Example: Major mines in Chile, Peru, and the United States.

10.2 Natural resources of Pakistan

Pakistan is blessed with a diverse array of natural resources that have profoundly influenced its landscapes, economy, and societal fabric (Figure 10.1). Situated amidst the majestic peaks of the Himalayas and the expansive deserts of Balochistan, Pakistan's geography encompasses fertile plains, meandering river valleys, extensive coastlines, and rugged mountain ranges. This geographical diversity, coupled with varied climatic conditions, has bestowed upon Pakistan a wealth of invaluable resources including fertile soils, abundant water bodies, mineral riches, and diverse biodiversity. These natural blessings have not only sustained generations through agriculture, fisheries, and livestock rearing but have also served as the cornerstone of industries, employment opportunities, and economic prosperity. As Pakistan endeavors to navigate the intricate challenges of modern development and environmental sustainability, a comprehensive understanding and judicious management of these natural resources assume paramount importance. Ensuring the prudent utilization and conservation of Pakistan's natural endowments are essential steps toward fostering a prosperous and harmonious future for both its citizens and its unique ecosystems on a global scale.

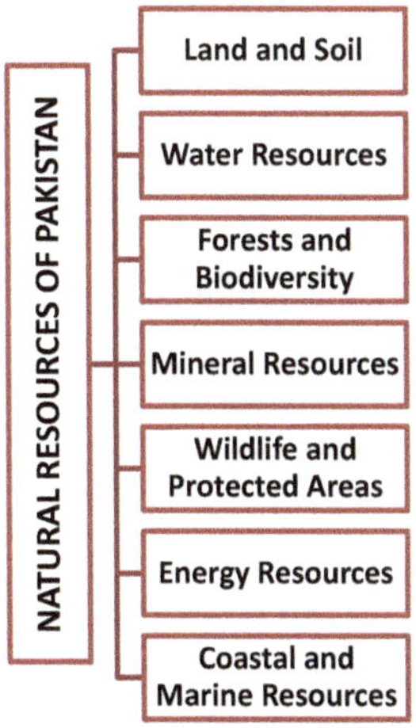

Figure 10.1: Natural resources of Pakistan.

10.2.1 Land and soil

Pakistan has a wide range of land types including fertile alluvial plains, arid and semi-arid regions, and mountainous areas. The agricultural sector heavily relies on land resources for crop cultivation and livestock grazing. Sustainable land management practices are essential to combat soil erosion, degradation, and land fragmentation.

10.2.2 Water resources

Water is a critical natural resource in Pakistan due to its agricultural, industrial, and domestic importance. The country's major rivers, including the Indus, Jhelum, Chenab, Ravi, and Sutlej, provide water for irrigation, hydropower generation, and domestic use. Efficient water management practices, such as reservoirs, canal systems, and water conservation measures, are vital for sustainable water resource management.

10.2.3 Forests and biodiversity

Pakistan is home to diverse forests including temperate forests in the north, mangroves along the coastal areas, and arid forests in the Balochistan region. Forests contribute to carbon sequestration, provide habitat for wildlife, protect watersheds, and support rural livelihoods. Conservation and sustainable management of forests are crucial for biodiversity conservation and environmental sustainability.

10.2.4 Mineral resources

Pakistan has significant mineral resources including coal, natural gas, oil, copper, gold, limestone, and salt. These resources contribute to energy production, industrial development, and export earnings. Responsible extraction and management of mineral resources are important for sustainable economic growth and minimizing environmental impacts.

10.2.5 Wildlife and protected areas

Pakistan is home to diverse wildlife species, including mammals, birds, reptiles, and marine life. The country has several protected areas, national parks, and wildlife sanctuaries aimed at conserving biodiversity and preserving endangered species. Effective management and conservation efforts are essential to protect wildlife habitats and promote ecotourism.

10.2.6 Energy resources

Pakistan possesses various energy resources including fossil fuels (coal, oil, and natural gas), hydropower potential, wind energy, solar energy, and biomass. Sustainable energy management and diversification of energy sources are crucial for meeting the growing energy demands, reducing dependency on fossil fuels, and mitigating climate change impacts.

10.2.7 Coastal and marine resources

Pakistan's coastal areas along the Arabian Sea harbor diverse marine ecosystems, mangrove forests, and fisheries resources. These resources contribute to seafood production, support coastal communities, and provide ecosystem services. Sustainable coastal and marine resource management is essential to protect marine biodiversity, ensure sustainable fishing practices, and maintain the ecological balance.

Managing and conserving these natural resources is of paramount importance to ensure sustainable development, mitigate environmental degradation, promote resilience to climate change, support livelihoods, and safeguard the well-being of present and future generations. These natural resources are critical for human survival, economic development, and technological progress. However, sustainable management and conservation efforts are necessary to ensure their availability for future generations.

10.3 Integrated natural resources management (INRM)

INRM encompasses a wide range of practices and strategies aimed at managing natural resources in a holistic and integrated manner. From land and water to forests, fisheries, and wildlife, INRM seeks to optimize the use of these resources while maintaining their ecological integrity and supporting the livelihoods of local communities. By integrating ecological, social, and economic considerations into decision-making processes, INRM offers a comprehensive approach to addressing pressing environmental issues such as deforestation, water scarcity, soil degradation, and biodiversity loss.

The key components of INRM are given as follows (Figure 10.2).

Figure 10.2: Key components of integrated natural resources management.

10.3.1 Holistic approach

Taking a holistic approach, INRM recognizes the intricate interplay between various components of the environment and the multifaceted needs of human societies. This approach acknowledges that natural resources are interconnected and that their management must address ecological, social, and economic dimensions to achieve sustainable development.

In understanding the ecological dimension, INRM recognizes the complex web of interactions within ecosystems. It considers factors such as biodiversity, habitat integrity, and ecosystem resilience when making decisions about resource management. For example, INRM might prioritize the conservation of critical habitats or the restoration of degraded ecosystems to maintain ecological balance and support the health of natural systems.

On the social front, INRM takes into account the diverse needs, values, and aspirations of communities that depend on natural resources for their livelihoods and well-being. This involves engaging with local stakeholders, including indigenous communities, farmers, fishers, and other resource users, to ensure that their voices are heard and their rights respected in decision-making processes. By incorporating social considerations, INRM aims to promote equity, social justice, and community empowerment in natural resource management.

Furthermore, INRM recognizes the economic dimension of natural resource management, acknowledging the importance of balancing economic development with environmental sustainability. This involves assessing the costs and benefits of resource use, exploring alternative livelihood opportunities, and promoting green economic growth. By integrating economic considerations, INRM seeks to foster prosperity while ensuring the long-term health and productivity of natural systems.

INRM seeks to strike a balance between ecological integrity, social equity, and economic prosperity. By addressing the complex interactions between humans and the environment and considering the needs of both present and future generations, INRM offers a comprehensive framework for achieving sustainable development and securing the health and resilience of natural resources for the benefit of all.

10.3.2 Multistakeholder collaboration

Multistakeholder collaboration lies at the heart of INRM, emphasizing the importance of involving diverse stakeholders in decision-making processes and action plans. This collaborative approach recognizes that effective natural resource management requires the collective efforts of various actors with different perspectives, interests, and expertise.

Government agencies play a crucial role in INRM by formulating policies, regulations, and strategies that guide natural resource management at the national and local levels. Their involvement ensures that INRM initiatives align with broader development goals, legal frameworks, and regulatory standards. Moreover, government agencies often provide the necessary institutional support, technical expertise, and financial resources to implement INRM projects and initiatives.

Local communities are essential stakeholders in INRM, as they are often the primary users and stewards of natural resources. Their traditional knowledge, practices, and perceptions of the environment contribute valuable insights to resource management decisions. Engaging local communities in INRM processes not only fosters ownership and empowerment but also enhances the effectiveness and sustainability of management interventions by ensuring they are culturally appropriate and socially acceptable.

Civil society organizations (CSOs) play a critical role in advocating for environmental protection, social justice, and community empowerment. CSOs often act as intermediaries between governments, communities, and other stakeholders, facilitating dialogue, building capacity, and promoting inclusive decision-making processes. Their involvement in INRM can help bridge gaps between different stakeholders, strengthen social cohesion, and mobilize resources for collective action.

The private sector, including businesses, industries, and corporations, also has a stake in INRM, as they rely on natural resources for raw materials, energy, and other inputs. By engaging with the private sector, INRM can leverage corporate resources, expertise, and innovation to promote sustainable production practices, responsible resource extraction, and corporate social responsibility initiatives. Collaboration with the private sector can unlock opportunities for investment, technology transfer, and market-based solutions to environmental challenges.

Multistakeholder collaboration is essential for the success of INRM initiatives, as it fosters shared ownership, builds trust, and mobilizes collective action toward achieving shared environmental goals. By bringing together government agencies, local communities, CSOs, and the private sector, INRM can harness the diverse perspectives, resources, and capacities needed to address complex environmental challenges and promote sustainable development.

10.3.3 Adaptive management

INRM embraces adaptive management principles, recognizing the dynamic nature of ecosystems and the inherent uncertainties associated with natural resource management. By adopting a flexible and iterative approach, INRM enables stakeholders to monitor environmental changes, adjust management practices, and respond effectively to emerging challenges and opportunities.

Adaptive management is a cornerstone of INRM, reflecting an understanding of the dynamic and often unpredictable nature of ecosystems as well as the inherent uncertainties associated with managing natural resources. This approach recognizes that ecosystems are complex, interconnected systems that can undergo rapid changes in response to various factors such as climate variability, human activities, and natural disturbances.

At its core, adaptive management involves a flexible and iterative process that allows stakeholders to continuously learn from their experiences, adjust their strategies, and respond effectively to emerging challenges and opportunities.

10.3.4 Monitoring and evaluation

INRM initiatives begin with the establishment of monitoring and evaluation mechanisms to assess the status and trends of natural resources as well as the effectiveness of management interventions. This may involve collecting data on ecological indicators, socioeconomic variables, and stakeholder feedback to inform decision-making processes.

10.3.5 Learning and knowledge sharing

Adaptive management emphasizes the importance of learning from both successes and failures. Stakeholders engage in regular knowledge sharing activities, such as workshops, seminars, and field visits, to exchange experiences, share best practices, and build collective understanding of ecosystem dynamics and management strategies.

10.3.6 Experimentation and innovation

Adaptive management encourages stakeholders to experiment with different management approaches and innovative solutions to address complex environmental challenges. This may involve piloting new techniques, conducting demonstration projects, or implementing small-scale interventions to test their feasibility and effectiveness before scaling up.

10.3.7 Flexibility and adaptability

Perhaps most importantly, adaptive management requires stakeholders to remain flexible and open to change. As new information becomes available or unexpected events occur, stakeholders must be willing to adjust their management practices, revise their strategies, and adapt to changing circumstances in real time.

By embracing adaptive management principles, INRM enables stakeholders to navigate the uncertainties and complexities of natural resource management more effectively. It empowers them to make informed decisions based on scientific evidence, local knowledge, and stakeholder input, while also fostering a culture of continuous learning, innovation, and resilience in the face of environmental change. Ultimately, adaptive management enhances the capacity of stakeholders to achieve their conservation and development objectives in a rapidly changing world.

10.3.8 Sustainable practices

INRM promotes the adoption of sustainable practices that enhance resource efficiency, minimize environmental impacts, and support long-term ecosystem resilience. From sustainable agriculture and agroforestry to integrated water management and habitat restoration, INRM emphasizes the importance of balancing human needs with ecological sustainability.

Sustainable practices are integral to INRM as they form the foundation for achieving long-term environmental sustainability, social equity, and economic prosperity. These practices are designed to enhance resource efficiency, minimize negative environmental impacts, and promote the resilience of ecosystems over time. INRM recognizes that human well-being depends on healthy and functioning ecosystems and therefore emphasizes the adoption of sustainable practices across various sectors and activities.

10.3.9 Sustainable agriculture

Sustainable agricultural practices focus on maximizing productivity while minimizing negative environmental impacts such as soil degradation, water pollution, and biodiversity loss. This may involve techniques such as crop rotation, organic farming, agroecology, and integrated pest management to improve soil health, conserve water, and reduce reliance on chemical inputs. By promoting sustainable agriculture, INRM seeks to ensure food security, enhance rural livelihoods, and protect natural resources for future generations.

10.3.10 Agroforestry

Agroforestry integrates trees and shrubs into agricultural landscapes to provide multiple benefits including soil conservation, carbon sequestration, and biodiversity conservation. Agroforestry systems can improve soil fertility, regulate water flow, and provide habitats for wildlife while also diversifying farm income and enhancing resilience to climate change. INRM encourages the adoption of agroforestry practices as a sustainable land-use strategy that balances agricultural production with environmental conservation.

10.3.11 Integrated water management

Integrated water management aims to optimize the use of water resources while safeguarding aquatic ecosystems and meeting the needs of various users including agriculture, industry, and urban populations. This may involve measures such as water harvesting, drip irrigation, water recycling, and watershed management to promote water efficiency, reduce pollution, and mitigate the impacts of droughts and floods. INRM recognizes the importance of managing water resources sustainably to ensure equitable access, ecosystem health, and water security for all.

10.3.12 Habitat restoration

Habitat restoration involves restoring degraded ecosystems, such as forests, wetlands, and mangroves, to their natural or near-natural condition to enhance biodiversity, ecosystem services, and resilience to environmental change. This may include reforestation, revegetation, and habitat enhancement projects aimed at recovering ecosystems that have been degraded by human activities such as deforestation, land degradation, and pollution. INRM supports habitat restoration as a means of conserving biodiversity, enhancing ecosystem functions, and promoting landscape resilience.

Sustainable practices are essential for achieving the goals of INRM by promoting the wise and responsible use of natural resources for the benefit of both current and future generations. By prioritizing sustainability in agriculture, forestry, water management, and habitat restoration, INRM seeks to ensure the long-term health and resilience of ecosystems while also supporting human well-being and socioeconomic development. By prioritizing environmental sustainability, social equity, and economic prosperity, INRM offers a pathway toward a more resilient and prosperous future for people and the planet.

10.4 Natural resource management (NRM)

Natural resource management (NRM) refers to the sustainable and responsible use, conservation, and management of natural resources to meet present and future needs while ensuring environmental integrity and social well-being. It involves the integrated management of various natural resources including land, water, forests, wildlife, minerals, and energy sources.

NRM is a holistic approach that encompasses the sustainable and responsible utilization, conservation, and governance of natural resources to meet the needs of present and future generations. At its core, NRM aims to ensure the balance between environmental integrity, social well-being, and economic development.

One of the fundamental principles of NRM is sustainability, which entails using natural resources in a manner that does not deplete or degrade them over time. This involves employing practices that maintain the productivity and resilience of ecosystems such as sustainable agriculture, forest management, and water conservation.

Moreover, NRM emphasizes the integration of various natural resources, recognizing their interconnectedness and interdependence. For instance, the management of water resources may affect both agricultural productivity and ecosystem health, highlighting the need for integrated approaches that consider multiple factors and stakeholders.

Furthermore, NRM involves conservation efforts aimed at protecting biodiversity, preserving habitats, and safeguarding ecosystem services. By conserving natural resources, NRM helps maintain essential ecological functions, such as pollination, water purification, and carbon sequestration, which are vital for human well-being and planetary health.

Effective NRM also requires sound governance mechanisms that promote transparency, accountability, and community participation. This entails developing and implementing policies, regulations, and management plans that balance the interests of various stakeholders while safeguarding the long-term sustainability of natural resources.

NRM is essential for addressing the complex challenges of environmental degradation, resource depletion, and climate change. By promoting sustainability, integration, conservation, and good governance, NRM contributes to building resilient and thriving ecosystems that support both human prosperity and ecological resilience.

10.5 Conservation agriculture and NRM

Conservation agriculture (CA) is a farming approach that emphasizes sustainable land management practices to enhance productivity while minimizing negative impacts on natural resources. It is closely related to natural resource management as it seeks to maintain and improve the health of the soil, water, and biodiversity, all of which are essential components of a healthy and productive agricultural ecosystem.

10.5.1 Conservation agriculture is connected to NRM

1. **Soil conservation:** One of the core principles of conservation agriculture is minimal soil disturbance. This means reducing or eliminating tillage, which helps pre-

vent soil erosion, improves soil structure, and retains moisture. By minimizing soil erosion, the fertility and productivity of the land are preserved, and sediment runoff into water bodies is reduced.

2. **Soil health:** Conservation agriculture encourages the use of cover crops and crop residues to protect the soil from erosion and maintain its organic matter content. This practice enhances soil structure, water-holding capacity, and nutrient retention, leading to healthier and more resilient soils.
3. **Erosion control:** CA mitigates soil erosion by protecting the soil surface with crop residues or cover crops. This helps prevent wind and water erosion, preserving the topsoil and minimizing nutrient loss. By reducing erosion, CA helps maintain soil productivity, prevents sedimentation in water bodies, and protects downstream ecosystems.
4. **Water management:** CA promotes practices like crop rotation and mulching, which help conserve soil moisture and reduce water evaporation. These strategies are essential for efficient water use, particularly in regions prone to water scarcity.
5. **Biodiversity:** Conservation agriculture promotes the use of diverse cover crops and crop rotations. These practices support biodiversity by providing habitat for beneficial insects and microorganisms, reducing the need for chemical inputs, and enhancing overall ecosystem health.

10.5.2 Integrated and sustainable biodiversity management

Integrated and sustainable biodiversity management refers to a holistic approach that aims to conserve and manage biodiversity while considering its interactions with ecosystems, human activities, and socioeconomic factors. Biodiversity encompasses the variety of life forms on Earth including plants, animals, microorganisms, and their genetic variations. This approach emphasizes the integration of biodiversity conservation into various sectors, such as agriculture, forestry, fisheries, and urban planning, to achieve both ecological and socioeconomic goals.

10.5.2.1 Importance of integrated and sustainable biodiversity management

Ecosystem services: Biodiversity provides essential ecosystem services such as pollination, soil fertility, water purification, and climate regulation. In Pakistan, a country with diverse ecosystems, these services contribute directly to agriculture, water management, and disaster risk reduction.

Food security: Biodiversity is the foundation of agriculture and fisheries. Integrated biodiversity management ensures genetic diversity in crops and livestock, making these systems more resilient to pests, diseases, and changing climatic conditions. It also supports sustainable fisheries management, crucial for food security and livelihoods.

Medicinal and traditional knowledge: Pakistan's biodiversity is a source of traditional medicines and indigenous knowledge. Sustainable management of biodiversity ensures that these resources are conserved and utilized in culturally sensitive ways, benefiting local communities.

Tourism and recreation: Pakistan's unique biodiversity, including rare species and diverse landscapes, attracts ecotourism and recreational activities. Integrated management safeguards these attractions while promoting sustainable tourism practices.

Climate change mitigation and adaptation: Biodiversity-rich ecosystems, such as forests and wetlands, play a role in carbon sequestration and reducing the impacts of climate change. Conserving these ecosystems enhances climate resilience.

Rural livelihoods: Many rural communities in Pakistan depend on biodiversity-related activities such as agriculture, forestry, and fishing. Integrated biodiversity management helps these communities adopt sustainable practices that ensure their long-term well-being.

Cultural and spiritual values: Biodiversity holds cultural and spiritual significance for many communities in Pakistan. Sustainable management respects these values and contributes to cultural preservation.

Conflict resolution: Biodiversity can be a source of conflicts over resource use. Integrated management approaches help balance competing interests and prevent conflicts.

International commitments: Pakistan is a signatory to international agreements like the Convention on Biological Diversity. Implementing integrated biodiversity management demonstrates the country's commitment to global conservation efforts.

Long-term sustainability: Integrating biodiversity management into various sectors ensures that economic development is not achieved at the expense of natural resources. It promotes sustainable practices that maintain biodiversity for future generations.

Integrated and sustainable biodiversity management is of paramount importance to Pakistan due to its rich natural heritage, ecological significance, and direct impact on people's lives. By adopting a holistic approach that considers biodiversity as an integral part of development, Pakistan can achieve both conservation goals and sustainable socioeconomic progress.

10.5.3 Wildlife conservation

CA encourages crop rotation or diversification, which promotes biodiversity in agricultural landscapes. A diverse range of crops provides habitats for beneficial insects, pollinators, and wildlife. By creating a more balanced and resilient ecosystem, CA contributes to the conservation of wildlife and ecological balance.

10.5.3.1 Integrated and sustainable wildlife management

Integrated and sustainable wildlife management involves the conservation and responsible use of wildlife resources while considering ecological, social, and economic factors. This approach seeks to balance the needs of wildlife conservation with human activities, ensuring the long-term survival of species and the well-being of local communities.

Integrated and sustainable wildlife management holds immense global significance in safeguarding biodiversity, preserving natural habitats, and promoting ecosystem resilience. By implementing holistic approaches that balance conservation objectives with socioeconomic needs, countries worldwide can address pressing challenges such as habitat loss, poaching, and human-wildlife conflicts.

Through collaborative efforts involving governments, local communities, conservation organizations, and other stakeholders, integrated wildlife management strategies can contribute to the protection of endangered species, the restoration of degraded ecosystems, and the promotion of sustainable livelihoods. This holistic approach not only benefits wildlife populations but also fosters a deeper appreciation for nature and enhances the well-being of both present and future generations.

10.5.3.2 Importance of integrated and sustainable wildlife management

1. **Biodiversity conservation:** Pakistan is home to a wide range of species including several endangered and threatened ones. Integrated wildlife management ensures the protection of these species, contributing to the overall biodiversity of the country.
2. **Ecosystem balance:** Wildlife plays a crucial role in maintaining ecosystem balance by controlling populations of certain species and participating in nutrient cycling. Proper management ensures that these roles are preserved.
3. **Tourism and recreation:** Pakistan's unique wildlife, such as the snow leopard, Markhor, and various bird species, attracts wildlife enthusiasts and tourists. Sustainable wildlife management supports ecotourism and recreational activities, generating income and employment opportunities.
4. **Research and education:** Wildlife provides valuable insights into ecosystems, behavior, and adaptations. Sustainable management facilitates research opportunities and contributes to environmental education and awareness.
5. **Cultural and traditional values:** Many communities in Pakistan have cultural and traditional ties to wildlife. Integrating these values into management plans fosters local support and participation.
6. **Livelihoods:** Local communities often depend on wildlife-related activities like hunting, fishing, and non-timber forest products. Integrated management ensures that these activities are sustainable and support livelihoods.

7. **Climate change resilience:** Wildlife-rich habitats contribute to climate resilience by storing carbon and providing buffer zones against extreme weather events. Proper management aids in adapting to changing climatic conditions.
8. **Conflict resolution:** Wildlife can come into conflict with human activities such as agriculture and infrastructure development. Integrated management seeks solutions that minimize such conflicts.
9. **Protection of endangered species:** Species like the snow leopard and Markhor are emblematic of Pakistan's wildlife heritage. Effective management helps protect these and other endangered species.
10. **Legal and illegal trade:** Integrated management includes measures to combat illegal wildlife trade, protecting species from exploitation and maintaining their populations.
11. **International commitments:** Pakistan is a signatory to international conservation agreements. Integrated and sustainable wildlife management aligns with these commitments.
12. **Economic diversification:** Managed wildlife resources contribute to economic diversification, offering sustainable income sources for local communities.

Integrated and sustainable wildlife management is essential for Pakistan's conservation efforts, ecosystem health, cultural preservation, and socioeconomic development. By adopting practices that safeguard wildlife and its habitats while involving local communities, Pakistan can achieve a balance between human needs and the well-being of its unique wildlife.

10.5.4 Pest and disease management

Through reduced soil disturbance and diverse cropping systems, CA creates an environment that's less conducive to the proliferation of pests and diseases. This minimizes the need for chemical pesticides, which can have negative effects on the environment.

Conservation agriculture (CA) has a significant impact on pest and disease management through its principles of reduced soil disturbance and diverse cropping systems. By minimizing soil disturbance, CA creates an environment that is less favorable for the proliferation of pests and diseases. When soil is left undisturbed, beneficial organisms like earthworms and beneficial bacteria thrive, helping to naturally suppress pest populations.

Additionally, diverse cropping systems in CA contribute to pest and disease management. Monoculture, or the continuous planting of the same crop in the same area, can create a haven for pests that specifically target that crop. In contrast, diverse cropping systems involve planting a variety of crops in the same area, which disrupts pest life cycles and reduces the buildup of pests and diseases.

The reduced need for chemical pesticides is a significant benefit of CA. Chemical pesticides can have detrimental effects on the environment including soil degradation, water contamination, harm to nontarget organisms, and resistance in pest populations. By relying less on these chemicals, CA promotes a more sustainable and environmentally friendly approach to pest and disease management.

10.5.4.1 Integrated Pest and Disease Management

Integrated pest and disease management (IPDM) is a holistic and sustainable approach to managing pests and diseases in agriculture. It involves the integration of various control methods and strategies to minimize the use of chemical pesticides while maximizing effectiveness and sustainability.

The key principles of IPDM include:

1. **Biological control:** This involves using natural enemies of pests, such as predators, parasites, and pathogens, to control pest populations. For example, introducing ladybugs to control aphids or using *Bacillus thuringiensis* to control certain insect pests.
2. **Cultural control:** This includes practices such as crop rotation, intercropping, and adjusting planting dates to disrupt pest life cycles and reduce pest pressure. Diverse cropping systems, as seen in CA, fall under this category.
3. **Mechanical and physical control:** These methods involve physically removing pests or creating barriers to prevent their entry. Examples include using traps, nets, barriers, or hand-picking pests.
4. **Chemical control as a last resort:** While minimizing chemical pesticide use is a goal of IPDM, when necessary, the judicious use of pesticides that are less harmful to nontarget organisms and the environment is recommended. This includes using selective pesticides that target specific pests while sparing beneficial insects.
5. **Monitoring and decision-making:** Regular monitoring of pest populations and crop health is crucial in IPDM. By monitoring, farmers can make informed decisions about when and how to implement control measures. This can include setting action thresholds based on pest population levels, weather conditions, and crop growth stage.
6. **Sustainable practices:** IPDM emphasizes practices that promote long-term sustainability. This includes promoting biodiversity, maintaining soil health, and minimizing the negative impacts of pest control measures on the environment.

Overall, IPDM aims to achieve effective pest control while minimizing risks to human health and the environment. By combining multiple approaches and strategies, farmers can manage pests and diseases in a sustainable manner that maintains ecosystem balance and agricultural productivity over the long term.

CA's practices of reduced soil disturbance and diverse cropping systems create an ecosystem that naturally suppresses pests and diseases. This reduces the reliance on chemical pesticides, which in turn mitigates environmental harm and promotes long-term agricultural sustainability.

10.5.4.2 Importance of integrated pest and disease management

Integrated and sustainable IPDM is crucial for several reasons:

IPDM is crucial for several reasons:

1. **Reduction in chemical pesticide use:** IPDM emphasizes the use of multiple strategies, such as biological control and cultural practices, to minimize reliance on chemical pesticides. This reduces the harmful impact of pesticides on the environment, nontarget organisms, and human health.
2. **Preservation of beneficial organisms:** By using biological control methods in IPDM, beneficial organisms like pollinators, natural predators, and soil microbes are preserved. This helps maintain biodiversity and ecosystem balance, which is essential for sustainable agriculture.
3. **Cost-effectiveness:** IPDM can be cost-effective in the long term. While initial investments in IPDM practices may be required, the reduction in pesticide use and potential crop losses from pests can lead to economic savings for farmers.
4. **Reduced pesticide resistance:** Continuous use of chemical pesticides can lead to the development of resistance in pest populations, rendering pesticides less effective over time. IPDM's diverse strategies make it more difficult for pests to develop resistance, prolonging the effectiveness of control measures.
5. **Improved crop health and yield:** Effective pest and disease management through IPDM leads to healthier crops with higher yields. By minimizing damage from pests and diseases, farmers can produce more food sustainably to meet the demands of a growing population.
6. **Environmental protection:** IPDM helps protect the environment by reducing pesticide runoff into water bodies, contamination of soil and air, and harm to nontarget organisms. It promotes a more balanced and sustainable agricultural ecosystem.
7. **Adaptability to climate change:** With the increasing unpredictability of weather patterns due to climate change, IPDM offers flexibility. Farmers can adjust their strategies based on changing conditions, such as altered pest pressures or shifts in crop growth cycles.
8. **Long-term sustainability:** Perhaps most importantly, IPDM supports the long-term sustainability of agriculture. By maintaining healthy soils, preserving biodiversity, and reducing environmental impacts, IPDM practices contribute to the resilience of farming systems against future challenges.

IPDM is essential for promoting environmentally friendly, economically viable, and socially acceptable pest and disease management practices in agriculture. It offers a balanced approach that considers not only short-term pest control but also long-term health of ecosystems and farming communities.

10.5.5 Carbon sequestration

By retaining crop residues and organic matter in the soil, CA contributes to carbon sequestration. This helps mitigate the effects of climate change by reducing greenhouse gas emissions and enhancing soil's ability to capture and store carbon.

10.5.5.1 Integrated and sustainable carbon sequestration management

Integrated and sustainable carbon sequestration management refers to strategies that enhance the capture and storage of carbon dioxide from the atmosphere, mitigating the effects of climate change. This approach involves various land-use practices and technologies that promote carbon storage in forests, soils, wetlands, and other ecosystems.

Integrated and sustainable carbon sequestration management plays a pivotal role in combating climate change and mitigating its adverse effects on the environment. As the concentration of greenhouse gases continues to rise, there is an urgent need to enhance carbon sequestration processes to offset emissions and stabilize atmospheric carbon dioxide levels. This chapter explores innovative approaches and strategies for integrating carbon sequestration into land-use practices, forestry management, and agricultural systems.

By adopting sustainable land management practices, promoting afforestation and reforestation efforts, and incentivizing carbon-friendly agricultural techniques, countries can enhance carbon sequestration capacities while simultaneously promoting environmental conservation and biodiversity preservation. Through a comprehensive examination of integrated and sustainable carbon sequestration management, this chapter aims to provide insights and recommendations for policymakers, practitioners, and stakeholders striving to address the global challenge of climate change.

10.5.5.2 Importance of integrated and sustainable carbon sequestration management

Climate change mitigation: Carbon sequestration reduces the concentration of greenhouse gases in the atmosphere, helping mitigate climate change effects such as rising temperatures and extreme weather events.

Ecosystem health: Sequestering carbon enhances the health and resilience of ecosystems, contributing to biodiversity conservation and habitat protection.

Natural resource management: Integrating carbon sequestration practices with sustainable land-use management improves soil fertility, water retention, and overall ecosystem productivity.

Agricultural productivity: Practices such as agroforestry and cover cropping not only sequester carbon but also enhance agricultural yields, benefiting food security and rural livelihoods.

Disaster risk reduction: Healthy ecosystems with enhanced carbon sequestration capacities can help buffer against floods, landslides, and other climate-induced disasters.

Renewable energy development: Carbon sequestration complements efforts to transition to renewable energy sources by offsetting carbon emissions.

Economic opportunities: Carbon trading and incentives for carbon sequestration projects can generate income and contribute to sustainable development.

Global commitments: Pakistan is a signatory to international agreements like the Paris Agreement. Implementing carbon sequestration aligns with these commitments.

Urban sustainability: Urban green spaces and tree planting contribute to carbon sequestration, improving air quality and enhancing the quality of life in cities.

Educational and awareness initiatives: Carbon sequestration management provides opportunities for education and awareness about climate change, sustainability, and ecosystem services.

Enhanced forest resources: Forests act as significant carbon sinks. Sustainable forest management and afforestation efforts contribute to carbon sequestration while supporting timber and non-timber forest products.

Local and indigenous knowledge: Integrating traditional knowledge and practices into carbon sequestration projects fosters community engagement and empowers local communities.

Integrated and sustainable carbon sequestration management is crucial for Pakistan's efforts to combat climate change, conserve natural resources, and promote sustainable development. By adopting practices that enhance carbon storage in various ecosystems, Pakistan can contribute to global climate goals while reaping local benefits in terms of ecosystem health, food security, and economic opportunities.

Energy efficiency: The reduced need for intensive plowing and tilling in CA leads to lower fuel consumption on farms. This not only reduces costs for farmers but also decreases the carbon footprint associated with agricultural practices.

10.5.6 Integrated and sustainable energy management

Integrated and sustainable energy management involves the harmonization of energy production, distribution, and consumption while considering environmental, social, and economic factors. This approach aims to meet energy demands efficiently, minimize environmental impacts, and ensure long-term energy security.

Integrated and sustainable energy management stands at the forefront of global efforts to transition toward a cleaner, more resilient energy future. In an era marked by growing energy demand, climate change concerns, and the imperative of sustainable development, effective energy management practices are crucial for achieving environmental, economic, and social objectives. The multifaceted realm of integrated and sustainable energy management, exploring strategies, technologies, and policies aimed at optimizing energy use, enhancing energy efficiency, and promoting renewable energy sources.

By integrating energy considerations into various sectors, fostering innovation, and fostering collaboration among stakeholders, integrated and sustainable energy management offers pathways toward a more secure, affordable, and environmentally responsible energy system. Through a comprehensive analysis of best practices, case studies, and policy frameworks, this chapter seeks to provide insights and guidance for policymakers, energy professionals, and stakeholders committed to advancing the global energy transition agenda.

10.5.6.1 Importance of integrated and sustainable energy management

Energy security: Pakistan's energy sector faces challenges of supply-demand gaps and dependency on imported fuels. Integrated energy management promotes diversified energy sources and reduces reliance on volatile international markets.

Climate change mitigation: Transitioning to cleaner and renewable energy sources reduces greenhouse gas emissions, contributing to global efforts to mitigate climate change.

Air quality improvement: Integrating renewable energy and energy-efficient technologies reduces air pollution, improving public health and quality of life.

Economic growth: Sustainable energy practices stimulate economic growth by creating jobs in renewable energy industries, reducing energy costs, and attracting investments.

Rural electrification: Integrated energy management includes decentralized renewable energy solutions that extend electricity access to remote and underserved areas.

Technology innovation: Sustainable energy management drives innovation in technologies such as solar, wind, and energy storage, fostering technological advancement and knowledge transfer.

Reduced energy poverty: By ensuring access to affordable and reliable energy services, integrated energy management helps alleviate energy poverty and improve living standards.

Energy efficiency: Implementing energy-efficient technologies and practices reduces energy waste and enhances overall energy productivity.

Local communities: Renewable energy projects can empower local communities by providing jobs, income opportunities, and infrastructure development.

Energy independence: Developing indigenous renewable energy sources reduces dependence on imported fossil fuels, enhancing energy security and reducing trade deficits.

Global commitments: Pakistan is a signatory to international climate agreements. Integrated energy management aligns with these commitments and enhances the country's global standing.

Community engagement: Involving communities in energy planning and decision-making fosters acceptance of projects and promotes sustainable energy practices.

Integrated and sustainable energy management is pivotal for Pakistan's development, addressing energy challenges, and achieving sustainable growth. By embracing renewable energy, enhancing energy efficiency, and adopting innovative technologies, Pakistan can ensure reliable, affordable, and environmentally friendly energy supplies while contributing to global efforts to combat climate change.

Economic sustainability: CA practices can lead to higher crop yields, reduced input costs (such as fuel and pesticides), and improved long-term land productivity. This contributes to the economic sustainability of farming operations.

10.5.7 Integrated economic sustainability management

Integrated and sustainable economic management involves strategies that promote economic growth, equity, and prosperity while considering social and environmental dimensions. This approach seeks to harmonize economic activities with the well-being of communities and the preservation of natural resources.

Integrated economic sustainability management is a comprehensive approach to achieving long-term economic prosperity while ensuring environmental protection, social equity, and resilience to global challenges. In today's interconnected world, economic activities have significant implications for the environment and society, necessitating integrated approaches that balance economic growth with environmental conservation and social well-being. This chapter explores the principles, strategies, and practices of integrated economic sustainability management, drawing on interdisciplinary insights from economics, environmental science, and social sciences.

By promoting sustainable production and consumption patterns, fostering inclusive growth, and harnessing the potential of innovation and technology, integrated economic sustainability management seeks to create economies that are both prosperous and environmentally responsible. Through case studies, best practices, and policy recommendations, this chapter aims to provide a roadmap for policymakers, businesses, and communities striving to achieve economic sustainability in the face of global challenges.

10.5.7.1 Importance of economic sustainability management

Equitable growth: Integrated economic sustainability management focuses on inclusive growth that benefits all segments of society, reducing income inequality and social disparities.

Resource management: Balancing economic growth with resource conservation ensures that economic development does not compromise the availability of natural resources for future generations.

Job creation: Sustainable economic practices, such as investing in renewable energy and eco-friendly industries, create employment opportunities, fostering economic resilience.

Innovation and entrepreneurship: Integrating sustainable business practices drives innovation, encourages entrepreneurship, and positions Pakistan as a hub for green technologies and solutions.

Resilience to shocks: Diversified and sustainable economic activities make Pakistan more resilient to economic shocks, market fluctuations, and global crises.

Local industries: Supporting local industries and value chains strengthens the domestic economy, reduces dependency on imports, and promotes self-sufficiency.

Tourism and cultural heritage: Integrated economic sustainability management includes cultural preservation and sustainable tourism practices, which contribute to economic growth and community development.

Reduced environmental impact: Sustainable economic practices minimize negative environmental impacts, such as pollution and resource depletion, enhancing overall ecological health.

Global competitiveness: Implementing sustainable business practices enhances Pakistan's competitiveness in international markets and attracts responsible investments.

Rural-urban linkages: Balanced economic development bridges the gap between rural and urban areas, promoting equitable opportunities and reducing rural-urban migration pressures.

Long-term prosperity: Integrated economic sustainability management focuses on long-term benefits, ensuring that economic gains are sustainable and do not come at the cost of future generations.

Social well-being: Prioritizing social indicators alongside economic growth ensures that economic development contributes to improved quality of life, education, and healthcare.

Global commitments: Pakistan is part of international agreements focusing on sustainable development. Integrated economic sustainability management aligns with these commitments.

Integrated and sustainable economic sustainability management is vital for Pakistan's progress, enabling economic growth while safeguarding natural resources and promoting equitable development. By adopting practices that balance economic, social, and environmental considerations, Pakistan can achieve lasting prosperity and resilience in the face of evolving global challenges.

10.5.8 Climate change adaptation and mitigation

CA practices help farmers adapt to the impacts of climate change and contribute to climate change mitigation. Improved soil health and moisture retention enhance crop resilience to droughts and extreme weather events. Additionally, CA reduces greenhouse gas emissions by sequestering carbon in the soil and reducing the use of machinery and agrochemicals. By improving soil health, water efficiency, and overall ecosystem balance, CA helps farmers adapt to the changing climate by making their systems more resilient to extreme weather events.

10.5.9 Sustainable agricultural productivity

CA supports sustainable agricultural productivity by preserving soil fertility, reducing reliance on synthetic fertilizers and pesticides, and improving crop resilience. It helps farmers optimize resource use, reduce production costs, and increase overall farm profitability, leading to sustainable agricultural practices.

10.5.10 Environmental protection

By minimizing soil erosion, conserving water, and reducing chemical inputs, CA contributes to environmental protection. It helps maintain water quality by preventing nutrient runoff and soil erosion that can carry pollutants into water bodies. It also reduces the risk of contamination of groundwater resources.

CA is closely intertwined with natural resource management. Its principles and practices work synergistically to safeguard soil, water, and biodiversity, ensuring that agriculture remains sustainable, productive, and environmentally responsible over the long term. Implementing CA practices requires farmer training, awareness campaigns, and policy support. Encouraging and incentivizing farmers to adopt CA can lead to more sustainable NRM practices, improved agricultural productivity, and better environmental outcomes in Pakistan.

10.6 Sustainable water resource management

Sustainable water resource management is crucial for agricultural, environmental, and societal well-being. It involves the responsible use, conservation, and allocation of water resources to ensure equitable access, ecosystem health, and long-term sustainability. By promoting water efficiency, reducing pollution, and safeguarding aquatic ecosystems, sustainable water resource management helps meet the needs of agriculture, industry, and urban populations while also preserving water for future generations. Additionally, it involves integrated approaches that consider the complex interactions between water availability, quality, and demand as well as the impacts of climate change and land-use practices. Sustainable water resource management is essential for addressing water scarcity, mitigating the effects of droughts and floods, and ensuring resilience to environmental change.

The key aspects of sustainable water resource management are as follows:

Water conservation and efficiency: Sustainable water management focuses on conserving and efficiently utilizing water resources. This includes adopting water-saving irrigation techniques such as drip irrigation and precision agriculture, improving on-farm water management practices, and promoting water-efficient technologies and practices.

Watershed management: Watershed management approaches help protect and enhance the quality and quantity of water resources. Implementing strategies such as reforestation, soil conservation measures, and wetland preservation contribute to sustainable water resource management. Integrated watershed management plans address both upstream and downstream impacts on water availability and quality.

Water governance and policies: Effective water governance and policies play a crucial role in sustainable water resource management. Developing water allocation systems, water pricing mechanisms, and regulatory frameworks ensure equitable access to water resources, promote efficient use, and prevent overexploitation.

Water recycling and reuse: Sustainable water management involves recycling and reusing water to minimize waste and maximize resource efficiency. Implementing technologies such as wastewater treatment systems and water recycling practices in agriculture help reduce freshwater demand and alleviate pressure on water resources.

Groundwater management: Managing groundwater resources sustainably is essential for long-term water security. Implementing measures such as groundwater monitoring, regulation of extraction practices, and promoting alternative water sources reduce the depletion and contamination of groundwater reserves.

10.7 Conservation and management of biodiversity

Conservation and management of biodiversity are critical for maintaining ecosystem health, resilience, and the sustainable use of natural resources. Biodiversity encompasses the variety of life forms, including plants, animals, and microorganisms, as well as the ecological processes and genetic diversity that sustain ecosystems. By conserving biodiversity, we can preserve essential ecosystem services such as pollination, soil fertility, and carbon sequestration, which are vital for human well-being and planetary health.

Moreover, managing biodiversity helps ecosystems adapt to environmental changes, resist disturbances, and recover from degradation. Sustainable management practices, such as protected area management, habitat restoration, and species conservation efforts, are essential for safeguarding biodiversity and promoting ecological resilience in the face of emerging threats such as habitat loss, climate change, and invasive species. Overall, conservation and management of biodiversity are integral components of sustainable development, ensuring the continued health and productivity of ecosystems for present and future generations.

The key aspects of biodiversity conservation and management are given as follows:

Protected areas and habitat conservation: Establishing protected areas, nature reserves, and wildlife sanctuaries help conserve and protect critical habitats and biodiversity hotspots. Effective management of these areas ensures the preservation of ecosystems, protects endangered species, and maintains biodiversity.

Sustainable land-use practices: Promoting sustainable land-use practices, such as agroforestry, organic farming, and sustainable forestry, minimizes habitat destruction and fragmentation. These practices enhance biodiversity conservation while supporting agricultural production and rural livelihoods.

Invasive species management: Invasive species pose a significant threat to biodiversity. Implementing strategies to prevent the introduction and control the spread of invasive species helps protect native ecosystems and species. Early detection, monitoring, and eradication programs are key components of invasive species management.

Conservation of endangered species: Conservation efforts focus on the protection and recovery of endangered species. This involves habitat restoration, captive breeding programs, and conservation breeding efforts to prevent the extinction of endangered species and restore their populations.

Education and awareness: Educating communities and raising awareness about the value of biodiversity are essential for its conservation. Promoting environmental education, community participation in conservation programs, and sustainable livelihood options help foster a sense of stewardship and collective responsibility for biodiversity.

10.8 Energy efficiency and renewable energy in agriculture

Energy efficiency and the utilization of renewable energy sources in agriculture are pivotal components of sustainable agricultural practices, offering multifaceted benefits ranging from environmental conservation to economic viability. By optimizing energy use and transitioning to renewable sources, agriculture can significantly reduce its carbon footprint and mitigate the impacts of climate change. Energy efficiency measures, such as upgrading equipment, improving irrigation systems, and adopting precision farming technologies, enhance productivity while minimizing energy consumption. These practices not only reduce greenhouse gas emissions but also lower operational costs and improve overall farm profitability.

Moreover, the integration of renewable energy sources such as solar, wind, and biomass into agricultural operations further enhances sustainability and resilience. Solar energy, for instance, can power irrigation pumps, lighting systems, and other farm machinery, providing a reliable and environmentally friendly alternative to fossil fuels. Similarly, wind turbines can generate electricity to meet on-farm energy needs or even contribute excess power to the grid, creating additional revenue streams for farmers. Biomass energy derived from agricultural residues, crop wastes, and animal manure offers another renewable energy option, providing a sustainable source of heat, electricity, and biofuels while reducing waste and pollution.

In addition to reducing reliance on finite fossil fuels and mitigating greenhouse gas emissions, energy efficiency and renewable energy technologies in agriculture foster innovation, job creation, and rural development. They empower farmers to become energy producers and managers, promoting energy independence and resilience in the face of energy price fluctuations and supply disruptions. Furthermore, by diversifying energy sources and reducing environmental impacts, agriculture can contribute to broader sustainability goals such as enhancing ecosystem health, conserving natural resources, and promoting social equity.

Overall, the integration of energy efficiency and renewable energy technologies in agriculture represents a transformative shift toward a more sustainable and resilient food system. By harnessing clean, renewable energy sources and embracing energy-efficient practices, agriculture can play a significant role in mitigating climate change, enhancing economic viability, and ensuring a prosperous and sustainable future for generations to come.

Here are some key aspects of energy efficiency and renewable energy in agriculture:

Efficient use of energy: Promoting energy-efficient practices in agriculture reduces energy consumption and greenhouse gas emissions. This includes using energy-efficient machinery and equipment, optimizing energy use in irrigation systems, and implementing energy-saving measures in farm operations.

On-farm renewable energy generation: Utilizing renewable energy sources on farms helps reduce dependence on fossil fuels and mitigates greenhouse gas emissions. Installing solar panels, wind turbines, or biogas digesters allows for on-site generation of clean energy to power farm operations and reduce carbon footprints.

Bioenergy production: Producing bioenergy from agricultural residues and dedicated energy crops provides a renewable source of energy. Biomass, such as crop residues, animal waste, and energy crops, can be converted into biofuels or used for biogas production. Bioenergy production helps diversify energy sources and supports sustainable agricultural practices.

Energy audits and management: Conducting energy audits and implementing energy management practices help identify energy-saving opportunities and optimize energy use on farms. Monitoring energy consumption, upgrading equipment, and adopting energy-efficient technologies contribute to energy efficiency in agriculture.

Knowledge transfer and capacity building: Providing farmers with information and training on energy-efficient practices and renewable energy technologies is crucial. Extension services, workshops, and training programs help farmers adopt energy-efficient practices, make informed decisions regarding renewable energy investments, and enhance their overall energy management skills.

10.9 Climate-smart agriculture and mitigation strategies

Climate-smart agriculture represents a holistic approach to farming that aims to address the challenges posed by climate change while promoting sustainable and resilient agricultural systems. At its core, climate-smart agriculture integrates three key pillars: adaptation, mitigation, and sustainable practices.

Adaptation strategies focus on enhancing the resilience of agricultural systems to the impacts of climate change such as shifting weather patterns, increased temperatures, and more frequent extreme weather events. These strategies may include implementing drought-resistant crop varieties, improving water management practices, diversifying cropping systems, and adopting agroforestry techniques to enhance ecosystem resilience and minimize vulnerability to climate-related risks.

Mitigation strategies, on the other hand, aim to reduce greenhouse gas emissions from agricultural activities, thereby mitigating the drivers of climate change. This may involve implementing practices such as conservation tillage, cover cropping, and integrated nutrient management to minimize emissions of carbon dioxide, methane, and nitrous oxide from agricultural soils. Additionally, adopting renewable energy sources, improving energy efficiency, and reducing emissions from livestock production can further contribute to mitigating climate change.

Sustainable agricultural practices serve as the foundation of climate-smart agriculture, providing the framework for implementing adaptation and mitigation strategies while promoting long-term environmental sustainability and food security. These practices emphasize optimizing resource use, conserving biodiversity, and enhancing ecosystem services to support agricultural productivity and resilience in the face of changing climatic conditions. By promoting soil health, water conservation, and biodiversity conservation, sustainable agricultural practices not only contribute to climate change adaptation and mitigation but also help to safeguard natural resources and improve livelihoods for farmers.

Climate-smart agriculture represents a proactive and integrated approach to addressing the complex challenges posed by climate change in agriculture. By combining adaptation, mitigation, and sustainable practices, climate-smart agriculture offers a pathway toward building resilient and sustainable agricultural systems that can thrive in a changing climate while contributing to global efforts to mitigate climate change and ensure food security for future generations.

Here are key aspects of climate-smart agriculture and mitigation strategies:

Adaptation to climate change: Climate-smart agriculture involves implementing practices that enhance resilience and adaptation to climate change impacts. This includes adopting drought-tolerant crop varieties, improving water management, implementing agroforestry practices, and diversifying farming systems to reduce vulnerability to climate-related risks.

Soil management and carbon sequestration: Sustainable soil management practices such as conservation tillage, cover cropping, and agroforestry, enhance soil health, fertility, and carbon sequestration. These practices help mitigate climate change by storing carbon in soils and reducing greenhouse gas emissions from agriculture.

Efficient nutrient management: Precision nutrient management practices minimize nutrient losses, reduce greenhouse gas emissions, and optimize fertilizer use. This includes adopting site-specific nutrient application, utilizing organic fertilizers, and implementing efficient irrigation practices to reduce nutrient runoff.

Water-smart agriculture: Water-smart agricultural practices enhance water-use efficiency, especially in water-stressed regions. These practices include implementing water-saving irrigation techniques, promoting water recycling and reuse, and adopting climate-resilient crop varieties that require less water.

Agroecological approaches: Agroecological approaches in agriculture such as organic farming, promote ecological balance, biodiversity conservation, and sustainable resource use. These practices contribute to climate change mitigation by reducing synthetic inputs, conserving natural resources, and enhancing ecosystem services.

Climate information and advisory services: Providing farmers with climate information, weather forecasts, and advisory services helps them make informed decisions and adapt their farming practices accordingly. Accessible and timely climate information supports climate-smart agriculture and improves farmers' resilience to climate change.

Beneath the canopy of natural resources management, the Earth's bounties unfurl in all their complexity and splendor. As we traverse the realms of land, water, and minerals, we unearth the essence of responsible custodianship. This chapter's narrative underscores the profound impact of judicious resource utilization – a journey that shapes the contours of sustainable development and echoes the sentiment that stewardship is the key to unlocking nature's enduring potential.

Rising from the depths of natural resources management, we transition to Chapter 11, where soil, water, and the environment converge into a harmonious tapestry of interconnectedness. Integrated soil, water, and environment management unveil the delicate balance between nurturing healthy soils, conserving water, and fostering ecological harmony. Within these pages, we embark on a journey through the pathways of sustainable agriculture, guided by the principles of conservation, regeneration, and the profound understanding that our choices reverberate through landscapes and generations.

References

Amanullah (2024). Harvesting Success: A Comprehensive Guide to Agronomy Courses. Punjab Books, Lahore, PK.

Amanullah., Ondrasek, G., Al –Tawaha, A. R. M. S., eds. (2023). Integrated nutrients management: An approach for sustainable crop production and food security in changing climates. Lausanne: Frontiers Media SA. doi: 10.3389/978-2-8325-3169-3

Amanullah, & Khalid, S. (2020). Agronomy – Climate Change & Food Security. IntechOpen, London, United Kingdom.

Amanullah, & Fahad, S. (2018). Nitrogen in Agriculture-Updates. InTechOpen, London, UK.

Amanullah, & Fahad. (2018). Corn – Production and Human Health in Changing Climate. InTech, London, UK.

Bawa, K. S., Seidler, R., & Raven, P. H. (2004). Reproductive biology of tropical forest plants: Historical perspectives. In: Tropical Forests and Global Atmospheric Change (pp. 53–61). Springer, Heidelberg, Germany.

Bekele, D., & Tilahun, S. A. (2016). Integrated soil and water conservation for sustainable livelihoods in Ethiopia: A synthesis. International Journal of Environmental Studies, 73(3), 388–400.

Custodio, M. R., Li, Y., Zhang, H., & Jin, L. (2015). Integrated natural resources management: Approaches and lessons from China. Journal of Environmental Management, 150, 166–176.

FAO. (2009). The State of the World's Land and Water Resources for Food and Agriculture (SOLAW) – Managing Systems at Risk. Food and Agriculture Organization of the United Nations. Rome, Italy.

Lal, R. (2014). Challenges and opportunities in soil organic matter research. European Journal of Soil Science, 65(1), 2–15.

Lambin, E. F., & Meyfroidt, P. (2011). Global land use change, economic globalization, and the looming land scarcity. Proceedings of the National Academy of Sciences, 108(9), 3465–3472.

MEA (Millennium Ecosystem Assessment). (2005). Ecosystems and Human Well-being: Synthesis. Island Press. Washington, D.C., USA.

Molden, D. (Ed.). (2007). Water for Food, Water for Life: A Comprehensive Assessment of Water Management in Agriculture. International Water Management Institute (IWMI). Colombo, Sri Lanka.

Oldeman, L. R., Hakkeling, R. T. A., & Sombroek, W. G. (1991). World Map of the Status of Human-induced Soil Degradation: An Explanatory Note. Wageningen Agricultural University. Wageningen, Netherlands.

Pretty, J., & Hine, R. (2001). Reducing food poverty by increasing agricultural sustainability in developing countries. Agriculture, Ecosystems & Environment, 86(3), 257–274.

Reenberg, A. (2007). Cultural landscapes of the world: Danish examples. GeoJournal, 68(3), 235–243.

Rockström, J., & Falkenmark, M. (2015). Agriculture: Increase water harvesting in Africa. Nature, 519(7543), 283–285.

Rockström, J., Steffen, W., Noone, K., Persson, Å., Chapin, F. S., Lambin, E., . . . & Foley, J. A. (2009). A safe operating space for humanity. Nature, 461(7263), 472–475.

Scoones, I. (1998). Sustainable rural livelihoods: A framework for analysis. IDS Working Paper, 72.

Simeonov, V., Stratis, J. A., Samara, C., Zachariadis, G., Voutsa, D., Anthemidis, A., . . . & Sofoniou, M. (2003). Assessment of the surface water quality in Northern Greece. Water Research, 37(17), 4119–4124.

Smith, P., & Smith, J. U. (1996). Description of the Rothamsted carbon model (RothC ver. 26.3). In: Evaluation of Soil Organic Matter Models (pp. 81–101). Springer. Heidelberg, Germany.

Steffen, W., Richardson, K., Rockström, J., Cornell, S. E., Fetzer, I., Bennett, E. M., . . . & Sörlin, S. (2015). Planetary boundaries: Guiding human development on a changing planet. Science, 347(6223), 1259855.

Sutton, M. A., Oenema, O., Erisman, J. W., Leip, A., van Grinsven, H., & Winiwarter, W. (2011). Too much of a good thing. Nature, 472(7342), 159–161.

United Nations. (2015). Transforming our world: The 2030 Agenda for Sustainable Development. United Nations.

World Bank. (2006). Environment Strategy 2001. World Bank. Washington, D.C., USA.

Chapter 11
Integrated soil, water, and environmental management: mitigating climate change impacts

Introduction

Integrated soil, water, and environment management (ISWEM) is an approach that recognizes the interconnections between soil, water, and the broader environment and seeks to manage these resources in a coordinated and sustainable manner. ISWEM integrates principles and practices from soil science, hydrology, ecology, and environmental management to promote the conservation, restoration, and sustainable use of soil and water resources while safeguarding environmental quality.

ISWEM is an approach that recognizes the interconnections between soil, water, and the broader environment and seeks to manage these resources in a coordinated and sustainable manner. ISWEM integrates principles and practices from soil science, hydrology, ecology, and environmental management to promote the conservation, restoration, and sustainable use of soil and water resources while safeguarding environmental quality.

One of the key aspects of ISWEM is soil conservation. This involves implementing practices such as contour plowing, terracing, and agroforestry to prevent soil erosion. By protecting the soil from erosion, ISWEM helps to maintain soil fertility, structure, and productivity. These practices also contribute to water conservation, as they reduce runoff and allow water to infiltrate into the soil, replenishing groundwater and supporting plant growth.

Additionally, ISWEM emphasizes the importance of managing water resources sustainably. This includes practices such as rainwater harvesting, drip irrigation, and water-efficient crop varieties. By optimizing water use efficiency, ISWEM helps farmers to cope with water scarcity and reduce their reliance on unsustainable water sources. Efficient water management also minimizes the risk of water pollution from agricultural runoff, preserving water quality for both agricultural and environmental purposes.

Furthermore, ISWEM incorporates principles of ecosystem management to enhance biodiversity and ecosystem services. This involves preserving natural habitats, creating buffer zones, and promoting the use of organic farming practices. These measures not only support diverse plant and animal species but also improve soil health and resilience to climate change.

ISWEM is a comprehensive approach that addresses the interconnectedness of soil, water, and environmental systems. By integrating various scientific disciplines and management practices, ISWEM promotes sustainable agriculture that balances productivity with environmental conservation. This approach is essential for ensuring the long-term viability of agricultural systems, protecting natural resources, and safeguarding the health of ecosystems for future generations.

https://doi.org/10.1515/9783111448077-011

11.1 Components of integrated soil, water, and environment management

Key components of ISWEM include:

Soil conservation and restoration: ISWEM focuses on protecting and restoring soil health and fertility through practices such as conservation tillage, agroforestry, cover cropping, and organic farming. These practices help to reduce soil erosion, improve soil structure and fertility, and enhance the resilience of agricultural systems to environmental stresses.

Water resource management: ISWEM emphasizes the sustainable management of water resources including surface water, groundwater, and rainwater. It promotes practices such as water harvesting, irrigation efficiency improvements, soil moisture conservation, and watershed management to optimize water use, minimize water pollution, and enhance water availability for agriculture, ecosystems, and human consumption.

Environmental conservation: ISWEM aims to minimize the negative environmental impacts of agricultural activities and land-use changes. It integrates environmental considerations into land management decisions to protect biodiversity, reduce greenhouse gas emissions, conserve natural habitats, and maintain ecosystem services. ISWEM also addresses pollution prevention and remediation efforts to safeguard air and water quality and mitigate the impacts of climate change.

Sustainable land-use planning: ISWEM advocates for integrated land-use planning approaches that balance competing land uses and promote sustainable development. It considers factors such as soil suitability, water availability, ecological sensitivity, and socioeconomic considerations to guide land management decisions and prevent land degradation, habitat loss, and conflicts over resource use.

Stakeholder engagement and capacity building: ISWEM encourages the active participation of stakeholders, including farmers, landowners, government agencies, NGOs, and local communities, in the planning, implementation, and monitoring of soil, water, and environmental management initiatives. It emphasizes capacity building, knowledge exchange, and awareness-raising activities to empower stakeholders to adopt sustainable land management practices and contribute to collective efforts for environmental conservation and resilience.

ISWEM is a holistic approach that promotes the sustainable management of natural resources by integrating soil and water management with broader environmental considerations. By adopting an integrated approach, ISWEM aims to address the complex challenges of soil degradation, water scarcity, and environmental degradation while fostering resilience, sustainability, and well-being for both present and future generations.

Soil, water, and the environment form the cornerstone of agricultural ecosystems, providing the foundation upon which food production, biodiversity conservation, and ecosystem services thrive. Integrated management approaches that harmonize these elements are essential for mitigating the adverse effects of climate change on agricultural productivity, water availability, and ecosystem health.

This chapter sheds light on the critical role of integrated approaches in mitigating climate change impacts. By adopting sustainable soil management practices, implementing water conservation measures, and promoting ecosystem restoration initiatives, agricultural stakeholders can enhance their capacity to adapt to climate variability and minimize greenhouse gas emissions. We examine the potential synergies between soil carbon sequestration, water resource management, and biodiversity conservation, showcasing how integrated strategies can yield multiple cobenefits for both agricultural productivity and environmental sustainability.

11.2 Sustainable soil, water, and environment management

Sustainable soil, water, and environment management hold global significance as the world grapples with the urgent need to address the challenges of climate change, food security, and environmental degradation. Across the globe, agricultural systems are facing increasing pressures from population growth, land degradation, water scarcity, and biodiversity loss. In light of these challenges, the sustainable management of soil, water, and the environment emerges as a critical imperative for ensuring the resilience, productivity, and sustainability of agricultural systems.

As we navigate through this chapter, we delve into the global relevance of sustainable soil, water, and environment management, recognizing their profound impact on the well-being of both current and future generations. In a world where agricultural expansion and intensification threaten to degrade soil fertility, deplete water resources, and degrade ecosystems, adopting sustainable management practices becomes essential for safeguarding natural resources, protecting biodiversity, and mitigating climate change impacts.

From the vast plains of the United States to the rice paddies of Southeast Asia, agricultural regions worldwide are confronted with similar challenges related to soil degradation, water scarcity, and environmental pollution. By embracing sustainable soil, water, and environment management practices, countries can enhance their capacity to adapt to climate change, improve agricultural productivity, and promote environmental sustainability. This chapter serves as a global call to action, urging policymakers, researchers, and stakeholders around the world to prioritize the sustainable management of soil, water, and the environment as a fundamental component of achieving the sustainable development goals and building a more resilient and prosperous future for all.

11.3 Understanding the intricate interplay: soil, water, and the environment

Understanding the intricate interplay between soil, water, and the environment is essential for sustainable land and resource management practices. These three components are interconnected in complex ways, influencing each other's health, productivity, and resilience. Soil provides the foundation for plant growth and plays a crucial role in regulating water flow and quality. Meanwhile, water is essential for soil health, supporting microbial activity and nutrient transport. The environment, including climate patterns, vegetation cover, and land use, further shapes the interactions between soil and water.

The triad of soil, water, and the environment forms an intricate web of interactions, each element intricately linked to the others. The health of the soil, characterized by its structure, composition, and fertility, lays the foundation for agricultural productivity. Water, a finite and invaluable resource, sustains life itself, nourishing crops and ecosystems alike. Meanwhile, the environment encompasses the broader ecological context within which agricultural systems operate, encompassing biodiversity, climate, and the delicate equilibrium of natural processes.

11.4 Strategies for sustainability: enabling sustainable agriculture

At the heart of this chapter lies the exploration of strategies and practices that usher in an era of sustainable agriculture. Soil health and management take center stage, with an examination of the dynamic factors that contribute to soil degradation, threatening agricultural productivity. Techniques for soil conservation, including contour farming, terracing, and the adoption of conservation tillage, come to the fore as solutions that counter erosion, retain soil structure, and prevent nutrient loss.

Moreover, the chapter delves into the importance of enhancing soil biodiversity and organic matter content. Soil biodiversity, which includes microorganisms, earthworms, and other soil organisms, plays a crucial role in nutrient cycling and soil fertility. Practices such as cover cropping and composting are discussed as methods to increase organic matter, improving soil structure and water-holding capacity. These practices not only promote healthier soils but also contribute to carbon sequestration, mitigating the effects of climate change.

Additionally, the chapter explores the integration of agroforestry practices into agricultural systems. Agroforestry involves planting trees and shrubs alongside crops or livestock, creating a diverse and productive landscape. Trees in agroforestry systems provide numerous benefits, such as shade for crops, windbreaks to reduce soil erosion, and additional sources of income through timber, fruits, or nuts. The symbiotic relationship between trees and agricultural activities enhances ecosystem services, promoting resilience and sustainability.

Furthermore, the role of precision agriculture technologies is highlighted as a pathway to sustainable farming. Precision agriculture utilizes tools like GPS, sensors, and drones to optimize resource use, minimize inputs, and improve efficiency. By precisely targeting irrigation, fertilization, and pesticide application, farmers can reduce waste, enhance yields, and protect environmental quality.

In essence, the chapter aims to provide a comprehensive overview of sustainable agricultural practices that prioritize soil health, conservation, and ecosystem resilience. By adopting these strategies, farmers can not only increase productivity and profitability but also contribute to broader environmental goals such as climate change mitigation and biodiversity conservation. This holistic approach to sustainable agriculture ensures the long-term viability of farming systems while promoting environmental stewardship and resilience in the face of global challenges.

11.5 Environmental conservation and climate resilience

Mitigating environmental degradation and fostering climate resilience stand as pillars of hope amidst the mounting challenges posed by climate change. In a world where agricultural systems are increasingly vulnerable to the impacts of shifting weather patterns and extreme events, it becomes imperative to embrace climate-smart approaches that adapt to these changes while promoting sustainability.

Within this chapter, we embark on a journey to explore innovative strategies for enhancing resilience and mitigating climate-induced risks. Crop diversification, agroforestry, and the integration of early warning systems emerge as key tools in the arsenal of climate-smart agriculture. By diversifying crop species and integrating trees into agricultural landscapes, farmers can buffer against climate extremes, enhance soil health, and safeguard ecosystem services.

Biodiversity conservation takes center stage as we delve into the intricate relationship between agricultural systems and the natural world. Through the lens of agroecological practices, we uncover the potential for harmonious coexistence, where agriculture and biodiversity thrive in symbiosis. By fostering biodiversity on farms, farmers can enhance pollination, pest control, and soil fertility, thereby reducing the reliance on external inputs and promoting ecological resilience.

Circular economy principles and waste management strategies offer additional pathways to minimize agricultural waste and harness byproducts for sustainable purposes. By closing the loop on resource use and minimizing waste generation, farmers can contribute to environmental stewardship while maximizing resource efficiency and reducing environmental impact. By embracing these principles and adopting innovative practices, farmers can not only adapt to a changing climate but also contribute to the conservation of biodiversity and the sustainability of agricultural landscapes for generations to come.

11.6 Forging a sustainable path forward

In today's rapidly changing world, the pursuit of sustainability has become paramount. As we grapple with pressing global challenges such as climate change, environmental degradation, and social inequality, the need for a sustainable path forward has never been more urgent. This imperative extends across all facets of human society from agriculture and industry to governance and culture.

Forging a sustainable path forward requires a concerted effort from individuals, communities, businesses, and governments worldwide. It entails reimagining our relationship with the planet and embracing practices that promote harmony between human activities and the natural world. Sustainability encompasses not only environmental stewardship but also social equity and economic prosperity, recognizing the interconnectedness of these dimensions in shaping our collective future.

In this endeavor, innovation, collaboration, and leadership are indispensable. By harnessing the power of technology, fostering partnerships, and adopting visionary policies, we can chart a course toward a more sustainable and resilient future. This journey will require bold action, transformative change, and a steadfast commitment to the well-being of current and future generations.

In this context, this document serves as a call to action and a roadmap for progress. It brings together insights, perspectives, and best practices from diverse fields and stakeholders, offering guidance and inspiration for those seeking to contribute to a more sustainable world. Through dialogue, education, and action, we can forge a sustainable path forward that preserves the planet's natural resources, fosters social justice, and ensures prosperity for all.

As this discussion draws to a close, the undeniable importance of sustainable soil, water, and environment management crystallizes. The intricate interdependence of these elements underscores the need for an integrated approach that transcends traditional boundaries. By adopting strategies that enhance soil health, optimize water usage, and prioritize environmental conservation, Pakistan's agricultural sector can usher in a resilient era of sustainable growth. The exploration of these strategies serves as a guiding light, illuminating a path toward a future where agricultural prosperity coexists harmoniously with the well-being of the planet.

11.7 Soil health and management

The foundation of agriculture lies in the very ground beneath our feet: the soil. A complex, living entity, soil sustains life, anchors ecosystems, and fuels the growth of crops that feed nations. In the intricate dance of soil dynamics, myriad factors converge to create a fertile canvas where plants thrive. Embarking on a journey into the realm of soil health and management, we delve into the multifaceted composition of soils and unravel the intricate relationship between soil fertility, productivity, and sustainable agricultural practices. As we explore the intricate balance between soil's hidden inhabitants, the menace of degradation, and the techniques that safeguard its vitality, we gain insights into how nurturing this precious resource is key to securing our agricultural future.

Furthermore, the critical role of soil conservation in preserving and enhancing soil health is illuminated. Sustainable land management practices, such as no-till farming and cover cropping, are discussed as effective methods to reduce erosion, improve soil structure, and increase water retention. These practices not only protect the soil from degradation but also contribute to long-term fertility and productivity. By adopting soil conservation techniques, farmers can ensure the resilience of their land against the impacts of climate change and intensive agricultural practices.

Moreover, exploring the concept of regenerative agriculture as a holistic approach to soil health and management reveals a transformative perspective. Regenerative agriculture goes beyond sustainability to actively restore and regenerate soil health and ecosystem function. Practices such as holistic grazing, agroforestry, and diversified cropping systems are highlighted for their ability to build soil organic matter, enhance biodiversity, and sequester carbon from the atmosphere. Through regenerative agriculture, farmers can transform their land into thriving ecosystems that not only produce food but also contribute to climate change mitigation and environmental conservation.

In essence, the discussion on soil health and management in agricultural practices is pivotal, emphasizing the foundational role of soil in both agriculture and the broader environment. Soil is not just a medium for plant growth; it is a complex ecosystem teeming with life and critical functions. Recognizing the intricate web of interactions within soil ecosystems, farmers gain insight into the vital role soil plays in nutrient cycling, water filtration, and carbon storage. Healthy soils are not only more productive but also contribute to climate change mitigation through carbon sequestration. This understanding drives home the point that soil health is not just a concern for farmers but for everyone, as healthy soils are essential for global food security and environmental stability.

Encouraging the adoption of sustainable practices that promote soil health, such as minimal tillage, cover cropping, and organic amendments, becomes imperative. These practices help to build soil organic matter, improve soil structure, and enhance biodiversity. By fostering healthy soils, farmers can mitigate the impacts of erosion, water runoff, and nutrient leaching, which are all threats to both agricultural productivity and environmental health.

Moreover, the long-term perspective necessary when considering soil management is highlighted. Soil degradation is a slow and often imperceptible process, but its effects can be devastating if left unchecked. By adopting sustainable practices now, farmers can protect and enhance soil fertility for future generations. This forward-thinking approach is not just about short-term gains but about stewardship of the land for the benefit of future agricultural productivity and environmental sustainability.

In essence, the discussion on soil health and management serves as a call to action for farmers, policymakers, and society as a whole. It emphasizes the need to move beyond conventional practices that degrade soil health and toward regenerative approaches that nurture and protect this vital resource. By doing so, we not only ensure the productivity and resilience of our agricultural systems but also safeguard the very foundation of life upon which all ecosystems, including our own, depend. It is a journey toward sustainability, resilience, and the preservation of our planet's most precious asset: the soil.

11.7.1 Understanding soil dynamics: nurturing earth's lifeblood

The soil beneath our feet is a dynamic ecosystem, teeming with life and complexity. As we delve into the intricate composition and structure of soils, we uncover a world where microorganisms, organic matter, and mineral content converge to create the foundation for agricultural fertility and productivity. Soil microbes, often unseen but immensely influential, play a pivotal role in nutrient cycling, breaking down organic matter into forms plants can readily absorb. Organic matter, composed of decaying plant and animal residues, enriches the soil with nutrients, improves its structure, and enhances water-holding capacity.

Mineral content, including essential elements like nitrogen, phosphorus, and potassium, dictates plant health and growth. However, this dynamic balance is susceptible to disruption, leading us to the concept of soil degradation. Erosion, driven by wind and water, strips away fertile topsoil, leaving behind barren land. Nutrient depletion occurs when continuous cultivation exhausts soil of essential elements, rendering it less conducive to productive growth. Salinity, exacerbated by improper irrigation practices, accumulates salts in the soil, rendering it inhospitable to many crops. The ramifications of soil degradation ripple through agricultural productivity, reducing yields, increasing input costs, and posing significant challenges to food security and sustainable livelihoods.

11.7.2 Soil conservation practices: safeguarding earth's skin

In response to the threats posed by soil degradation, a suite of soil conservation practices emerges as a shield against erosion, nutrient loss, and diminished productivity. Contour farming, a technique where crops are planted along the contours of the land, forms a natural barrier against runoff, preventing soil erosion while maximizing water absorption. Terracing takes this approach further, constructing level platforms on slopes to trap rainwater and minimize erosion.

Cover cropping, a method involving the planting of crops like legumes between main crops, shields the soil from direct exposure to the elements. These cover crops act as living mulch, preventing soil erosion, improving soil structure, and adding organic matter as they decompose.

Conservation tillage techniques, like no-till and reduced tillage, revolutionize traditional plowing methods. By minimizing soil disturbance, these techniques maintain the integrity of soil structure, enhancing water retention and promoting the growth of beneficial soil microorganisms. Conservation tillage also fosters carbon sequestration, where carbon dioxide from the atmosphere is stored in the soil as organic matter, mitigating the impacts of climate change.

11.7.3 Nutrient management: balancing nature's account

Efficient nutrient management lies at the heart of sustainable soil practices. Balanced fertilization involves applying fertilizers that match the specific nutrient requirements of crops, optimizing growth while reducing the risk of nutrient runoff. Organic amendments, such as compost and manure, enrich the soil with organic matter and nutrients, enhancing its fertility and structure. Precision agriculture technologies, including soil sensors and GPS-guided equipment, provide real-time data to tailor nutrient applications, minimizing waste and environmental impact.

Soil testing, a cornerstone of nutrient management, analyzes soil composition and offers tailored recommendations for nutrient inputs. Nutrient-specific application methods, like banding fertilizers below the soil surface, minimize contact with air and water, reducing the risk of nutrient leaching and pollution.

In embracing these strategies, we unearth the power of nurturing soil health as the bedrock of agricultural sustainability. With an understanding of soil dynamics, the implementation of conservation practices, and meticulous nutrient management, we embark on a journey toward cultivating resilient landscapes and securing a nourished future.

11.7.4 Integrated soil management (ISM)

Integrated soil management (ISM) is a holistic approach to soil conservation and improvement that considers the interactions between various soil components, land management practices, and socioeconomic factors. It aims to maintain or enhance soil productivity while preserving the soil's quality, fertility, and long-term sustainability. ISM recognizes that soils are dynamic ecosystems influenced by both natural processes and human activities.

Key principles of ISM include (Figure 11.1).

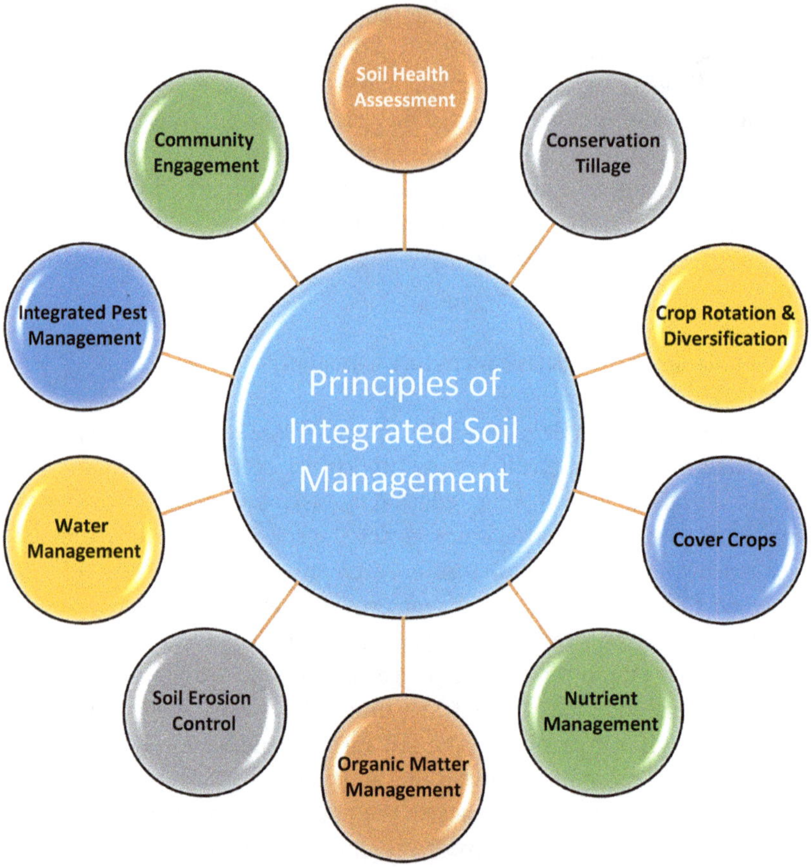

Figure 11.1: Key principles of integrated soil management.

Soil health assessment: ISM begins with assessing the current state of soil health. This involves analyzing soil properties such as texture, structure, nutrient content, pH, organic matter, and microbial activity. Understanding the soil's characteristics helps in making informed decisions about management practices.

Conservation tillage: Traditional tillage practices can lead to soil erosion, compaction, and loss of organic matter. ISM promotes reduced or no-till techniques that disturb the soil less, helping to retain its structure, reduce erosion, and improve water infiltration.

Crop rotation and diversification: Growing a variety of crops in a sequence helps break pest and disease cycles, improves soil fertility, and reduces the risk of nutrient depletion. Different crops have varying root structures and nutrient needs, contributing to improved soil health.

Cover crops: Cover crops are grown during periods when the main crop isn't cultivated. They protect the soil from erosion, improve organic matter content, and enhance microbial activity. When cover crops are eventually incorporated into the soil, they contribute to soil structure and nutrient availability.

Nutrient management: ISM emphasizes balanced nutrient application based on soil testing and crop requirements. This reduces the risk of nutrient imbalances, excessive use of fertilizers, and subsequent pollution of water bodies.

Organic matter management: Adding organic matter through compost, manure, or crop residues improves soil structure, water-holding capacity, and nutrient availability. Organic matter also enhances microbial activity, which contributes to nutrient cycling.

Soil erosion control: Implementing practices like contour farming, terracing, and planting buffer strips along water bodies helps control soil erosion. These measures prevent sediment and nutrient runoff into rivers and lakes.

Water management: Proper water management, including efficient irrigation practices and maintaining proper drainage, prevents waterlogging and salinity, which can degrade soil quality.

Integrated pest management (IPM): Managing pests through a combination of techniques, such as biological control, crop rotation, and resistant crop varieties reduces the need for excessive pesticide use, which can harm soil and the environment.

Community engagement: ISM often involves local communities, as they have valuable knowledge about the land and its challenges. Collaborative efforts can lead to the adoption of sustainable practices that suit local conditions.

ISM recognizes that soil health is interconnected with various aspects of agriculture from crop production to environmental sustainability. By considering these interconnected elements, ISM aims to create a balanced and resilient agricultural system that not only supports current productivity but also preserves soil health for future generations.

11.8 Water resource management

In the vast agricultural expanses of Pakistan and beyond, water emerges as the lifeblood that sustains growth and prosperity. However, this precious resource is not limitless, and its scarcity poses a formidable challenge to agricultural sustainability worldwide. Water resource management navigates through the intricate labyrinth of water management, unveiling the complexities of water scarcity and the innovative solutions that promise to ensure its efficient utilization.

As we delve into the delicate balance between the demands of agriculture and the preservation of water sources, we uncover strategies that have global significance. Water resource management is not just about meeting the immediate needs of agriculture; it is about ensuring the long-term availability of water for future generations. In a world where water scarcity is becoming increasingly prevalent due to factors such as population growth and climate change, the importance of efficient water management cannot be overstated.

Exploring techniques such as drip irrigation, rainwater harvesting, and water-efficient crop varieties reveals ways to enhance water efficiency in agriculture. These practices not only reduce water wastage but also improve crop yields, making agriculture more sustainable in the face of water scarcity. Additionally, there is an emphasis on integrated water management (IWM) approaches that consider the entire water cycle, from source to consumption to reuse. By adopting these strategies, nations can optimize water use across sectors from agriculture to industry to domestic use.

Furthermore, the global significance of water resource management extends beyond agricultural productivity. Water is a fundamental resource for ecosystems, supporting biodiversity, aquatic life, and natural habitats. The section emphasizes the need to protect and restore water sources to maintain ecosystem health and resilience. This includes measures such as river restoration, wetland conservation, and sustainable groundwater extraction.

In essence, water resource management highlights the critical importance of water for global sustainability. By exploring innovative techniques and integrated approaches, nations can ensure the efficient and equitable use of water resources for agricultural production, ecosystem preservation, and human well-being. It is a journey toward balancing the needs of today with the preservation of water for the generations to come, securing a sustainable and resilient future for all.

11.8.1 Water scarcity and efficiency: unveiling the drought's impact

Water scarcity and efficiency are pressing concerns that extend far beyond the arid landscapes of Pakistan, resonating with agricultural regions worldwide. Across the globe, the intersection of population growth, climate variability, and unsustainable water management practices exacerbates the challenge of ensuring water security for agricultural production.

In many regions, inefficient irrigation practices, overextraction from water sources, and inadequate water infrastructure contribute to water scarcity and pose significant threats to agricultural sustainability. Traditional flood irrigation, though widely practiced, is known to result in substantial water wastage due to evaporation, runoff, and deep percolation, further exacerbating the strain on limited water resources.

As agricultural systems worldwide grapple with the impacts of water scarcity, there is an urgent need to adopt water-efficient technologies and management practices that optimize water use, minimize waste, and enhance water productivity. By embracing drip irrigation, precision agriculture techniques, and water-saving technologies, farmers can mitigate the impacts of water scarcity, improve crop yields, and ensure the long-term sustainability of agricultural production systems.

Moreover, addressing water scarcity requires a multifaceted approach that integrates water management strategies with broader efforts to promote water conservation, watershed management, and sustainable land-use practices. Collaboration between government agencies, local communities, and stakeholders is essential to develop and implement effective water governance frameworks that prioritize equitable access to water resources while safeguarding environmental sustainability.

In essence, the issue of water scarcity and efficiency transcends geographical boundaries, highlighting the need for global cooperation and concerted action to address this critical challenge. By prioritizing water efficiency and sustainability in agricultural practices, nations can build resilience against water scarcity, safeguard food security, and promote the sustainable use of water resources for generations to come.

In the arid landscape of Pakistan, water scarcity emerges as a formidable challenge that casts a shadow over agricultural viability. The intersection of population growth, inefficient irrigation practices, and overextraction from water sources amplifies the issue, threatening the sustainability of agricultural endeavors. Traditional flood irrigation, while widespread, often results in excessive water wastage through evaporation, runoff, and deep percolation.

Amidst this challenge, precision irrigation techniques offer a beacon of hope. Drip and sprinkler systems emerge as solutions that meticulously deliver water directly to plant roots, minimizing wastage and maximizing utilization. Drip irrigation, characterized by its slow and targeted water delivery, prevents surface evaporation and waterlogging, while sprinkler systems disperse water like rain, reducing contact with the soil surface and mitigating runoff.

11.8.2 Watershed management: sustaining the lifeline

Integrated watershed management strategies play a pivotal role in safeguarding the lifeline of water resources. The approach revolves around protecting water catchment areas, mitigating soil erosion, and preserving water quality. Afforestation and reforestation emerge as formidable tools within this framework, where the planting of trees on watershed lands fortifies the landscape against degradation.

Afforestation, the establishment of forests in previously barren areas, enhances soil structure and reduces surface runoff, allowing water to infiltrate and recharge groundwater reservoirs. Reforestation, the restoration of degraded forests, amplifies these benefits, mitigating soil erosion, increasing water retention, and fostering habitat for biodiversity. As trees transpire, releasing water vapor into the atmosphere, they contribute to local precipitation patterns, creating a harmonious cycle of water regeneration.

11.8.3 Rainwater harvesting: quenching thirst from the skies

Amidst water scarcity, rainwater harvesting emerges as a simple yet transformative technique that bridges the gap between precipitation and agricultural needs. This practice involves capturing and storing rainwater for later use, mitigating reliance on external sources and minimizing pressure on dwindling groundwater reserves. From rooftop systems that channel rainwater into storage tanks to ground-level storage structures that capture runoff from fields, rainwater harvesting manifests in diverse forms.

The benefits are manifold. Rainwater harvesting supplements irrigation needs, serving as a backup during dry periods and augmenting water availability. It also contributes to groundwater recharge, replenishing aquifers that sustain agriculture and communities. By reducing runoff and erosion, rainwater harvesting fosters soil health, improving the water-holding capacity of fields and enhancing overall agricultural resilience.

In navigating the complex landscape of water resource management, Pakistan's agriculture faces a dual challenge: to address water scarcity while enhancing efficiency. Precision irrigation, watershed management, and rainwater harvesting emerge as pillars of sustainable practice, each offering unique solutions to a shared dilemma. As the nation endeavors to secure its water future, these strategies serve as guiding lights, illuminating a path toward equilibrium where water, the essence of life, meets the needs of agriculture and sustains the generations to come.

11.8.4 Integrated water management (IWM)

IWM is a comprehensive approach that seeks to balance the competing demands for water resources while considering social, economic, and environmental factors. It involves the coordinated development and management of water, land, and related resources to ensure sustainable use, efficient allocation, and equitable distribution of water for various purposes. IWM aims to address challenges such as water scarcity, pollution, and ecosystem degradation by integrating the diverse needs of users and stakeholders.

Key components of IWM include (Figure 11.2).

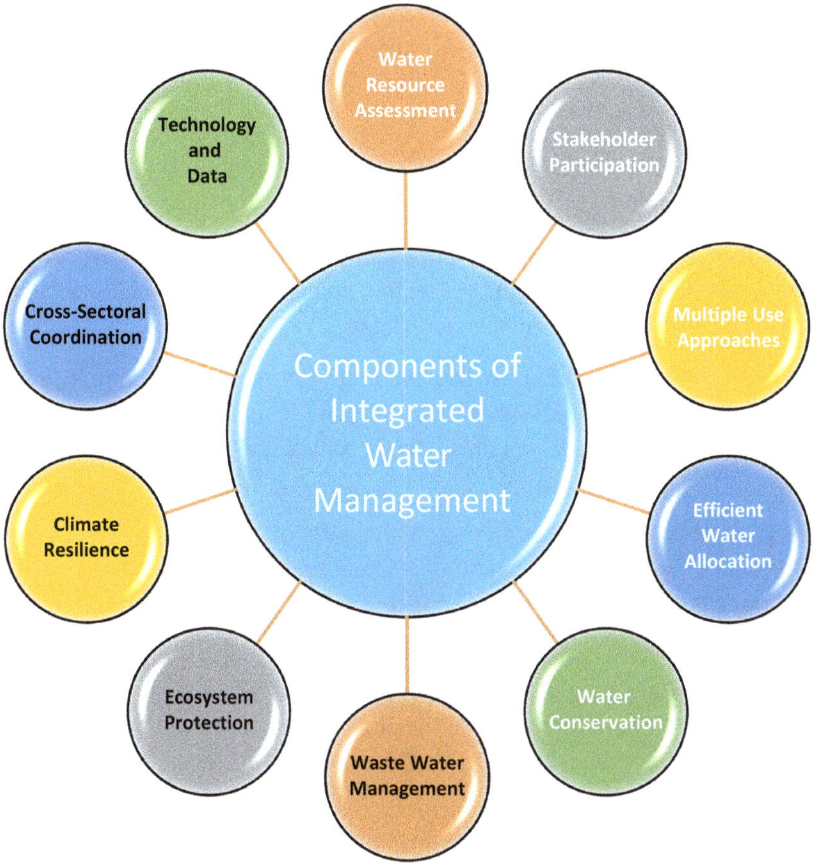

Figure 11.2: Key components of integrated water management.

Water resource assessment: IWM begins with a thorough assessment of available water resources, both surface water and groundwater. This includes evaluating water availability, quality, and potential uses in a specific region.

Stakeholder participation: Effective IWM involves engaging various stakeholders, including government agencies, communities, industries, and agriculture, in the decision-making process. Local knowledge and perspectives are valuable for understanding water needs and challenges.

Multiple use approaches: IWM encourages considering various uses of water such as domestic, agricultural, industrial, and environmental needs. Balancing these uses helps prevent conflicts and ensures equitable distribution.

Efficient water allocation: IWM focuses on optimizing water allocation to achieve the best possible outcomes for both humans and ecosystems. This might involve prioritizing water for drinking, food production, and ecological health.

Water conservation: IWM promotes efficient water use through measures like modern irrigation techniques, leak detection, and water-saving technologies. Water conservation helps alleviate scarcity issues and minimizes wastage.

Wastewater management: Proper treatment and management of wastewater are essential components of IWM. This includes treating and reusing wastewater for non-potable purposes, reducing pollution of water bodies.

Ecosystem protection: IWM acknowledges the importance of maintaining healthy aquatic ecosystems. It includes provisions for maintaining minimum environmental flows in rivers and preserving wetlands, which contribute to water quality and biodiversity.

Climate resilience: IWM takes into account the impacts of climate change on water resources. Strategies might include building resilience against droughts and floods, enhancing water storage capacity, and adjusting water management practices to changing precipitation patterns.

Cross-sectoral coordination: Collaboration between different sectors such as agriculture, industry, and urban planning is crucial to avoid overuse and pollution of water resources. IWM promotes integrated approaches that consider the impacts of one sector on others.

Technology and data: Modern technologies, such as remote sensing, geographic information systems, and data modeling, play a vital role in monitoring water resources, predicting water availability, and making informed decisions.

IWM recognizes that water is a finite and shared resource that requires careful management to meet the needs of present and future generations. By incorporating diverse perspectives, optimizing water use, and safeguarding ecosystems, IWM aims to foster sustainable water resource management and mitigate conflicts over water use.

11.9 Environmental conservation and climate resilience

The harmonious coexistence of agriculture and the environment lies at the heart of sustainable progress. In a world marked by shifting climate patterns, ecosystem degradation, and the pressing need for resource stewardship, the integration of environmental conservation and climate resilience becomes paramount. This chapter embarks on an exploration of these intertwined concepts, unraveling the profound impact of climate change on agriculture and the imperative to preserve the natural world that sustains it.

As we journey through the dynamic landscape of climate-smart practices, biodiversity protection, and circular economy principles, we delve into the transformative potential of aligning agricultural pursuits with the health of the planet.

11.9.1 Climate-smart agriculture: navigating the changing climate

The looming specter of climate change casts its shadow on Pakistan's agriculture, presenting a formidable challenge. Shifting weather patterns, erratic rainfall, and extreme events disrupt traditional agricultural calendars, threatening crop yields and livelihoods. Climate-smart agricultural practices emerge as a beacon of resilience in the face of this uncertainty. These practices adapt to climate-induced changes while mitigating their impacts.

Examining the impact of climate change on agriculture underscores the urgency of action. Embracing crop diversification, where a variety of crops are cultivated, reduces vulnerability to climate shocks. Agroforestry, the integration of trees into agricultural landscapes, creates microclimates, provides windbreaks, and enhances carbon sequestration. Early warning systems, enabled by technology, empower farmers to anticipate and prepare for impending weather extremes.

11.9.2 Biodiversity conservation: nature's unsung allies

Biodiversity, often overlooked in the context of agriculture, emerges as a silent champion of sustainability. Preserving biodiversity within agricultural landscapes unlocks a host of benefits. Diverse plant species act as natural pest deterrents, reducing the reliance on chemical pesticides. A mosaic of plant types enhances ecosystem stability, bolstering resilience against disease outbreaks and extreme weather events.

Agroecological approaches celebrate biodiversity by emulating natural ecosystems. Polyculture, the cultivation of multiple crops in proximity, fosters harmonious coexistence, mimicking the diversity found in nature. This approach enhances the availability of habitats for beneficial insects, pollinators, and soil microorganisms, contributing to sustainable agricultural productivity.

11.9.3 Circular economy and waste management: closing the loop

Within the realm of environmental conservation, the concept of a circular economy takes center stage. Strategies to reduce agricultural waste and promote circular economies not only minimize the ecological footprint but also offer economic opportunities. The circular economy envisions utilizing agricultural byproducts as resources rather than waste. Crop residues, often discarded, can be harnessed for bioenergy production, offering an eco-friendly alternative to fossil fuels.

Composting, the natural decomposition of organic matter, transforms agricultural waste into nutrient-rich compost that enriches soil fertility. This practice closes the loop, returning organic matter to the land while reducing the reliance on synthetic fertilizers.

As this chapter concludes, the essence of environmental conservation and climate resilience is woven into a tapestry that binds the fate of Pakistan's agriculture. The challenges of a changing climate and the imperative to conserve nature find resonance in climate-smart practices, biodiversity stewardship, and circular economy principles. By embracing these practices, Pakistan's agricultural sector charts a course toward a sustainable future. This chapter stands as a testament to the power of conscious choices and collective action in forging a resilient path that ensures both agricultural prosperity and environmental well-being.

11.9.4 Integrated environmental management (IEM)

Integrated environmental management (IEM) is a comprehensive approach that aims to achieve sustainable development by considering the interconnections between environmental, social, and economic factors. It involves the coordinated planning, management, and decision-making processes that address environmental challenges while promoting human well-being and prosperity. IEM recognizes that environmental issues are complex and often intertwined with social and economic considerations.

The key components of IEM include

Holistic approach: IEM acknowledges that environmental challenges cannot be effectively addressed in isolation. It considers the broader context of economic development, social equity, and cultural values to find solutions that are both environmentally sound and socially beneficial.

Interdisciplinary collaboration: IEM involves collaboration among various disciplines such as ecology, economics, sociology, and policy-making. This interdisciplinary approach allows for a more comprehensive understanding of complex environmental issues.

Stakeholder engagement: Engaging a diverse range of stakeholders, including government agencies, communities, businesses, and NGOs, is essential in IEM. These stakeholders bring different perspectives and expertise to the decision-making process.

Environmental impact assessment (EIA): IEM often includes conducting environmental impact assessments before initiating development projects. EIAs identify potential environmental impacts and suggest mitigation measures to minimize negative effects.

Sustainable resource management: IEM focuses on the sustainable use and management of natural resources including water, forests, land, and minerals. It seeks to balance resource extraction with preservation to ensure long-term availability.

Biodiversity conservation: IEM promotes the conservation of biodiversity and ecosystems. This includes protecting habitats, preventing species extinction, and restoring degraded ecosystems.

Pollution prevention and control: IEM emphasizes the prevention and reduction of pollution through regulations, technology upgrades, and cleaner production practices.

Climate change mitigation and adaptation: IEM considers the impacts of climate change and seeks to mitigate its effects through reducing greenhouse gas emissions and adapting to changing conditions.

Waste management: Proper waste management is integral to IEM. This involves waste reduction, recycling, safe disposal, and the promotion of a circular economy to minimize waste generation.

Education and awareness: IEM aims to raise public awareness about environmental issues and promote environmentally responsible behavior through education and information campaigns.

Policy and regulation: IEM involves the development and implementation of environmental policies, regulations, and standards to guide sustainable development practices.

Long-term planning: IEM emphasizes long-term planning that considers the needs of current and future generations. It strives to balance short-term economic gains with the preservation of environmental quality for the long term.

IEM recognizes that a healthy environment is crucial for human well-being and economic prosperity. By integrating environmental considerations into various sectors and decision-making processes, IEM seeks to create a sustainable future where the environment is preserved, social equity is promoted, and economic development is balanced with ecological integrity.

In the grand tapestry of sustainable agriculture, the chapters of soil health, water resource management, and environmental conservation intertwine to forge a future that harmonizes productivity with ecological integrity. The journey through these essential elements has illuminated the profound interconnectedness that underpins agricultural success. As the chapter on soil health drew us into the intricate realm beneath our feet, we learned that nurturing this living entity is not just an option but a necessity. It became evident that understanding the nuances of soil composition, guarding against degradation, and embracing practices that sustain its vitality are pivotal to ensuring bountiful harvests for generations to come.

Our exploration of water resource management revealed that water's scarcity calls for not only vigilance but also innovation. The delicate balance between water supply and agricultural demands precision, and the techniques unveiled in this chapter offer a roadmap for equitable distribution and efficient utilization. We've come to appreciate that precision irrigation, watershed management, and rainwater harvesting are not just strategies; they are lifelines that bridge the gap between water scarcity and agricultural vitality.

Finally, as we ventured into the world of environmental conservation and climate resilience, the imperative to safeguard the planet became palpable. The profound impacts of climate change on agriculture awakened us to the urgency of adopting climate-smart practices that bolster resilience and mitigate risks. The significance of biodiversity conservation stood as a testament to nature's role as an unsung ally in agriculture. Circular economy principles underscored the potential to align agricultural production with resource preservation, closing loops and minimizing waste.

In uniting these chapters, we weave a narrative of agricultural sustainability that transcends fields and ecosystems. The intricate dance of soil, water, and environmental management culminates in a symphony of resilience, productivity, and harmony. As we navigate the complex challenges and opportunities of modern agriculture, the lessons learned from these chapters echo with a resounding truth: sustainable practices are not just an aspiration but an imperative, a path that leads us toward a future where prosperity, nourishment, and ecological well-being are intertwined. By embracing these principles, we pave the way for a world where agriculture thrives, nature flourishes, and generations to come inherit a planet poised for enduring prosperity.

Amidst the symphony of earth and water, Chapter 11 invites us to explore the profound interplay between soil, water, and the environment. ISWEM transcends boundaries, advocating for the preservation of essential resources while nurturing thriving ecosystems. As we delve into the practices of sustainable land use, watershed management, and environmental conservation, this chapter underscores the inseparable connection between healthy landscapes, vibrant communities, and a flourishing planet.

From the embrace of soil and water stewardship, we step into Chapter 12 a realm where the heartbeat of rural communities pulses through the veins of cottage industries. Cottage industry and rural development shine a light on local craftsmanship, innovation, and economic growth. Here, we witness the fusion of tradition and modernity, as small-scale enterprises and sustainable livelihoods converge to sculpt the landscape of vibrant rural futures.

References

Amanullah, Yar, M., Khalid, S., et al. (2021). Phenology, growth, productivity, and profitability of mungbean as affected by potassium and organic matter under water stress vs. no water stress conditions. Journal of Plant Nutrition, 45(9), 629–650.

Amanullah, & Khalid, S. (2020). Agronomy – Climate Change & Food Security IntechOpen, London, United Kingdom.

Amanullah, Ullah, H., Soliman Elshikh, M., Alwahibi, M. S., Alkahtani, J., Muhammad, A., & Khalid, S. (2020). Nitrogen contents in soil, grains, and straw of hybrid rice differ when applied with different organic nitrogen sources. Agriculture, 10, 386.

Amanullah, Inamullah, Alkahtani, J., Elshikh, M. S., Alwahibi, M. S., Muhammad, A., Ahmad, M., & Khalid, S. (2020). Phosphorus and zinc fertilization influence crop growth rates and total biomass of coarse vs. fine types rice cultivars. Agronomy, 10, 1356.

Amanullah, Inamullah, Alwahibi, M. S., Elshikh, M. S., Alkahtani, J., Muhammad, A., Khalid, S., Ahmad, M., Khan, N., Ullah, S., & Ali, I. (2020). Phosphorus and zinc fertilization improve zinc biofortification in grains and straw of coarse vs. fine rice genotypes. Agronomy, 10, 1155.

Amanullah, Hidayatullah, A. J., Shah, Z., Khan, M. J., Parmar, B., & Fahad, S. (2019). Organic carbon sources and nitrogen management improve biomass of hybrid rice (*Oryza sativa* L.) under nitrogen deficient condition. In: M. Hasanuzzaman, M. Fujita, K. Nahar & J. K. Biswas (Eds.) Advances in Rice Research for Abiotic Stress Tolerance (pp. 447–468). Woodhead Publishing, Chennai.

Amanullah, & Fahad, S. (2018). Nitrogen in Agriculture-Updates InTechOpen, London, UK.

Amanullah, & Fahad, S. (2018). Corn – Production and Human Health in Changing Climate InTech, London, UK.

Bawa, K. S., Seidler, R., & Raven, P. H. (2004). Reproductive biology of tropical forest plants: Historical perspectives. In: Tropical Forests and Global Atmospheric Change (pp. 53–61). Springer, Heidelberg, Germany.

Bekele, D., & Tilahun, S. A. (2016). Integrated soil and water conservation for sustainable livelihoods in Ethiopia: A synthesis. International Journal of Environmental Studies, 73(3), 388–400.

Custodio, M. R., Li, Y., Zhang, H., & Jin, L. (2015). Integrated natural resources management: Approaches and lessons from China. Journal of Environmental Management, 150, 166–176.

FAO. (2019). The International Code of Conduct for the Sustainable Use and Management of Fertilizers (pp. 30). FAO, Rome.

FAO. (2018). Soil Pollution: A Hidden Reality FAO, Rome-Italy.

FAO. (2017). Unlocking the Potential of Soil Organic Carbon FAO, Rome-Italy.

FAO. (2016). Soil and Pulses: Symbiosis for Life FAO, Rome-Italy.

FAO and ITPS. (2016). Voluntary Guidelines for Sustainable Soil Management (VGSSM) Rome, Italy.

FAO. (2009). The State of the World's Land and Water Resources for Food and Agriculture (SOLAW) – Managing Systems at Risk Food and Agriculture Organization of the United Nations.

Lal, R. (2014). Challenges and opportunities in soil organic matter research. European Journal of Soil Science, 65(1), 2–15.

Lambin, E. F., & Meyfroidt, P. (2011). Global land use change, economic globalization, and the looming land scarcity. Proceedings of the National Academy of Sciences, 108(9), 3465–3472.

MEA (Millennium Ecosystem Assessment). (2005). Ecosystems and Human Well-being: Synthesis Island Press, Washington, D.C., USA.

Molden, D. (Ed.). (2007). Water for Food, Water for Life: A Comprehensive Assessment of Water Management in Agriculture International Water Management Institute (IWMI), Colombo, Sri Lanka.

Oldeman, L. R., Hakkeling, R. T. A., & Sombroek, W. G. (1991). World Map of the Status of Human-induced Soil Degradation: An Explanatory Note Wageningen Agricultural University, Wageningen, Netherlands.

Pretty, J., & Hine, R. (2001). Reducing food poverty by increasing agricultural sustainability in developing countries. Agriculture, Ecosystems & Environment, 86(3), 257–274.

Reenberg, A. (2007). Cultural landscapes of the world: Danish examples. GeoJournal, 68(3), 235–243.

Rockström, J., & Falkenmark, M. (2015). Agriculture: Increase water harvesting in Africa. Nature, 519(7543), 283–285.

Rockström, J., Steffen, W., Noone, K., Persson, Å., Chapin, F. S., Lambin, E., . . . & Foley, J. A. (2009). A safe operating space for humanity. Nature, 461(7263), 472–475.

Scoones, I. (1998). Sustainable rural livelihoods: A framework for analysis. IDS Working Paper, 72.

Simeonov, V., Stratis, J. A., Samara, C., Zachariadis, G., Voutsa, D., Anthemidis, A., . . . & Sofoniou, M. (2003). Assessment of the surface water quality in Northern Greece. Water Research, 37(17), 4119–4124.

Smith, P., & Smith, J. U. (1996). Description of the Rothamsted carbon model (RothC ver. 26.3). In: Evaluation of Soil Organic Matter Models (pp. 81–101). Springer, Heidelberg, Germany.

Steffen, W., Richardson, K., Rockström, J., Cornell, S. E., Fetzer, I., Bennett, E. M., . . . & Sörlin, S. (2015). Planetary boundaries: Guiding human development on a changing planet. Science, 347(6223), 1259855.

Sutton, M. A., Oenema, O., Erisman, J. W., Leip, A., van Grinsven, H., & Winiwarter, W. (2011). Too much of a good thing. Nature, 472(7342), 159–161.

United Nations. (2015). Transforming Our World: The 2030 Agenda for Sustainable Development United Nations, United Nations, New York, USA.

World Bank. (2006). Environment Strategy 2001 World Bank, Washington, D.C., USA.

Chapter 12
Cottage industry and rural development: global initiatives

Introduction

A cottage industry refers to a small-scale manufacturing operation, typically run from a person's home or a small workshop, where products are produced manually or with limited mechanization. These industries often involve traditional craftsmanship and utilize locally available resources and skills to produce goods such as textiles, handicrafts, food products, and artisanal items.

The global significance of cottage industries lies in their multifaceted contributions to various aspects of society and economy. Firstly, they play a vital role in economic development by providing avenues for entrepreneurship and small business growth. Cottage industries offer individuals the opportunity to start their own businesses with relatively low initial capital, fostering a spirit of innovation and self-reliance.

Secondly, cottage industries are important for employment generation, especially in rural and marginalized communities where job opportunities may be limited. By engaging in cottage industries, individuals can earn a livelihood without having to migrate to urban areas in search of work. This not only helps to alleviate poverty but also contributes to the overall economic stability of rural regions.

Moreover, cottage industries are significant for cultural preservation. Many of these industries involve traditional crafts and skills that have been passed down through generations. By supporting cottage industries, communities can preserve their cultural heritage and traditions, ensuring that valuable skills and techniques are not lost to modernization.

Additionally, cottage industries often promote sustainable practices and the use of locally sourced materials. This can have positive environmental impacts by reducing the carbon footprint associated with production and transportation. Furthermore, these industries often prioritize quality over quantity, producing handmade goods that are unique and have a story behind them, appealing to consumers who value authenticity and craftsmanship.

In summary, cottage industries play a vital role in economic development, employment generation, cultural preservation, and sustainable production practices. Their significance extends beyond just the products they create; they are integral to the social fabric of communities, particularly in rural areas. By supporting cottage industries, societies can promote inclusive growth, preserve cultural heritage, and foster a sense of community pride and identity.

https://doi.org/10.1515/9783111448077-012

12.1 Key aspects of cottage industry

Some key aspects of their global significance are listed in Figure 12.1.

Figure 12.1: Key aspects of cottage industry in Pakistan.

Economic contribution: Cottage industries play a vital role in the economy by providing employment and income opportunities for millions of people worldwide, especially in regions where formal employment opportunities are limited. These industries contribute to household income generation, poverty reduction, and economic diversification, particularly in rural areas where agriculture may not provide sufficient livelihood opportunities.

Cultural preservation: Cottage industries often involve traditional craftsmanship and production techniques that have been passed down through generations. These industries help preserve cultural heritage, traditional knowledge, and artisanal skills, ensuring the continuation of cultural practices and identities. They also contribute to the promotion of cultural tourism and the preservation of cultural diversity.

Sustainable development: Cottage industries are typically based on small-scale production, local resources, and environmentally friendly practices. They often have a lower environmental footprint compared to large-scale industrial operations, as they utilize natural materials, reduce energy consumption, and minimize waste generation. By promoting sustainable production and consumption patterns, cottage industries contribute to the achievement of sustainable development goals.

Community empowerment: Cottage industries empower local communities by providing opportunities for entrepreneurship, skill development, and social cohesion. They enable individuals, particularly women and marginalized groups, to participate in economic activities, gain financial independence, and contribute to community development initiatives. Cottage industries also foster social networks, collective decision-making, and community resilience.

Global trade and market access: Cottage industries produce a diverse range of artisanal products and handicrafts that are valued both locally and internationally. They contribute to global trade and cultural exchange, allowing artisans and producers to access wider markets and generate income from exports. Platforms such as e-commerce and fair trade initiatives further facilitate market access for cottage industry products, enabling artisans to reach global consumers while promoting fair wages and ethical production practices.

Cottage industries play a crucial role in promoting inclusive and sustainable economic development, preserving cultural heritage, and empowering communities worldwide. Their global significance extends beyond economic considerations to encompass social, cultural, and environmental dimensions, making them valuable contributors to a more equitable and resilient global economy.

In the realm of cottage industry and rural development, we delve into the pivotal role of small-scale enterprises in fostering economic growth, empowering communities, and promoting sustainable development on a global scale. Cottage industries, characterized by small-scale production units often operated from homes or small workshops, play a significant role in generating employment, preserving traditional crafts, and fostering economic diversity in rural areas.

Across the globe, these industries contribute to poverty alleviation, women's empowerment, and community resilience by providing livelihood opportunities to marginalized populations and preserving cultural heritage. The significance of cottage industries extends beyond economic benefits; they are vital for social and cultural sustainability. Through these small-scale operations, communities can maintain their traditional crafts and skills, ensuring that valuable heritage is passed down to future generations.

Moreover, cottage industries often serve as a catalyst for innovation and entrepreneurship. They provide a platform for individuals to showcase their creativity and turn their ideas into tangible products. This not only boosts local economies but also encourages a sense of self-reliance and empowerment within communities.

Additionally, cottage industries are inherently sustainable, often utilizing locally available resources and promoting eco-friendly practices. By reducing the need for extensive transportation and mass production, they contribute to lower carbon footprints and promote a more environmentally friendly approach to manufacturing.

In essence, the role of cottage industries in rural development is multifaceted. They are not just about producing goods; they are about creating opportunities, preserving culture, fostering innovation, and promoting sustainability. By recognizing and supporting these small-scale enterprises, we can contribute to the well-being of rural communities, empower individuals, and build a more inclusive and sustainable global economy.

Furthermore, rural development initiatives encompass a wide range of efforts aimed at improving living standards, infrastructure, and access to essential services in rural areas. From agricultural extension programs that enhance farmers' productivity to microfinance initiatives that support rural entrepreneurs, these initiatives strive to unlock the potential of rural communities and bridge the urban-rural divide.

The global initiatives and best practices aimed at promoting cottage industry and rural development underscore the transformative potential of inclusive economic policies, capacity-building programs, and enhanced market linkages. These initiatives prioritize the empowerment of rural communities, recognizing them as vital contributors to sustainable development and economic growth. By investing in entrepreneurship training, vocational skills development, and access to finance, nations can cultivate a conducive environment for cottage industries to thrive.

Moreover, fostering collaboration between governments, civil society organizations, and private sector stakeholders facilitates the creation of supportive ecosystems that enable cottage industries to flourish. Through targeted interventions that address infrastructural challenges, provide technical assistance, and facilitate market access, countries can harness the full potential of cottage industries as engines of inclusive growth and poverty alleviation on a global scale.

12.2 Rural development

Rural development is a multifaceted process aimed at improving the quality of life and economic well-being of people living in rural areas. It encompasses various strategies and initiatives designed to address the unique challenges faced by rural communities, such as limited access to basic services, infrastructure, employment opportunities, and markets.

Central to rural development efforts are programs focused on agriculture, which often serves as the backbone of rural economies. These programs aim to enhance agricultural productivity, promote sustainable farming practices, provide access to modern technologies, and improve market linkages for farmers. By boosting agricultural output and income, rural development initiatives contribute to poverty reduction and food security.

Moreover, rural development extends beyond agriculture to include initiatives related to education, healthcare, sanitation, housing, and infrastructure development. Improving access to quality education and healthcare services in rural areas is essential for enhancing human capital and overall well-being. Infrastructure projects such as roads, bridges, and electricity not only connect rural communities to urban centers but also stimulate economic activities and create employment opportunities locally.

Rural development is not just about economic growth but also about fostering social inclusion and empowerment. It involves empowering marginalized groups, including women and youth, through skill development programs, entrepreneurship training, and access to financial services. By promoting social cohesion and equitable opportunities, rural development aims to create vibrant and sustainable rural communities that can thrive in the face of challenges and contribute to the overall progress of the nation.

Cottage industries play a vital role in rural development by providing employment opportunities, preserving traditional skills, and fostering local economies. These small-scale enterprises are typically operated from homes or small workshops, often using locally available raw materials and simple tools. In rural areas where large industries may be scarce, cottage industries serve as a significant source of livelihood for many households.

One of the key benefits of cottage industries is their ability to generate employment locally. They offer opportunities for rural residents, including women and artisans, to engage in productive work without the need to migrate to urban areas in search of jobs. This not only reduces rural-to-urban migration but also helps in maintaining the social fabric of rural communities.

Cottage industries also play a role in preserving traditional craftsmanship and skills that are often passed down through generations. Whether it's handloom weaving, pottery, handicrafts, or artisanal food production, these industries contribute to the cultural heritage of a region. By supporting cottage industries, rural areas can retain their unique identity and traditions while creating products that have a market both locally and globally.

Furthermore, cottage industries contribute to local economies by creating value-added products from locally available resources. This can lead to increased income for rural households and improved standards of living. Additionally, the growth of cottage industries can stimulate related businesses such as transportation, packaging, and marketing services, creating a ripple effect of economic development in the area.

In essence, cottage industries are a cornerstone of rural development, offering a path to economic empowerment, cultural preservation, and community resilience. Supporting and promoting these small-scale enterprises not only boosts rural economies but also helps in achieving sustainable development goals by fostering inclusive growth and preserving the rich heritage of rural areas.

12.3 Cottage industry and its role in rural development

Cottage industry plays a pivotal role in rural development by serving as a catalyst for economic growth and community empowerment. These small-scale enterprises, often run from homes or small workshops, contribute significantly to the economic diversity of rural areas. By providing employment opportunities locally, cottage industries help to reduce migration to urban centers in search of work, thus bolstering the economic stability of rural communities. Moreover, cottage industries preserve traditional crafts and skills, ensuring that cultural heritage is passed down to future generations. Through these enterprises, rural communities can tap into their unique strengths, promote entrepreneurship, and foster a sense of self-reliance. Additionally, cottage industries are often sustainable, utilizing locally available resources and promoting eco-friendly practices, which further contributes to the overall sustainability of rural development. In essence, cottage industry stands as a cornerstone of rural development, intertwining economic progress, cultural preservation, and community resilience.

Key aspects of cottage industry and its role in rural development:

Employment generation: Cottage industries provide employment opportunities, particularly in rural areas where alternative job opportunities may be limited. They offer livelihood options for individuals and families, helping to reduce unemployment and poverty.

Skill development and capacity building: Cottage industries contribute to skill development and capacity building at the local level. Traditional crafts, artisanal skills, and specialized knowledge are often preserved and passed down through generations. Training programs and skill enhancement initiatives support the development of new skills and techniques.

Value addition to local resources: Cottage industries add value to local resources and raw materials. They transform locally available materials into finished products, creating economic value and marketable goods. This encourages the sustainable utilization of local resources and promotes local entrepreneurship.

Income generation and poverty alleviation: Cottage industries provide income-generating opportunities, particularly for marginalized communities and women. They enable individuals to earn a livelihood and improve their standard of living, contributing to poverty alleviation and economic empowerment.

Cultural preservation: Cottage industries often involve traditional crafts and artisanal practices that are deeply rooted in local culture and heritage. By supporting and promoting cottage industries, cultural traditions, craftsmanship, and traditional knowledge can be preserved and celebrated.

Community development: Cottage industries contribute to community development by fostering social cohesion, local pride, and community engagement. They often serve as hubs for social interaction, knowledge exchange, and collective initiatives. Cottage industries can contribute to the overall well-being and resilience of rural communities.

Promoting cottage industries entails providing comprehensive support mechanisms that encompass access to finance, specialized training, infrastructure development, and effective market linkages. These initiatives serve as catalysts for rural development, fostering economic growth and vitality in remote areas. By empowering local artisans and entrepreneurs, cottage industries not only generate employment opportunities but also preserve cultural heritage and traditional craftsmanship.

Moreover, investments in infrastructure such as roads, electricity, and telecommunications bolster the competitiveness of cottage industries by facilitating efficient production and distribution networks. Equally important are efforts to strengthen market linkages, enabling cottage industries to access broader consumer markets and command fair prices for their products. Through these multifaceted interventions, governments and development agencies can unlock the untapped potential of cottage industries as engines of sustainable rural development and poverty alleviation.

12.4 Role of cottage industries in rural economies

Cottage industries play a significant role in rural economies, serving as engines of economic growth, employment generation, and poverty alleviation. These small-scale enterprises contribute to the diversification of rural economies by providing alternative sources of income beyond agriculture. They often leverage local resources and traditional skills, preserving cultural heritage and fostering community cohesion. Moreover, cottage industries empower marginalized groups, including women and artisans, by creating opportunities for entrepreneurship and skill development. By stimulating local production and consumption, cottage industries contribute to the revitalization of rural communities and the promotion of sustainable development.

Here are key aspects of their role:

Employment generation: Cottage industries provide employment opportunities for local communities, particularly in rural areas where formal employment may be limited. They create jobs for artisans, craftsmen, and small-scale entrepreneurs, helping to reduce rural unemployment rates and contributing to income generation.

Economic diversification: Cottage industries contribute to economic diversification in rural areas. They promote the development of nonagricultural sectors and provide alternative sources of income beyond traditional farming activities. Economic diversification helps reduce dependence on agriculture, enhances resilience, and supports overall economic growth.

Value addition and market access: Cottage industries add value to raw materials and agricultural products by processing them into finished goods. This value addition enhances the marketability and profitability of products. Cottage industries provide a means for farmers to access higher-value markets by transforming their raw materials into value-added products.

Preservation of traditional skills and cultural heritage: Cottage industries often involve traditional crafts and artisanal practices that are deeply rooted in local culture and heritage. They preserve traditional skills, knowledge, and cultural practices, promoting cultural identity and heritage conservation. This contributes to the overall well-being and resilience of rural communities.

Small-scale entrepreneurship: Cottage industries offer opportunities for small-scale entrepreneurship in rural areas. They empower individuals to start their own businesses, create their own livelihoods, and exercise their entrepreneurial skills. Cottage industries foster entrepreneurship, innovation, and creativity, leading to local economic growth and self-sufficiency.

12.5 Value addition and processing of agricultural products

Value addition and processing of agricultural products through cottage industries bring several benefits to both rural economies and consumers alike. By transforming raw agricultural commodities into value-added products such as jams, pickles, handicrafts, and textiles, cottage industries create opportunities for higher profitability and market diversification. Additionally, value-added processing extends the shelf life of perishable goods, reducing postharvest losses and ensuring food security.

Moreover, cottage industries often employ traditional techniques and local ingredients, adding unique flavors and characteristics to the final products. This not only preserves cultural heritage but also enhances the marketability of rural products, attracting discerning consumers seeking authenticity and sustainability. Overall, value addition and processing play a crucial role in maximizing the economic potential of agricultural resources, promoting rural entrepreneurship, and contributing to the overall development of rural communities.

Key aspects of value addition and processing:

Increased market value: Value addition through processing enhances the market value of agricultural products. By transforming raw agricultural commodities into processed goods, such as jams, pickles, dairy products, or textiles, their value and desirability increase. This enables farmers and rural entrepreneurs to fetch higher prices and generate more income.

Preservation and storage: Processing agricultural products extends their shelf life and helps preserve their quality. Value-added products, such as canned fruits and vegetables, dried herbs, or frozen meats, can be stored and sold throughout the year. This reduces postharvest losses and ensures a more consistent income stream for farmers.

Market diversification: Value-added agricultural products open up opportunities to access diverse markets. Processed goods often have wider consumer appeal and can target niche markets or cater to specific consumer preferences. This diversification helps reduce reliance on traditional commodity markets and increases market resilience for farmers.

Job creation: Value addition and processing require additional labor, leading to job creation in rural areas. Cottage industries that process agricultural products generate employment opportunities for individuals involved in sorting, cleaning, processing, packaging, and marketing. This contributes to rural employment, income generation, and poverty reduction.

Quality control and branding: Processing agricultural products allows for quality control measures and the establishment of branded products. Through value addition, farmers and entrepreneurs can ensure consistent product quality, adhere to food safety standards, and build a brand reputation for their products. Quality control and branding enhance consumer trust and support market competitiveness.

Small-scale entrepreneurship in rural areas: Small-scale entrepreneurship holds a pivotal role in rural areas, serving as a cornerstone of economic activity and community development. In these regions, where traditional industries often dominate the landscape, small-scale enterprises inject vitality into local economies by providing diverse goods and services tailored to the needs of rural populations. These enterprises encompass a wide range of activities, from agricultural processing and handicraft production to retail and service-oriented ventures. Often rooted in local knowledge and cultural traditions, small-scale entrepreneurship fosters innovation and resilience, enabling rural communities to adapt to changing economic conditions and market demands.

Moreover, small-scale entrepreneurship serves as a potent engine for job creation and income generation in rural areas, offering livelihood opportunities to a broad spectrum of individuals, including farmers, artisans, women, and youth. By harnessing their entrepreneurial spirit and ingenuity, individuals can transform locally available resources into value-added products and services, thus unlocking economic potential and enhancing household incomes. Additionally, small-scale entrepreneurship empowers marginalized groups, such as women and indigenous communities, by providing avenues for economic participation and social inclusion. This inclusive approach to entrepreneurship not only fosters economic empowerment but also strengthens social cohesion and community resilience in rural areas.

Key aspects of small-scale entrepreneurship are:

Local economic development: Small-scale entrepreneurship drives local economic development by creating employment opportunities and income generation within rural communities. It stimulates economic growth, improves livelihoods, and contributes to poverty reduction.

Innovation and creativity: Small-scale entrepreneurs often bring innovation and creativity to their businesses. They identify niche markets, develop unique products or services, and find innovative solutions to local challenges. This entrepreneurial spirit fosters resilience, adaptability, and competitiveness in rural economies.

Community engagement and empowerment: Small-scale entrepreneurship promotes community engagement and empowerment. It encourages individuals to take ownership of their economic activities, fosters a sense of self-reliance, and strengthens social ties within communities. Small-scale entrepreneurs often engage in community initiatives, collaborate with other stakeholders, and contribute to local development.

Local resource utilization: Small-scale entrepreneurship leverages local resources and assets, including natural resources, cultural heritage, and traditional knowledge. It promotes sustainable use of resources, encourages local production, and supports local supply chains. This strengthens the resilience of rural economies and reduces dependency on external resources.

Youth and women empowerment: Small-scale entrepreneurship offers opportunities for youth and women in rural areas. It provides avenues for economic participation, skill development, and empowerment, addressing issues of youth unemployment and gender inequality. Small-scale entrepreneurship can be a pathway for young entrepreneurs and women to contribute to local development.

12.6 Rural development policies and programs: enhancing rural prosperity

Rural development policies and programs form a crucial component of a nation's quest for equitable and sustainable growth. In the context of Pakistan, where a significant portion of the population resides in rural areas, these policies become instrumental in narrowing the urban-rural development gap and fostering socioeconomic progress in remote and underserved regions. Within this framework, several key aspects of rural development policies and programs emerge, each geared toward enhancing infrastructure, access to finance, and overall well-being.

12.6.1 Infrastructure development: paving the path for connectivity and progress

Infrastructure development constitutes a cornerstone of rural development initiatives. These programs channel resources into creating and improving essential facilities that form the backbone of rural economies. Roads, often referred to as rural arteries, play a pivotal role in connecting rural areas to urban centers and markets. Improved road networks facilitate the transportation of agricultural products, reduce transportation costs, and open doors to socioeconomic opportunities.

Beyond roads, infrastructure development encompasses other critical elements such as electricity, water supply, and telecommunications. Reliable access to electricity empowers rural communities by supporting agricultural processing, small-scale industries, and household activities. Clean water supply ensures improved health and sanitation, while telecommunications bridge information gaps and enable digital connectivity, allowing rural residents to access market information, financial services, and educational resources.

12.6.2 Access to finance and credit: empowering rural entrepreneurs

Access to finance and credit is a fundamental catalyst for rural development, enabling entrepreneurs and smallholder farmers to invest in their businesses, seize opportunities, and enhance their economic status. Rural development programs in Pakistan emphasize the provision of financial resources to uplift rural communities. Microfinance schemes, designed to cater to the specific needs of rural entrepreneurs, offer accessible credit facilities for agricultural activities, cottage industries, and small businesses.

In addition to microfinance, subsidies and grants play a pivotal role in encouraging rural development. These mechanisms alleviate the financial burden on farmers and entrepreneurs, making essential resources such as fertilizers, seeds, and modern agricultural technologies more affordable. Subsidies can also incentivize the adoption of sustainable practices, such as water-efficient irrigation techniques and organic farming methods.

12.6.3 Education and skill development: nurturing human capital

A comprehensive approach to rural development extends beyond physical infrastructure and financial support. Education and skill development programs play a pivotal role in nurturing human capital in rural areas. Accessible and quality education equips individuals with knowledge and skills necessary to engage in diverse economic activities. Vocational training initiatives empower rural youth with practical skills that cater to local demands, reducing unemployment and enhancing income-generating opportunities.

12.6.4 Women empowerment and social inclusion: inclusive development

Rural development policies prioritize gender equality and social inclusion. Recognizing the significant contributions of women in agriculture and rural economies, these programs promote women's participation in decision-making, entrepreneurship, and income-generating activities. By addressing gender disparities and fostering women's economic empowerment, rural development initiatives contribute to more inclusive and holistic progress.

Rural development policies and programs are pivotal in shaping the trajectory of Pakistan's rural areas. By investing in infrastructure, improving access to finance, nurturing human capital, and promoting inclusive growth, these initiatives pave the way for vibrant rural economies that contribute to the overall socioeconomic development of the nation.

Within the tapestry of cottage industry and rural development, the delicate threads of tradition, innovation, and economic empowerment intertwine. This chapter unveils the transformative power of local craftsmanship, echoing the stories of artisans who shape rural landscapes through creativity and ingenuity. As we immerse ourselves in the world of cottage industries, we recognize their pivotal role in nurturing resilient communities, preserving cultural heritage, and igniting the flames of prosperity.

As the mosaic of rural life comes alive, the curtain rises on **Chapter 13** an exploration of present scenario and future prospects globally. This chapter offers a panoramic view of the nation's agricultural journey, shedding light on the challenges, successes, and aspirations that define its trajectory. By delving into the dynamics of evolving policies, market trends, and technological advancements, we gain insight into the canvas upon which the future of Pakistan's agriculture is painted.

References

Amanullah, & Khalid, S. (2020). Agronomy – Climate Change & Food Security. IntechOpen, London, United Kingdom.

Amanullah, & Fahad, S. (2018). Nitrogen in Agriculture-Updates. InTechOpen, London, UK.

Amanullah, & Fahad. 2018). Corn – Production and Human Health in Changing Climate. InTech, London, UK.

Anand, G. (2014). Cottage industry in India: Past, present, and future. International Journal of Business and Administration Research Review, 1(5), 144–149.

Biggs, S. D. (2017). The rise of the cottage industry in India: De-industrialization, manufacturing revival, and labour regulations. Indian Journal of Labour Economics, 60(3), 475–504.

Borse, R. B., & Sonar, R. M. (2013). Cottage industry: A boom for rural development. Journal of Emerging Technologies and Innovative Research, 1(4), 199–204.

Das, B., & Das, S. (2016). Cottage industry in India: An economic analysis. International Journal of Humanities, Social Sciences and Education, 3(11), 61–67.

Duarte, C. L. (2012). Micro and small cottage industries in West Bengal, India: A potential to rural development. International Journal of Research in Economics and Social Sciences, 2(2), 125–144.

Gopalan, M. R., & Chakrabarti, B. (2014). Cottage industry in rural Tamil Nadu: An analysis. International Journal of Social and Economic Research, 4(4), 379–390.

Government of India. (2008). Report of the National Commission for Enterprises in the Unorganized Sector. Ministry of Micro, Small and Medium Enterprises.

Kalamkar, S. S., & Devkate, D. V. (2019). Cottage industry and its role in rural development: A case study. International Journal of Research in Management, Science & Technology, 7(1), 136–140.

Kumar, A. (2017). Cottage industry: A potential for rural development in India. International Journal of Business Management & Research, 7(4), 1–10.

Prajapati, P. (2018). Cottage industry: A tool for rural development. International Journal of Recent Research Aspects, 5(2), 13–19.

Chapter 13
Global agriculture: current trends and future outlook

Introduction

Global agriculture, spanning the production, distribution, and trade of agricultural products across continents, stands as a cornerstone of human civilization. This multifaceted system encompasses the cultivation of crops, husbandry of livestock, forestry practices, and fisheries activities, collectively contributing to the sustenance of billions and the economic vitality of nations. In essence, global agriculture represents a complex interplay of factors, stakeholders, and challenges, with far-reaching implications for food security, poverty alleviation, and livelihood sustainability worldwide. As we navigate the intricacies of this global system, understanding its dynamics and addressing its challenges are imperative for ensuring a resilient and equitable future for all.

Inviting us to explore the ever-evolving panorama of global agriculture, we delve into its dynamic trends and illuminate the path forward for agricultural systems worldwide. Within these pages, we embark on a voyage through the intricate tapestry of modern agriculture, where technological innovation, market dynamics, and environmental imperatives converge to shape the future of food production. As we traverse this terrain, we unravel the multifaceted dimensions of agricultural development, from the burgeoning advances in agricultural technology to the profound shifts in consumer preferences and global trade patterns.

This exploration serves as a beacon, guiding our understanding of the current trends that are reshaping the global agricultural landscape and offering invaluable insights into the future trajectory of agricultural systems. By navigating through the complexities of modern agriculture, we gain a deeper understanding of the challenges and opportunities that lie ahead, empowering us to chart a course toward a more sustainable, resilient, and equitable agricultural future.

The current state of global agriculture presents a complex landscape shaped by various factors such as technological advancements, climate change, market dynamics, and socioeconomic considerations. Innovations in agriculture, including precision farming, biotechnology, and digitalization, have revolutionized the sector, increasing productivity, efficiency, and sustainability. These advancements enable farmers to optimize resource use, manage risks, and adapt to changing environmental conditions. Climate change poses significant challenges to global agriculture, affecting crop yields, water availability, pest and disease patterns, and overall food security. Extreme weather events, such as droughts, floods, and heatwaves, disrupt agricultural production systems, leading to yield losses and economic instability.

https://doi.org/10.1515/9783111448077-013

Global agricultural markets are influenced by trade policies, consumer preferences, and supply chain disruptions. Volatility in commodity prices, trade tensions, and shifts in dietary patterns impact farmers' incomes, market access, and food prices, influencing agricultural practices and investments. Socioeconomic factors, including population growth, urbanization, income inequality, and rural development, shape agricultural landscapes and food systems. Addressing issues of poverty, access to land and resources, gender equity, and labor rights is essential for achieving sustainable agriculture and rural development goals. The United Nations' Sustainable Development Goals (SDGs), particularly Goal 2 (Zero Hunger) and Goal 12 (Responsible Consumption and Production), provide a framework for addressing global agriculture challenges. Achieving these goals requires concerted efforts to promote sustainable agriculture practices, ensure food security, and reduce food waste and environmental degradation.

The current state of global agriculture is characterized by a complex interplay of technological, environmental, economic, and social factors. Addressing the challenges and opportunities presented by this dynamic landscape requires collaborative efforts from governments, international organizations, farmers, researchers, and other stakeholders to promote sustainable agricultural practices, enhance resilience, and ensure food security for present and future generations.

Here are some key details covered in this chapter (Figure 13.1).

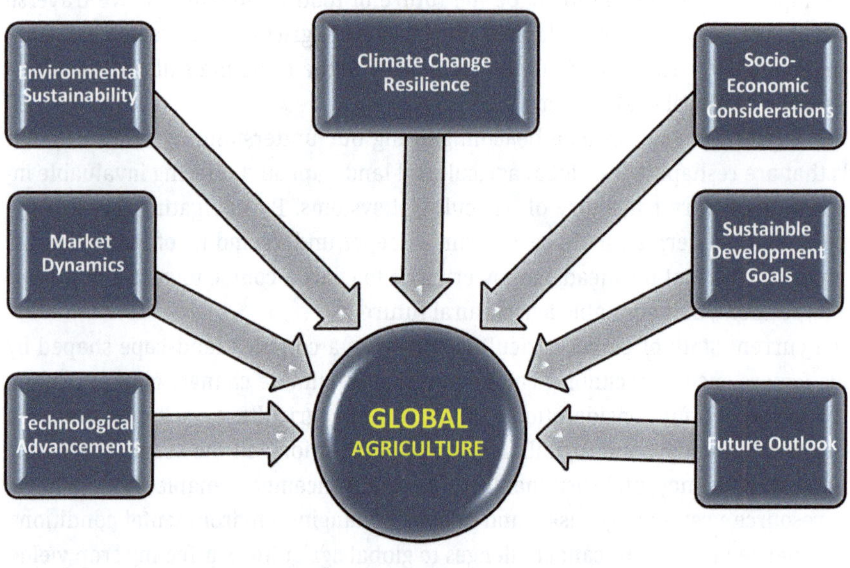

Figure 13.1: Key aspects of global agriculture.

13.1 Technological advancements

The chapter explores the role of technology in revolutionizing agriculture, from precision farming and remote sensing to digital agriculture and artificial intelligence (AI). By harnessing innovative technologies, farmers can enhance productivity, optimize resource use, and mitigate environmental impacts, paving the way for a more efficient and sustainable agricultural sector.

Technological advancements have emerged as transformative forces in modern agriculture, revolutionizing traditional farming practices and unlocking new avenues for productivity, sustainability, and resilience. In Chapter 13, we delve into the profound role of technology in shaping the agricultural landscape, exploring a myriad of innovations that are reshaping the way farmers cultivate crops, manage resources, and adapt to changing environmental conditions.

Precision farming stands at the forefront of technological innovation, offering farmers precision-guided tools and techniques to optimize crop management practices. Through the use of GPS-enabled machinery, sensors, and data analytics, precision farming enables farmers to precisely monitor and manage factors such as soil moisture, nutrient levels, and crop health. By tailoring inputs such as irrigation, fertilization, and pest control to the specific needs of each field or even individual plants, farmers can maximize yields while minimizing resource waste and environmental impacts.

Remote sensing technologies, including satellite imagery and drones, provide farmers with unprecedented insights into their fields, allowing for real-time monitoring of crop growth, soil conditions, and pest infestations. By harnessing these data-driven insights, farmers can make informed decisions regarding crop management practices, enabling proactive interventions to optimize yields and mitigate risks.

Digital agriculture platforms leverage the power of information technology and connectivity to enhance farm management and decision-making processes. From cloud-based farm management software to mobile applications for weather forecasting and market analysis, digital agriculture tools empower farmers with access to real-time information and analytics, enabling them to streamline operations, reduce costs, and improve productivity.

AI represents the cutting edge of agricultural innovation, offering the potential to revolutionize farm management practices through autonomous machinery, predictive analytics, and machine learning (ML) algorithms. By harnessing AI-driven solutions, farmers can automate routine tasks, optimize resource allocation, and anticipate future challenges, thereby enhancing efficiency, sustainability, and resilience in agriculture.

Overall, technological advancements hold immense promise for revolutionizing agriculture, enabling farmers to overcome challenges, seize opportunities, and navigate the complexities of an ever-changing agricultural landscape. By embracing innovation and leveraging the power of technology, farmers can unlock new frontiers in productivity, sustainability, and resilience, paving the way for a more efficient and sustainable agricultural sector.

13.2 Market dynamics

Market trends and consumer preferences are examined to understand the changing demands and challenges facing agricultural producers and stakeholders. Topics such as global trade patterns, market volatility, and supply chain disruptions are analyzed to provide insights into the factors influencing agricultural markets and shaping future opportunities.

Within the realm of Chapter 13, the exploration of market dynamics unravels a complex tapestry of factors that influence the agricultural sector, shaping its present landscape and guiding its future trajectory. By delving into the intricacies of market trends and consumer preferences, we gain a deeper understanding of the dynamic forces at play, illuminating the challenges and opportunities that agricultural producers and stakeholders face in the global marketplace.

At the heart of market dynamics lies a nuanced examination of shifting consumer preferences and demands. As consumer tastes evolve and diversify, driven by factors such as health consciousness, sustainability concerns, and cultural influences, agricultural producers must adapt to meet changing market expectations. By analyzing consumer trends and behaviors, we gain insights into the demand for organic, locally sourced, and ethically produced food products, as well as emerging opportunities in niche markets such as plant-based alternatives and functional foods.

Global trade patterns emerge as another focal point of analysis, offering a window into the interconnected nature of agricultural markets and the forces shaping international trade flows. From the impacts of trade agreements and tariffs to geopolitical tensions and currency fluctuations, the dynamics of global trade exert profound influences on agricultural producers and stakeholders worldwide. By examining trade data and market trends, we gain insights into the opportunities and challenges presented by international trade, as well as the strategies needed to navigate the complexities of the global marketplace.

Market volatility stands as a pervasive challenge facing agricultural producers, characterized by fluctuations in commodity prices, input costs, and currency exchange rates. These fluctuations, driven by factors such as weather events, geopolitical developments, and macroeconomic trends, pose risks to agricultural producers and stakeholders, impacting profitability, investment decisions, and supply chain management. By analyzing market volatility and its underlying drivers, we gain a deeper understanding of the factors influencing price movements and the strategies needed to manage risk and uncertainty in agricultural markets.

Supply chain disruptions emerge as a critical consideration in the context of market dynamics, highlighting the vulnerabilities inherent in global supply chains and the need for resilience and adaptability. From disruptions caused by natural disasters and pandemics to logistical challenges and trade disruptions, supply chain disruptions can have far-reaching impacts on agricultural producers, affecting production, distribution, and market access. By examining the resilience of supply chains and identifying vulnerabilities, we gain insights into the strategies needed to enhance supply chain resilience and mitigate the impacts of disruptions on agricultural markets.

In essence, the exploration of market dynamics within Chapter 13 offers a comprehensive analysis of the forces shaping the global agricultural sector, providing valuable insights into the challenges and opportunities that agricultural producers and stakeholders face in the ever-evolving marketplace. By understanding the complexities of market dynamics, stakeholders can better navigate the challenges and seize the opportunities presented by the dynamic landscape of global agriculture.

13.3 Environmental sustainability

The chapter highlights the growing importance of sustainability in agriculture and examines efforts to promote conservation practices, reduce greenhouse gas emissions, and enhance ecosystem resilience. By adopting sustainable agricultural practices, farmers can address environmental challenges such as soil degradation, water scarcity, and biodiversity loss while ensuring the long-term viability of agricultural systems.

The emphasis on environmental sustainability underscores a pivotal shift toward more conscientious and responsible agricultural practices. This section delves into the imperative of sustainability within agriculture, shedding light on the critical efforts aimed at preserving and enhancing the health of our ecosystems.

A central focus of this discussion revolves around the promotion of conservation practices designed to mitigate the adverse impacts of agricultural activities on the environment. From soil erosion control measures and reforestation initiatives to the establishment of protected areas and wildlife corridors, these conservation efforts seek to safeguard the integrity of natural habitats and protect biodiversity.

Furthermore, the chapter delves into strategies aimed at reducing greenhouse gas emissions associated with agricultural production. By exploring innovative approaches such as carbon sequestration through agroforestry, methane capture from livestock operations, and the adoption of renewable energy sources, the chapter underscores the pivotal role of agriculture in mitigating climate change and reducing its environmental footprint.

Efforts to enhance ecosystem resilience are also examined within this section, highlighting the importance of fostering robust and resilient ecosystems capable of withstanding environmental stresses and disturbances. Through practices such as ecosystem restoration, habitat conservation, and sustainable land management, farmers can contribute to the preservation of ecosystem services and the maintenance of ecological balance.

Overall, the chapter offers a comprehensive analysis of the multifaceted efforts underway to promote environmental sustainability within agriculture. By adopting sustainable agricultural practices, farmers can not only mitigate environmental challenges such as soil degradation, water scarcity, and biodiversity loss but also ensure the long-term viability and resilience of agricultural systems in the face of environmental change.

13.4 Climate change resilience

Climate change poses significant challenges to global agriculture, impacting crop yields, water availability, and pest and disease patterns. The need for climate resilience in agricultural systems has never been more urgent.

To address this, strategies for building climate resilience in agriculture are explored. These strategies encompass a wide array of approaches, ranging from the adoption of climate-smart practices to the development and utilization of resilient crop varieties.

Climate-smart practices involve the implementation of agricultural techniques and technologies that optimize resource use, minimize environmental impact, and enhance resilience to climate variability and extremes. Examples include precision irrigation methods, conservation agriculture techniques, and the integration of agroforestry into farming systems.

Resilient crop varieties play a crucial role in enhancing climate resilience by possessing traits that enable them to withstand and adapt to changing climatic conditions. These traits may include heat and drought tolerance, resistance to pests and diseases, and the ability to thrive in varying soil and moisture conditions.

Adaptive management approaches are also key components of climate resilience in agriculture. These approaches involve continually monitoring environmental conditions, assessing risks, and adjusting management practices accordingly. By remaining flexible and responsive to changing climatic conditions, farmers can better mitigate the impacts of climate change on their crops and livelihoods.

Furthermore, mitigating greenhouse gas emissions from agricultural activities is essential for building climate resilience. By implementing practices that reduce emissions, such as improving nutrient management, minimizing tillage, and promoting renewable energy use, agricultural systems can contribute to global efforts to mitigate climate change while also enhancing their own resilience to its impacts.

Building climate resilience in agriculture is imperative for ensuring food security, maintaining rural livelihoods, and safeguarding the environment in the face of climate change. By embracing climate-smart practices, resilient crop varieties, adaptive management approaches, and emission reduction strategies, agricultural systems can become more resilient, sustainable, and capable of thriving in a changing climate.

13.5 Socioeconomic considerations

Socioeconomic factors, including population growth, urbanization, income inequality, and rural development, shape agricultural landscapes and food systems. Addressing issues of poverty, access to land and resources, gender equity, and labor rights is essential for achieving sustainable agriculture and rural development goals. Socioeconomic considerations play a crucial role in shaping agricultural landscapes and food systems globally.

Here's a detailed overview along with a few examples:

1. **Population growth:** The increasing global population puts pressure on agricultural systems to produce more food to meet rising demand. This necessitates innovations in farming practices, increased productivity, and sustainable land use to ensure food security for growing populations. For example, countries like India and China, with large and growing populations, implement agricultural policies and programs to boost productivity and improve food access.
2. **Urbanization:** Urbanization leads to the conversion of agricultural land into urban areas, impacting food production and land availability for farming. It also affects labor availability, market access, and agricultural practices. Urban agriculture initiatives, such as rooftop gardens, vertical farming, and community gardens, are emerging as solutions to produce food within urban areas, reduce food miles, and promote local food systems.
3. **Income inequality:** Disparities in income distribution affect access to resources, including land, credit, technology, and markets, which in turn influence agricultural productivity and rural livelihoods. Addressing income inequality through policies that promote equitable access to resources and market opportunities can improve agricultural productivity and alleviate poverty. For instance, Brazil's Zero Hunger program focuses on reducing poverty and hunger through social welfare programs and support for smallholder farmers.
4. **Rural development:** Rural areas are often characterized by limited infrastructure, inadequate services, and lack of economic opportunities, leading to migration to urban centers. Rural development initiatives aim to improve living conditions, livelihoods, and economic opportunities in rural areas through investments in infrastructure, education, healthcare, and agriculture. For example, the European Union's Common Agricultural Policy supports rural development projects and subsidies for farmers to enhance competitiveness and sustainability.
5. **Gender equity:** Gender disparities in access to resources, decision-making power, and participation in agricultural activities are prevalent in many societies. Promoting gender equity in agriculture by ensuring women's access to land, education, credit, and technology can improve agricultural productivity, household food security, and rural development. Projects like the World Bank's Gender in Agriculture Support Program work to empower women in agriculture by providing training, resources, and access to markets.
6. **Labor rights:** Agricultural laborers, including farm workers and migrant workers, often face exploitation, unsafe working conditions, and inadequate wages. Ensuring labor rights and decent working conditions for agricultural workers is essential for sustainable agriculture and social justice. International labor standards, such as those set by the International Labour Organization, advocate for fair wages, safe working environments, and social protections for agricultural workers worldwide.

Addressing socioeconomic considerations in agriculture is crucial for achieving sustainable development goals, ensuring food security, and promoting inclusive growth. Policies and programs that address population growth, urbanization, income inequality, rural development, gender equity, and labor rights are essential for building resilient and equitable food systems globally.

13.6 The Sustainable Development Goals (SDGs)

The SDGs established by the United Nations provide a comprehensive framework for addressing global agriculture challenges and promoting sustainable development. Goal 2, "Zero Hunger," aims to end hunger, achieve food security, improve nutrition, and promote sustainable agriculture. Goal 12, "Responsible Consumption and Production," focuses on ensuring sustainable consumption and production patterns including within the agricultural sector.

Here's a detailed explanation with a few examples:

13.6.1 Goal 2: Zero Hunger

Objective: This goal aims to end hunger, achieve food security, improve nutrition, and promote sustainable agriculture.

Importance: Ensuring food security and adequate nutrition is essential for the well-being and development of individuals and communities worldwide.

Examples

Promotion of sustainable agriculture: Implementing sustainable agriculture practices, such as agroecology, organic farming, and conservation agriculture, can enhance food production while minimizing environmental impact. For example, the Food and Agriculture Organization (FAO) supports projects promoting sustainable farming methods in countries like Kenya and India to improve food security and livelihoods.

Agricultural research and innovation: Investing in agricultural research and innovation helps develop resilient crop varieties, improve farming techniques, and adapt to climate change. For instance, research institutions and organizations collaborate to develop drought-resistant crops, such as heat-tolerant wheat varieties in South Asia, to mitigate the impact of climate change on food production.

Strengthening food systems: Enhancing food systems' resilience, accessibility, and affordability is crucial for ensuring food security. Initiatives like the World Food Programme provide food assistance, nutrition support, and capacity-building programs in countries affected by conflict, natural disasters, and food insecurity.

13.6.2 Goal 12: Responsible Consumption and Production

Objective: This goal aims to ensure sustainable consumption and production patterns, including within the agricultural sector, to minimize waste, reduce environmental impact, and promote resource efficiency.

Importance: Promoting responsible consumption and production practices is essential for conserving natural resources, reducing pollution, and achieving sustainable development goals.

Examples

Reducing food waste: Addressing food loss and waste throughout the food supply chain, from production to consumption, is critical for sustainable agriculture and food security. Initiatives like the Save Food campaign by the FAO raise awareness about food waste and encourage actions to minimize waste at all levels.

Sustainable packaging and distribution: Adopting eco-friendly packaging materials and distribution practices helps reduce the environmental footprint of agricultural products. For instance, companies and organizations promote biodegradable packaging, reusable containers, and efficient transportation methods to minimize packaging waste and carbon emissions.

Sustainable consumption practices: Encouraging consumers to make informed choices and adopt sustainable eating habits, such as consuming locally sourced, seasonal, and plant-based foods, contributes to responsible consumption patterns. Educational campaigns, food labeling initiatives, and sustainable food certification programs help raise awareness and promote sustainable consumption behaviors.

Achieving the SDG, particularly Goal 2 (Zero Hunger) and Goal 12 (Responsible Consumption and Production), requires collaborative efforts from governments, organizations, businesses, and individuals to promote sustainable agriculture practices, ensure food security, and reduce environmental impact throughout the food supply chain. By addressing these goals, countries can work toward building resilient and equitable food systems that support the well-being of present and future generations.

13.7 Future outlook

Drawing on current trends and projections, the chapter offers insights into the future of global agriculture, highlighting emerging opportunities and challenges. From the adoption of precision agriculture and biotechnology to the rise of alternative protein sources and sustainable food systems, the chapter provides a forward-looking perspective on the direction of agricultural development and the potential pathways for achieving food security and sustainability in the years to come.

One key aspect of the future outlook is the increasing adoption of advanced technologies in agriculture such as precision agriculture and biotechnology. Precision agriculture involves the use of data-driven technologies, including GPS, sensors, and drones, to optimize farming practices and resource management. Biotechnology, on the other hand, encompasses genetic engineering techniques aimed at enhancing crop traits, increasing yields, and improving resistance to pests, diseases, and environmental stressors. The widespread adoption of these technologies is expected to revolutionize farming practices, enabling more efficient use of resources, higher productivity, and greater resilience to environmental challenges.

Another trend shaping the future of agriculture is the rise of alternative protein sources and sustainable food systems. As global demand for protein continues to increase, there is growing interest in alternative protein sources, such as plant-based proteins, cultured meat, and insect-based foods. These alternatives offer potential solutions to address concerns related to environmental sustainability, animal welfare, and food security. Additionally, there is a growing emphasis on developing sustainable food systems that prioritize the health of people and the planet, integrating principles of agroecology, biodiversity conservation, and social equity into food production and distribution processes.

Overall, this chapter " provides a forward-looking perspective on the evolving landscape of global agriculture, highlighting the transformative potential of advanced technologies, alternative protein sources, and sustainable food systems. By anticipating and preparing for emerging opportunities and challenges, stakeholders in the agricultural sector can better navigate the complexities of the future and work toward achieving food security, environmental sustainability, and resilience in agricultural systems worldwide.

13.8 Current scenario and future prospects of Pakistan

The current scenario and future prospects of Pakistan are discussed in the following paragraphs:

Present scenario: Pakistan's present scenario is characterized by a combination of challenges and opportunities across various sectors. Here are some key aspects of the present scenario in Pakistan:

Economy: Pakistan's economy is facing challenges such as fiscal deficits, high inflation, and a high debt burden. However, efforts are being made to stabilize the economy through structural reforms, attracting foreign investment, and promoting exports. The country has also entered into an Extended Fund Facility with the International Monetary Fund to address economic challenges.

Agriculture: Agriculture plays a vital role in Pakistan's economy, employing a significant portion of the population. However, the sector faces challenges such as water scarcity, land degradation, outdated farming practices, and limited access to resources. Efforts are underway to promote sustainable agriculture, improve water management, enhance productivity, and modernize farming techniques.

Energy sector: Pakistan faces energy shortages, leading to frequent power outages and hindering industrial growth. The government is working on expanding the energy mix, including investments in renewable energy sources such as wind and solar power. Efforts to improve energy infrastructure and attract private investment are being made to address the energy crisis.

Infrastructure development: Infrastructure development is a key focus area in Pakistan. Projects such as the China-Pakistan Economic Corridor (CPEC) aim to improve transportation networks, energy infrastructure, and industrial development. These initiatives have the potential to boost economic growth and create employment opportunities.

Education and skills development: The education sector in Pakistan faces challenges of low literacy rates, inadequate access to quality education, and a mismatch between education and market needs. Efforts are being made to improve access to education, enhance the quality of teaching, and promote skill development programs to address the skills gap and improve human capital.

Future prospects: Despite the challenges, Pakistan has several future prospects that can contribute to its development and progress. Here are some key areas with future prospects in Pakistan:

CPEC and economic development: The CPEC has the potential to transform Pakistan's economy by attracting foreign investment, developing infrastructure, and boosting industrial growth. CPEC projects are expected to create employment opportunities, enhance connectivity, and contribute to regional trade and economic integration.

Agriculture and food security: Pakistan's agricultural sector has immense potential to contribute to food security and economic growth. Promoting sustainable agriculture practices, improving water management, enhancing access to credit and technology, and value addition in the agriculture sector can lead to increased productivity, better market linkages, and improved livelihoods for farmers.

Information technology and digital economy: The IT sector in Pakistan is growing rapidly, with a young and tech-savvy population. The government is focusing on promoting the digital economy, e-commerce, and entrepreneurship. The expansion of broadband connectivity and initiatives to enhance digital skills can drive innovation, job creation, and economic growth.

Renewable energy: Pakistan has abundant renewable energy resources including solar, wind, and hydropower. Investments in renewable energy can address the energy crisis, improve access to electricity, and reduce reliance on fossil fuels. The

promotion of renewable energy projects and the development of a renewable energy policy framework can pave the way for a sustainable and clean energy future.
Human capital development: Investing in human capital development through improved education and skills training can enhance productivity, innovation, and economic growth. Focusing on quality education, vocational training programs, and promoting research and development can strengthen the country's human resource potential.
Regional trade and connectivity: Pakistan's strategic location makes it a potential hub for regional trade and connectivity. Strengthening regional cooperation, improving trade infrastructure, and enhancing connectivity with neighboring countries can open up new markets, promote cross-border trade, and drive economic integration.

While Pakistan faces various challenges, it also has promising future prospects. With concerted efforts and effective implementation of strategies, Pakistan can address its challenges and leverage its potential in areas such as infrastructure development, agriculture, renewable energy, digital economy, and human capital development to achieve sustainable and inclusive growth.

13.9 Analysis of the current state of Pakistan's agriculture

The current state of Pakistan's agriculture sector presents a mix of challenges and opportunities. While agriculture remains a significant contributor to the country's economy and employs a large portion of the population, it faces various issues that hinder its full potential. Here is an analysis of the current state of Pakistan's agriculture:

Productivity challenges: Pakistan's agricultural productivity is relatively low compared to its potential. Outdated farming practices, limited access to modern inputs, and inadequate use of technology contribute to lower yields and productivity levels. There is a need to promote the adoption of improved farming techniques, mechanization, and precision agriculture to enhance productivity.

Water scarcity: Water scarcity is a critical challenge in Pakistan's agriculture sector. The country heavily relies on irrigation, but water resources are limited and depleting. Inefficient water management, outdated irrigation infrastructure, and unsustainable water practices contribute to water scarcity issues. Implementing efficient water management strategies, such as drip irrigation and water-efficient practices, can help address this challenge.

Land degradation: Land degradation, including soil erosion, salinization, and waterlogging, poses a significant threat to agricultural productivity. Unsustainable agricultural practices, inadequate soil conservation measures, and deforestation contribute to land degradation. Implementing sustainable land management practices, promoting agroforestry, and encouraging soil conservation techniques can help address land degradation challenges.

Climate change impacts: Pakistan is highly vulnerable to climate change, with increasing temperatures, irregular rainfall patterns, and more frequent extreme weather events. These climate change impacts disrupt agricultural production, affect crop yields, and pose risks to livestock health. Implementing climate-smart agriculture practices, promoting drought-tolerant crop varieties, and enhancing resilience through improved water management and conservation are essential to address climate change challenges.

Market constraints: Pakistan's agricultural sector faces challenges related to market access, price volatility, and value chain development. Limited market infrastructure, middlemen dominance, and inadequate cold chain facilities hinder farmers' ability to access fair prices and effectively market their produce. There is a need for developing efficient market linkages, improving postharvest infrastructure, and promoting value addition to enhance market opportunities for farmers.

13.10 Identification of key opportunities and potential for growth

Despite the challenges, Pakistan's agriculture sector offers significant opportunities and growth potential. Here are some key areas of opportunity:

Value addition and processing: The agriculture sector can benefit from value addition and processing activities. Promoting agro-processing industries, developing supply chains for value-added products, and enhancing linkages between farmers and processors can create new economic opportunities and higher value for agricultural produce.

Organic and high-value crops: The demand for organic and high-value crops, such as fruits, vegetables, and spices, is increasing both domestically and internationally. Pakistan has favorable agro-climatic conditions for growing these crops. Encouraging the adoption of organic farming practices, promoting export-oriented high-value crops, and supporting farmers in meeting quality and certification standards can open new markets and generate higher incomes.

Export potential: Pakistan has the potential to increase its agricultural exports. By focusing on improving the quality and safety of agricultural produce, complying with international standards, and exploring new export markets, the country can enhance its export earnings. Targeting high-demand products, such as rice, fruits, vegetables, and seafood, can contribute to export growth.

Technology adoption: Embracing technological advancements and innovations in agriculture can significantly enhance productivity and efficiency. Promoting the use of modern technologies, such as precision agriculture, remote sensing, and data analytics, can optimize resource utilization, improve decision-making, and increase yields.

Farm mechanization: There is ample opportunity for farm mechanization in Pakistan. Increasing access to modern agricultural machinery, promoting the use of farm equipment for efficient land preparation, planting, and harvesting, and establishing machinery rental services can improve productivity, reduce labor requirements, and enhance agricultural efficiency.

13.11 Emerging trends and technologies in agriculture

The agriculture sector is undergoing a significant transformation driven by emerging trends and technologies that hold the promise of revolutionizing traditional farming practices. One such trend is the adoption of precision agriculture, which leverages advanced technologies such as GPS, drones, and sensors to optimize crop management and resource use. Precision agriculture enables farmers to monitor field conditions in real-time, apply inputs more efficiently, and make data-driven decisions to maximize yields while minimizing environmental impacts.

Furthermore, the integration of AI and ML algorithms into agricultural systems is revolutionizing how farmers manage crops, predict pests and diseases, and optimize supply chains. AI-powered agricultural robots and autonomous vehicles are increasingly used for tasks such as planting, harvesting, and weeding, reducing labor costs and increasing efficiency. Additionally, ML algorithms analyze vast amounts of data to provide insights into crop health, soil conditions, and weather patterns, enabling farmers to optimize their operations and improve productivity.

Another emerging trend in agriculture is the adoption of vertical farming and controlled environment agriculture (CEA) techniques. These innovative farming methods involve growing crops indoors, often in stacked layers or controlled environments such as greenhouses or hydroponic systems. Vertical farming and CEA offer several advantages including year-round production, higher crop yields, reduced water usage, and minimal reliance on pesticides. These techniques are particularly well-suited for urban areas where land availability is limited and consumers demand locally grown, fresh produce.

Moreover, the rise of blockchain technology is transforming agricultural supply chains by enhancing transparency, traceability, and trust. Blockchain enables secure and immutable recording of transactions, allowing stakeholders to track the journey of agricultural products from farm to fork. By providing visibility into the origin, quality, and sustainability of food products, blockchain technology empowers consumers to make informed choices and promotes fair trade practices. Additionally, blockchain-based smart contracts facilitate faster and more transparent transactions, benefiting farmers, traders, and consumers alike.

Emerging trends and technologies in agriculture are reshaping the way food is produced, distributed, and consumed on a global scale. By harnessing the power of precision agriculture, AI, vertical farming, and blockchain technology, farmers can enhance productivity, sustainability, and resilience in the face of evolving challenges such as climate change, resource scarcity, and food security. Embracing these innovations is essential for building a more efficient, equitable, and sustainable agricultural system to meet the needs of future generations.

Here are some notable examples (Figure 13.2):

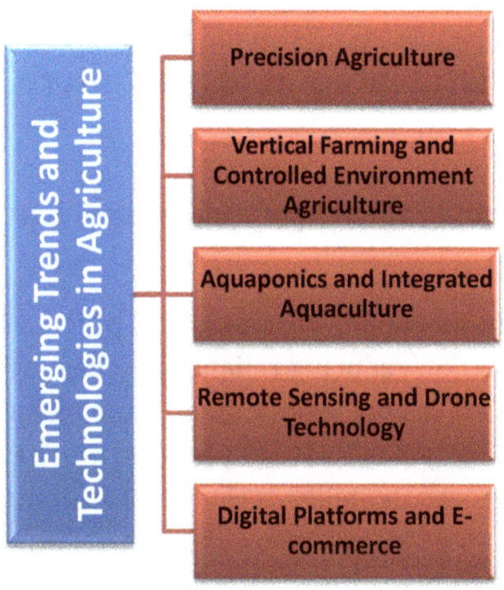

Figure 13.2: Emerging trends and technologies in agriculture.

Precision agriculture: Precision agriculture involves the use of technologies such as GPS, remote sensing, and data analytics to optimize resource management, improve crop monitoring, and enhance decision-making. Precision agriculture allows for targeted application of inputs, such as fertilizers and pesticides, based on crop needs, resulting in improved productivity and resource efficiency.

Vertical farming and controlled environment agriculture (CEA): Vertical farming and CEA involve the cultivation of crops in controlled environments such as vertical stacks, greenhouses, or hydroponic systems. These technologies offer opportunities to maximize land use, optimize water and nutrient management, and produce high-value crops throughout the year, irrespective of external climatic conditions.

Aquaponics and integrated aquaculture: Aquaponics combines aquaculture (fish farming) and hydroponics (soilless plant cultivation) in a symbiotic system. Fish waste provides nutrients for plants, and plants filter and purify water for the fish. Integrated aquaculture systems allow for efficient resource utilization, reduced water consumption, and simultaneous production of fish and crops.

Remote sensing and drone technology: Remote sensing techniques, including satellite imagery and aerial drone surveys, enable monitoring and analysis of crop health, water stress, and pest outbreaks. These technologies aid in timely interventions, resource management, and precision farming practices.

Digital platforms and e-commerce: Digital platforms and e-commerce solutions provide farmers with access to market information, weather forecasts, agricultural advisory services, and online marketplaces. These platforms enable direct linkages between farmers and buyers, facilitating transparent and efficient trade.

13.12 Case studies highlighting successful integrated agriculture models in Pakistan

Integrated agriculture models have been successful in various parts of Pakistan. Here are two notable case studies:

Integrated pest management (IPM) in cotton: In Punjab province, IPM practices have been adopted in cotton farming. IPM focuses on reducing reliance on chemical pesticides by promoting the use of biological control agents, crop rotation, monitoring of pests, and timely intervention. This approach has led to reduced pesticide use, improved pest control, and enhanced environmental sustainability in cotton cultivation.

Agroforestry in Khyber Pakhtunkhwa: Agroforestry practices have been successfully implemented in the Khyber Pakhtunkhwa province. Farmers have integrated fruit-bearing trees, such as apple and walnut, with their agricultural fields. This not only provides additional income through fruit production but also contributes to soil conservation, biodiversity preservation, and improved microclimatic conditions.

Within the corridors of Chapter 13, we stand at the crossroads of history and vision as we unravel the tapestry of Pakistan's agricultural landscape. Here, the chapters of the past, present, and future intersect, unveiling a narrative of resilience, adaptability, and forward-thinking. As we reflect on the present scenario and chart the course for future prospects, we recognize the intricate dance between policy formulation, technological innovation, and the aspirations of a nation striving for sustenance, prosperity, and enduring growth.

From the panorama of present scenarios and future horizons, we step into the realm of Chapter 14. This chapter casts a spotlight on the pillars that support the edifice of agricultural progress, shedding light on the frameworks that shape the nation's agricultural trajectory. By navigating through policy landscapes and institutional frameworks, we come to understand the profound influence that well-crafted policies and dynamic institutions wield in sculpting the contours of Pakistan's agricultural destiny.

References

Abbasi, S. A., & Butt, A. R. (2010). Pakistan's agriculture: A backward and forward linkage analysis. Pakistan Economic and Social Review, 48(1), 37–58.

Ahmed, M., Ashfaq, M., Khan, A. A., & Ahmed, Z. (2011). An overview of agricultural trends and policies in Pakistan. The Lahore Journal of Economics, 16, 43–62.

Amanullah. (2024). Harvesting Success: A Comprehensive Guide to Agronomy Courses Punjab Books, Lahore, PK.

Amanullah., Ondrasek, G., Al –Tawaha, A. R. M. S., eds. (2023). Integrated nutrients management: An approach for sustainable crop production and food security in changing climates. Lausanne: Frontiers Media SA. doi: 10.3389/978-2-8325-3169-3

Amanullah, & Khalid, S. (2020). Agronomy – Climate Change & Food Security IntechOpen, London, United Kingdom.

Amanullah, & Fahad, S. (2018). Nitrogen in Agriculture-Updates InTechOpen, London, UK.

Amanullah, & Fahad, S. (2018). Corn – Production and Human Health in Changing Climate InTech, London, UK.

FAO. (2018a). The Future of Food and Agriculture: Alternative Pathways to 2050 Food and Agriculture Organization of the United Nations, Rome, Italy.

FAO. (2018b). Soil Pollution: A Hidden Reality. FAO, Rome-Italy.

FAO. (2019). The International Code of Conduct for the Sustainable Use and Management of Fertilizers (pp. 30). FAO, Rome.

FAO. (2017). Unlocking the Potential of Soil Organic Carbon. FAO, Rome-Italy.

FAO. (2016). Soil and Pulses: Symbiosis for Life. FAO, Rome-Italy.

FAO and ITPS. (2016). Voluntary Guidelines for Sustainable Soil Management (VGSSM), Rome, Italy.

Government of Pakistan. (2019a). Agriculture Policy 2018 Ministry of National Food Security and Research, Islamabad, Pakistan.

Government of Pakistan. (2019b). Pakistan Economic Survey 2018–2019 Ministry of Finance, Islamabad, Pakistan.

Hafeez, M. M., & Imtiaz, M. (2019). Agriculture sector of Pakistan: The past, present, and future. Journal of Agricultural Science and Technology, 21(3), 625–636.

Hasan, M. F., & Shahbaz, M. (2020). Agricultural productivity and rural-urban inequality nexus in Pakistan: New evidence using district-level data. Sustainability, 12(1), 270.

Iftikhar, M., Hassan, S., & Hussain, A. (2012). Role of institutions in development of agriculture in Pakistan: A review. Pakistan Journal of Agricultural Sciences, 49(3), 299–309.

Janjua, P. Z., & Haider, S. J. (2019). Institutional dimensions of agricultural extension system in Punjab, Pakistan. Journal of Agricultural Extension and Rural Development, 11(8), 175–182.

Javed, M., & Bakhsh, K. (2019). Agriculture and the economy of Pakistan: A macroeconomic overview. International Journal of Financial Research, 10(2), 116–122.

Khan, M. A., & Iqbal, J. (2018). Role of institutions in agricultural development: A case study of Pakistan. European Journal of Agriculture and Forestry Research, 6(3), 25–32.

Malik, M. A., Ashraf, S., & Ahmad, Z. (2019). Institutions and policies in Pakistan's agriculture sector: An overview. Journal of Agriculture and Environmental Sciences, 2(1), 38–49.

Qureshi, M. E., & Mahmood, A. (2015). Pakistan's agricultural growth: An analysis of recent trends. International Journal of Economics, Commerce and Management, 3(7), 51–61.

Qureshi, S. E., & Ahmed, M. (2011). The agricultural sector of Pakistan: An overview. Munich Personal RePEc Archive (MPRA) Paper No. 28858.

Rashid, A., Hameed, S., & Abid, M. (2019). Role of agriculture in the economy of Pakistan. South Asian Studies, 34(2), 383–397.

Rehman, T., Dorward, A., & Balcombe, K. (2007). Constraints on development of the Punjab non-farm sector: The case of agricultural machinery services and trading in Pakistan. World Development, 35(10), 1746–1766.

Siddiqui, R., & Kemal, A. R. (2005). Environment, growth and development in Pakistan: A case of missing linkages. The Pakistan Development Review, 44(4), 711–723.

Siddiqui, R., & Kemal, A. R. (2006). Macro and micro policy environment and farm production in Pakistan. The Pakistan Development Review, 45(4), 739–761.

Tariq, S. A. (2013). Agriculture sector in Pakistan: An analysis of growth trends and prospects. Pakistan Journal of Social Sciences (PJSS), 33(1), 119–130.

Zafar, H. M., & Ali, S. (2019). An evaluation of agriculture policies and their impacts on the agricultural sector of Pakistan. Journal of Agriculture and Environment for International Development, 113(1), 185–200.

Chapter 14
Agricultural institutions and policies

Introduction

In the ever-evolving landscape of global agriculture, the role of institutions and policies cannot be overstated. As nations grapple with the complexities of food security, sustainability, and rural development, the frameworks and strategies established by agricultural institutions and policies play a pivotal role in shaping the trajectory of agricultural systems worldwide. This chapter delves into the realm of agricultural institutions and policies from an international perspective, offering insights into the diverse approaches, challenges, and best practices observed across different regions and countries.

At the heart of this exploration lies a profound recognition of the interconnectedness of agricultural policies and institutions with broader socioeconomic, environmental, and political dynamics. From trade agreements and regulatory frameworks to subsidy programs and land tenure systems, the decisions made within the realm of agricultural policy have far-reaching implications for food production, rural livelihoods, and environmental sustainability on a global scale.

Against the backdrop of rapid globalization, climate change, and technological advancements, the need for international cooperation and coordination in agricultural governance has become increasingly apparent. This chapter seeks to unravel the complexities of this dynamic landscape, shedding light on the roles played by international organizations, bilateral agreements, and multilateral platforms in shaping agricultural policies and institutions across borders.

Moreover, as the world grapples with emerging challenges such as climate change, resource scarcity, and shifting consumer preferences, the chapter explores how agricultural institutions and policies are evolving to address these pressing issues. From the promotion of climate-smart agriculture to the integration of sustainability principles into policy frameworks, the chapter highlights innovative approaches and initiatives aimed at fostering agricultural resilience and adaptation in the face of global challenges.

Through a comparative analysis of agricultural policies and institutions in diverse contexts, this chapter aims to provide valuable insights and lessons learned that can inform decision-making processes and contribute to the development of more effective and equitable agricultural policies on a global scale. By examining international perspectives on agricultural institutions and policies, we gain a deeper understanding of the complex interplay between governance, economics, and sustainability in the realm of global agriculture.

Agricultural institutions and policies play a crucial role in shaping the landscape of global agriculture. This chapter provides an in-depth exploration of agricultural institutions and policies from an international perspective, offering insights into the diverse approaches, challenges, and best practices observed across different regions and countries.

https://doi.org/10.1515/9783111448077-014

14.1 International organizations and agricultural governance

International organizations such as the Food and Agriculture Organization (FAO), World Bank, and World Trade Organization (WTO) play pivotal roles in agricultural governance at the global level. These organizations have mandates, functions, and initiatives aimed at promoting food security, rural development, and sustainable agriculture practices worldwide.

The FAO is a specialized agency of the United Nations that leads international efforts to defeat hunger and improve nutrition. Its mandate includes providing technical assistance, policy advice, and capacity-building support to member countries in areas such as agricultural production, food safety, and natural resource management. The FAO also coordinates global initiatives to address food insecurity, malnutrition, and agricultural emergencies.

The World Bank is a multilateral development bank that provides financial and technical assistance to developing countries for projects and programs aimed at reducing poverty and promoting sustainable development. In the agricultural sector, the World Bank supports initiatives to enhance agricultural productivity, improve market access for smallholder farmers, and strengthen rural infrastructure. It also works to build resilience to climate change and promote sustainable natural resource management practices.

The WTO is an international organization that regulates international trade and sets rules for trade negotiations among its member countries. In the agricultural sector, the WTO oversees trade agreements related to agriculture, including the Agreement on Agriculture (AoA), which aims to promote fair and market-oriented agricultural trade while ensuring food security and protecting the environment. The WTO also provides a forum for member countries to negotiate trade policies and resolve disputes related to agricultural trade.

Case studies and examples serve as valuable tools to demonstrate the collaborative efforts of international organizations, governments, NGOs, and other stakeholders in addressing key agricultural challenges. Take, for instance, the FAO, which frequently collaborates with national governments to devise policies and initiatives aimed at enhancing food security and nutrition. Through partnerships with FAO, governments can leverage expertise and resources to develop effective strategies for improving agricultural practices, enhancing productivity, and ensuring access to nutritious food for their populations.

Similarly, the **World Bank** plays a pivotal role in financing agricultural projects that aim to alleviate poverty and promote sustainable development. By providing funding and technical assistance to agricultural initiatives, the World Bank supports efforts to improve infrastructure, modernize farming practices, and enhance market access for smallholder farmers. These projects not only contribute to poverty reduction but also foster economic growth and resilience within rural communities.

The WTO also plays a significant role in shaping global agricultural governance by facilitating negotiations among member countries to address trade barriers and promote agricultural exports. Through trade agreements and dispute resolution mechanisms, the WTO helps to create a more open and predictable trading environment for agricultural products, thereby enabling farmers and agribusinesses to access international markets more efficiently.

Overall, these international organizations serve as catalysts for collaboration and coordination among diverse stakeholders in the agricultural sector. By pooling resources, sharing knowledge, and fostering partnerships, they help to address the multifaceted challenges facing agriculture, from food insecurity and poverty to environmental degradation and trade barriers. Through case studies and examples, the effectiveness of these collaborative efforts can be demonstrated, highlighting the collective impact of international cooperation on global agricultural governance and development.

14.2 Bilateral and multilateral agreements

Bilateral and multilateral agreements play significant roles in shaping agricultural trade, investment, and cooperation among nations. These agreements encompass various types, including free trade agreements (FTAs), regional trade blocs, and bilateral investment treaties (BITs), each impacting agricultural policies and practices in distinct ways.

FTAs are agreements between two or more countries that facilitate trade by reducing or eliminating tariffs, quotas, and other trade barriers. These agreements aim to promote economic growth, increase market access, and enhance competitiveness. In the agricultural sector, FTAs can lead to increased exports and imports of agricultural products as well as changes in domestic agricultural policies to align with international trade rules.

Regional trade blocs, such as the European Union (EU), ASEAN (Association of Southeast Asian Nations), and Mercosur (Southern Common Market), are groups of countries that form economic unions to promote trade and economic integration within a specific geographic region. These blocs often have their own trade agreements and regulations governing agricultural trade, investment, and cooperation among member countries.

BITs are agreements between two countries that aim to promote and protect foreign investment by providing legal protections and guarantees to investors. These treaties can influence agricultural investment decisions, technology transfer, and market access for agricultural products.

Case studies are commonly used to highlight the opportunities and challenges associated with bilateral and multilateral agreements in the agricultural sector. For example, an FTA between two countries may lead to increased market access for agricultural exports but could also result in increased competition from imported agricultural products. Similarly, regional trade blocs may create opportunities for regional cooperation and integration but may also pose challenges related to differing agricultural policies and regulations among member countries.

Overall, bilateral and multilateral agreements play critical roles in shaping agricultural trade, investment, and cooperation among nations, with implications for agricultural development, market access, and technology transfer. Understanding the impact of these agreements is essential for policymakers, agricultural stakeholders, and researchers seeking to navigate the complex landscape of international agricultural trade and investment.

14.3 Policy frameworks and regulatory mechanisms

Policy frameworks and regulatory mechanisms are crucial components of agricultural governance, governing various aspects of agricultural production, marketing, and land tenure. These frameworks provide the structure and guidelines for how agricultural activities are managed and regulated within a country.

Government agencies, legislative bodies, and regulatory authorities play central roles in formulating and implementing agricultural policies. These entities are responsible for developing policies, laws, and regulations that govern agricultural practices, ensuring compliance with national standards and objectives.

Within policy frameworks, countries address a wide range of issues related to agriculture, including land reform, agricultural subsidies, environmental regulations, and food safety standards. For example, land reform policies may aim to address issues of land ownership, distribution, and tenure to promote equitable access to land for agricultural purposes. Agricultural subsidies may be implemented to support farmers, enhance agricultural productivity, or stabilize food prices. Environmental regulations may focus on mitigating the environmental impacts of agriculture such as soil erosion, water pollution, and biodiversity loss.

Case studies are often used to illustrate how countries navigate these complex policy issues within their unique contexts. For instance, a case study may examine how a country implements land reform policies to address historical inequities in land ownership and promote sustainable agricultural development. Another case study may analyze the impact of agricultural subsidies on farmer livelihoods, agricultural productivity, and market dynamics.

Overall, policy frameworks and regulatory mechanisms are essential for governing agricultural activities, ensuring sustainability, and addressing societal needs and priorities. Understanding the roles and functions of government agencies, legislative bodies, and regulatory authorities is crucial for effective agricultural governance and the development of policies that promote agricultural sustainability, food security, and rural development.

14.4 Some international policy frameworks and regulatory mechanisms in agriculture

International policy frameworks and regulatory mechanisms in agriculture refer to the various agreements, treaties, and regulations established by countries and international organizations to govern agricultural production, trade, and sustainability on a global scale. These frameworks aim to address key issues such as food security, environmental conservation, trade liberalization, and rural development.

Some key aspects include:

Trade agreements: Trade agreements, both bilateral and multilateral, play a significant role in shaping agricultural policies and practices. These agreements aim to promote trade liberalization, reduce tariffs and trade barriers, and facilitate market access for agricultural products. Examples include regional trade blocs like the EU's Common Agricultural Policy (CAP) and multilateral agreements under the WTO.

Trade agreements are pivotal instruments that not only facilitate the exchange of agricultural goods but also influence the regulatory frameworks governing agricultural practices. Bilateral trade agreements, negotiated between two countries, often entail provisions related to tariffs, quotas, and sanitary and phytosanitary measures, which directly impact agricultural trade flows. These agreements aim to foster economic cooperation, enhance market access, and promote mutual trade benefits.

On a larger scale, multilateral trade agreements negotiated under international organizations like the **WTO** establish overarching principles and rules governing global trade. The **AoA** set standards for agricultural trade including market access, domestic support, and export subsidies. They provide a framework for resolving trade disputes and promoting fair competition among member countries.

These agreements, like the AoA established under the WTO, serve as regulatory frameworks governing international agricultural trade. They delineate standards and guidelines pertaining to various aspects of agricultural commerce such as market accessibility, domestic support programs, and the regulation of export subsidies. Essentially, these agreements aim to create a level playing field for agricultural producers and traders by standardizing trade practices, ensuring fair competition, and mitigating distortions in global agricultural markets.

Regional trade blocs, such as the EU's **CAP**, go a step further by integrating agricultural policies across member states. The CAP, for instance, harmonizes agricultural regulations, allocates subsidies, and establishes market mechanisms to ensure the smooth functioning of agricultural trade within the EU. Such regional agreements aim to create a unified market, promote agricultural competitiveness, and facilitate economic integration among participating countries.

Moreover, trade agreements often include provisions related to nontariff barriers, intellectual property rights, and environmental and labor standards, which have implications for agricultural production practices. For instance, agreements may require adherence to specific environmental regulations or the protection of intellectual property rights related to agricultural innovations. These provisions can influence farmers' practices and incentivize the adoption of sustainable agricultural methods to comply with international standards.

Overall, trade agreements significantly shape agricultural policies and practices by influencing market access, trade regulations, and production standards. By promoting trade liberalization and fostering economic cooperation, these agreements contribute to agricultural development, food security, and economic growth on a global scale.

Environmental regulations: International agreements and conventions address environmental concerns related to agriculture such as deforestation, water pollution, and biodiversity loss. For instance, the United Nations Framework Convention on Climate Change (UNFCCC) and the Convention on Biological Diversity (CBD) establish guidelines and frameworks for mitigating climate change and conserving biodiversity, which are essential for sustainable agricultural practices.

International agreements and conventions play a crucial role in addressing environmental challenges associated with agriculture, providing frameworks and guidelines for sustainable resource management and conservation. One notable example is the UNFCCC, which aims to stabilize greenhouse gas concentrations in the atmosphere to prevent dangerous anthropogenic interference with the climate system. Through mechanisms such as the Kyoto Protocol and the Paris Agreement, countries commit to reducing emissions and enhancing resilience to climate change, including in the agricultural sector.

Within the realm of agriculture, climate change poses significant threats including shifts in precipitation patterns, increased frequency of extreme weather events, and alterations in temperature regimes. These impacts can disrupt agricultural productivity, water availability, and ecosystem stability. Therefore, international efforts under the UNFCCC focus on promoting climate-smart agriculture practices that enhance resilience, sequester carbon, and mitigate greenhouse gas emissions from agricultural activities.

Similarly, the CBD addresses biodiversity conservation, recognizing the vital role of ecosystems in supporting agricultural productivity, genetic diversity, and ecosystem services. The CBD emphasizes the conservation and sustainable use of biological resources including genetic resources for food and agriculture. Through initiatives such as the Nagoya Protocol and the Global Strategy for Plant Conservation, the CBD promotes sustainable agriculture practices that safeguard biodiversity and ecosystem integrity.

Moreover, international agreements often intersect with regional initiatives and national policies to address environmental challenges holistically. For instance, the sustainable development goals adopted by the United Nations provide a comprehensive framework for sustainable development, integrating environmental, social, and economic dimensions. Goal 2, which aims to end hunger, achieve food security, improve nutrition, and promote sustainable agriculture, underscores the importance of environmental sustainability in agricultural development.

International environmental agreements and conventions provide crucial frameworks and mechanisms for addressing environmental challenges in agriculture including climate change mitigation, biodiversity conservation, and sustainable resource management. By promoting cooperation, knowledge exchange, and collective action, these agreements contribute to the development and implementation of sustainable agricultural practices on a global scale.

Food security policies: International organizations like the FAO and the World Food Programme (WFP) work to develop policies and initiatives to enhance global food security. These efforts include promoting sustainable agriculture, supporting smallholder farmers, and addressing hunger and malnutrition in vulnerable populations.

Food security policies are critical components of global efforts to ensure that all people have access to safe, nutritious, and sufficient food to meet their dietary needs and lead healthy lives. International organizations such as the FAO and the **WFP** play pivotal roles in developing and implementing policies and initiatives aimed at enhancing food security worldwide.

One of the primary objectives of food security policies is to promote sustainable agriculture practices that increase agricultural productivity while minimizing environmental impacts. This involves supporting smallholder farmers, who produce a significant portion of the world's food, by providing access to resources such as land, seeds, credit, and agricultural extension services. By investing in agricultural research and innovation, food security policies aim to develop resilient crop varieties, improve soil fertility, and enhance water management practices to adapt to climate change and variability.

Furthermore, food security policies address issues of hunger and malnutrition, particularly in vulnerable populations such as children, women, and the elderly. The FAO and WFP collaborate with governments, nongovernmental organizations (NGOs), and other stakeholders to implement nutrition-sensitive interventions that improve access to diverse and nutritious foods, promote breastfeeding and complementary feeding practices, and prevent and treat malnutrition. Additionally, social protection programs, including cash transfers, food vouchers, and school feeding programs, help vulnerable households cope with food insecurity and provide a safety net during times of crisis.

International cooperation and coordination are essential for effective food security policies, as food systems are interconnected and global in nature. The FAO's Committee on **World Food Security (CFS)** serves as a platform for dialogue and decision-making on food security issues, bringing together governments, civil society organizations, and the private sector to exchange knowledge, share best practices, and develop consensus-based policy recommendations. Similarly, the WFP works closely with governments and partners to deliver food assistance and support emergency response efforts in crisis-affected regions.

Food security policies developed and implemented by international organizations like the FAO and WFP are crucial for addressing the complex challenges of hunger, malnutrition, and food insecurity on a global scale. By promoting sustainable agriculture, supporting smallholder farmers, and addressing the nutritional needs of vulnerable populations, these policies contribute to building more resilient and equitable food systems that can withstand shocks and ensure food security for all.

Genetically modified organisms (GMOs) regulations: The use of GMOs in agriculture is regulated by international agreements such as the Cartagena Protocol on Biosafety, which governs the transboundary movement of GMOs and aims to ensure their safe handling, transport, and use.

The regulation of GMOs in agriculture is a subject of considerable debate and concern globally. International agreements such as the Cartagena Protocol on Biosafety play a crucial role in establishing guidelines and protocols for the safe handling, transportation, and use of GMOs, particularly in cross-border contexts.

The Cartagena Protocol on Biosafety, which is a supplementary agreement to the **CBD**, aims to ensure the safe transfer, handling, and use of living modified organisms **(LMOs)** resulting from modern biotechnology that may have adverse effects on biodiversity and human health. It establishes procedures for the advance informed agreement prior to the intentional transboundary movement of LMOs and requires countries to develop national biosafety frameworks to regulate the import, export, and use of GMOs within their jurisdictions.

In addition to international agreements, many countries have established their own regulatory frameworks for the assessment and approval of GMOs in agriculture. These frameworks typically involve rigorous risk assessment procedures to evaluate the potential environmental, health, and socioeconomic impacts of GMOs before they are approved for commercial release or importation. Regulatory agencies, such as the United States Food and Drug Administration and the European Food Safety Authority, play a central role in overseeing the safety of GMOs and ensuring compliance with regulatory requirements.

However, the regulation of GMOs remains a contentious issue, with proponents arguing that GMOs have the potential to increase agricultural productivity, improve crop resilience, and reduce the need for chemical inputs, while opponents raise concerns about the potential risks to human health, biodiversity, and socioeconomic equity. As a result, the regulation of GMOs continues to evolve in response to scientific advancements, public opinion, and political pressures, highlighting the complexity of balancing innovation with safety and sustainability in agricultural biotechnology.

International standards and certification: International organizations and initiatives set standards and certification schemes to ensure the quality, safety, and sustainability of agricultural products. Examples include the International Organization for

Standardization (ISO) standards and certification programs like Fair Trade and Organic certification.

International standards and certification schemes play a crucial role in ensuring the quality, safety, and sustainability of agricultural products in global trade. These standards are established by international organizations and initiatives to provide guidelines and criteria that producers, processors, and traders must adhere to in order to meet certain quality and sustainability benchmarks.

One prominent organization involved in setting international standards is the **ISO**. ISO develops voluntary standards that cover a wide range of industries including agriculture. These standards address various aspects of agricultural production, processing, and distribution, with the aim of promoting efficiency, quality, and safety throughout the supply chain. For example, ISO standards may specify requirements for food safety management systems, traceability, and quality control measures to ensure that agricultural products meet established criteria for safety and quality.

In addition to ISO standards, certification programs like Fair Trade and Organic certification provide consumers with assurance that agricultural products have been produced in accordance with certain social and environmental criteria. Fair Trade certification, for instance, ensures that producers receive fair prices for their products, adhere to labor rights and environmental standards, and invest in community development initiatives. Similarly, organic certification verifies that agricultural products have been produced using organic farming practices that minimize synthetic inputs, promote soil health, and protect biodiversity.

By adhering to international standards and certification schemes, agricultural producers can access global markets, differentiate their products, and meet consumer demand for safe, sustainable, and ethically produced goods. These standards also help to facilitate trade by harmonizing regulations and ensuring transparency and consistency in the global marketplace. However, compliance with international standards and certification requirements may entail additional costs and administrative burdens for producers, particularly smallholders and farmers in developing countries. Therefore, efforts to support capacity building, technical assistance, and market access for producers can help to enhance the benefits of international standards and certification for sustainable agriculture and economic development.

Research and development (R&D) collaboration: International collaboration in agricultural R&D is facilitated through initiatives like the Consultative Group on International Agricultural Research (CGIAR) and partnerships between national research institutions and universities. These collaborations aim to promote innovation, technology transfer, and knowledge sharing to address global agricultural challenges.

International collaboration in agricultural R&D is essential for advancing scientific knowledge, fostering innovation, and addressing the complex challenges facing global agriculture. Initiatives like the **CGIAR** serve as platforms for coordinating and facilitating research efforts among a network of national and international research institutions, universities, and NGOs. The CGIAR is a global partnership that aims to reduce poverty, improve food security, and enhance natural resource management through agricultural research, innovation, and capacity building.

Here are the details and roles of CGIAR and its related international institutes:

Coordination and collaboration: CGIAR serves as a platform for coordinating and facilitating collaborative research efforts among a network of national and international research institutions, universities, and NGOs. It brings together diverse expertise and resources to address complex agricultural challenges on a global scale.

Research and innovation: CGIAR supports research programs and projects focused on enhancing agricultural productivity, resilience, and sustainability. It conducts research on a wide range of topics including crop breeding, soil health, water management, pest and disease control, livestock production, and climate change adaptation. Through its research initiatives, CGIAR develops innovative solutions and technologies to improve agricultural practices and livelihoods.

Capacity building: CGIAR invests in building the capacity of researchers, scientists, extension workers, policymakers, and farmers to promote knowledge sharing, technology transfer, and skill development. It provides training, workshops, fellowships, and other capacity-building activities to empower individuals and institutions to contribute effectively to agricultural R&D efforts.

Policy influence: CGIAR engages with policymakers, government agencies, international organizations, and other stakeholders to advocate for evidence-based policies and investments that support sustainable agriculture, food security, and rural development. It provides scientific evidence, policy recommendations, and technical expertise to inform decision-making processes at the national, regional, and global levels.

International research centers: CGIAR comprises a network of international agricultural research centers (IARCs) located around the world. These centers specialize in specific crops, livestock, and natural resource management areas and conduct research to address the unique agricultural challenges facing different regions and agroecological zones.

Some notable CGIAR centers include (Figure 14.1):
- International Rice Research Institute (IRRI)
- International Maize and Wheat Improvement Center (CIMMYT)
- International Center for Tropical Agriculture (CIAT)
- International Livestock Research Institute (ILRI)
- International Water Management Institute (IWMI)
- International Food Policy Research Institute (IFPRI)

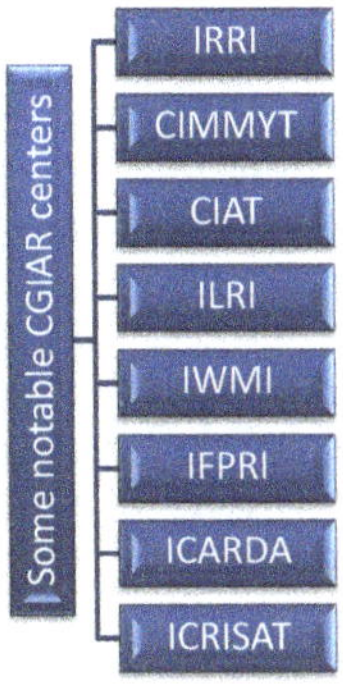

Figure 14.1: Some notable CGIAR centers.

The **International Food Policy Research Institute (IFPRI)** holds significant importance in global agriculture for several reasons:

Policy research and analysis: IFPRI conducts research and analysis on food policy issues, agricultural development, and nutrition to provide evidence-based policy recommendations to governments, international organizations, and other stakeholders. Its research helps policymakers design and implement effective policies and programs to address food security, poverty, and malnutrition.

Capacity building: IFPRI invests in building the capacity of researchers, policymakers, and institutions to conduct policy-relevant research, analyze data, and formulate evidence-based policies and programs. It offers training, workshops, fellowships, and technical assistance to empower individuals and institutions to contribute to agricultural development and food policy initiatives.

Food security and nutrition: IFPRI's research focuses on improving food security, nutrition, and livelihoods, particularly for vulnerable populations in developing countries. It analyzes the drivers of food insecurity, identifies policy interventions to enhance food access and availability, and evaluates the impact of nutrition-sensitive agricultural interventions on human health and well-being.

Global collaboration and partnerships: IFPRI collaborates with governments, international organizations, research institutions, civil society organizations, and the private sector to address food security challenges and promote sustainable agricultural development. It engages in global initiatives, networks, and partnerships to foster knowledge sharing, innovation, and collective action to achieve food security goals.

Evidence-based policy advocacy: IFPRI advocates for evidence-based policies and investments that support agricultural development, poverty reduction, and food security at the national, regional, and global levels. It disseminates research findings, policy briefs, and recommendations to policymakers, media, and the public to raise awareness and mobilize support for policy reforms and initiatives.

Monitoring and evaluation: IFPRI monitors and evaluates the implementation and impact of agricultural policies, programs, and investments to assess their effectiveness and identify areas for improvement. It conducts rigorous impact evaluations, policy assessments, and performance monitoring to inform decision-making and accountability in food and agriculture systems.

Overall, IFPRI plays a critical role in generating knowledge, building capacity, influencing policies, and mobilizing action to address global food security challenges, promote sustainable agriculture, and improve the well-being of millions of people around the world.

CGIAR impact and results: CGIAR's research and innovations have had significant impacts on global agriculture, contributing to increased crop yields, improved livelihoods, and enhanced food security for millions of people, particularly in developing countries. Its work has led to the development of new crop varieties, sustainable farming practices, and resilient agricultural systems that help farmers adapt to climate change and other challenges.

CGIAR and its related international institutes play a crucial role in advancing agricultural research, innovation, and capacity building to address global food security challenges, promote sustainable development, and improve the livelihoods of smallholder farmers and rural communities around the world.

International R&D: International R&D collaboration refers to cooperative efforts among countries, institutions, and organizations to conduct research and innovation activities aimed at addressing agricultural challenges and promoting sustainable development. This collaboration involves pooling resources, expertise, and knowledge to tackle common agricultural issues, share best practices, and develop innovative solutions.

The primary objectives of international R&D collaboration in agriculture include:

Promoting innovation: International R&D collaboration facilitates the exchange of ideas, expertise, and technologies to drive innovation in agriculture. By leveraging diverse perspectives and resources from different countries and institutions, collaborative research efforts can lead to the development of new technologies, practices, and solutions to enhance agricultural productivity, resilience, and sustainability.

One of the primary objectives of international R&D collaboration is to promote innovation in agriculture by developing new technologies, practices, and solutions to improve productivity, resilience, and sustainability. Through collaborative research projects, scientists and experts from different countries can leverage their expertise and resources to tackle common agricultural problems such as pests and diseases, climate change, and soil degradation. By pooling their knowledge and resources, researchers can accelerate the pace of innovation and generate novel solutions that benefit farmers and communities worldwide.

Addressing global challenges: Collaborative research initiatives focus on addressing pressing global challenges facing agriculture such as climate change, food insecurity, soil degradation, water scarcity, and pest and disease outbreaks. By pooling resources and expertise, researchers can develop holistic and context-specific solutions to these challenges that can be applied across different regions and agroecosystems.

Sharing knowledge and best practices: International R&D collaboration facilitates the sharing of knowledge, data, and best practices among researchers, policymakers, and practitioners worldwide. Through collaborative research projects, workshops, training programs, and knowledge exchange platforms, stakeholders can learn from each other's experiences, successes, and failures, leading to more effective and evidence-based decision-making.

Another important aspect of international R&D collaboration is technology transfer, which involves sharing scientific knowledge, best practices, and technologies between countries and regions. Collaborative research projects often result in the development of new agricultural technologies and practices that can be adapted and adopted by farmers in different contexts. For example, drought-resistant crop varieties developed through international research partnerships can help farmers mitigate the impacts of water scarcity and climate change in regions prone to drought.

Building capacity: Collaborative R&D initiatives contribute to building the capacity of researchers, institutions, and policymakers in participating countries. By providing training, technical assistance, and mentorship opportunities, international collaborations empower individuals and organizations to conduct high-quality research, analyze data, and implement evidence-based policies and practices.

International R&D collaboration facilitates knowledge sharing and capacity building among researchers, extension agents, and policymakers. By participating in collaborative research projects, scientists gain exposure to diverse perspectives, methodologies, and approaches, which can enhance their skills and expertise. Training programs, workshops, and exchange opportunities provided through international partnerships enable researchers to learn from each other, build networks, and strengthen research capabilities in their respective countries.

Fostering networking and partnerships: International R&D collaboration fosters networking and partnerships among researchers, institutions, governments, and NGOs across borders. By establishing collaborative networks and partnerships, stakeholders can leverage complementary expertise, resources, and infrastructure to achieve common research goals and maximize impact.

International collaboration in agricultural R&D contributes to advancing the frontiers of science, promoting innovation diffusion, and driving sustainable agricultural development worldwide. By fostering partnerships, sharing knowledge, and leveraging collective expertise, researchers can address the complex challenges facing global agriculture and contribute to achieving food security, poverty reduction, and environmental sustainability on a global scale. Overall, international R&D collaboration plays a crucial role in driving innovation, addressing global agricultural challenges, promoting knowledge sharing, building capacity, and fostering partnerships to advance sustainable agriculture and food security worldwide.

International policy frameworks: International policy frameworks and regulatory mechanisms in agriculture play a crucial role in promoting sustainable development, ensuring food security, and addressing global challenges such as climate change, biodiversity loss, and poverty alleviation. International policy frameworks and regulatory mechanisms in agriculture serve as essential instruments for promoting sustainable development, ensuring food security, and addressing pressing global challenges. These frameworks provide guidelines, standards, and regulations that guide the actions of governments, organizations, and stakeholders in the agricultural sector. By establishing common principles and objectives, international policy frameworks facilitate cooperation and coordination among countries to achieve shared goals and aspirations.

One of the key functions of international policy frameworks is to promote sustainable agriculture practices that minimize environmental impact, conserve natural resources, and enhance resilience to climate change. These frameworks encourage the adoption of agroecological approaches, organic farming methods, and integrated pest management systems to promote biodiversity, soil health, and ecosystem services. Additionally, they advocate for the sustainable use of water resources, the conservation of genetic diversity, and the reduction of greenhouse gas emissions from agricultural activities.

Moreover, international policy frameworks play a crucial role in ensuring food security and nutrition for all people, particularly in vulnerable populations. These frameworks prioritize the equitable distribution of food, the eradication of hunger and malnutrition, and the promotion of sustainable food systems that provide safe, nutritious, and culturally appropriate diets. They also address issues such as food waste and loss, food safety, and access to markets, ensuring that food reaches those who need it most.

Furthermore, international policy frameworks are instrumental in addressing global challenges such as climate change, biodiversity loss, and poverty alleviation. By promoting climate-smart agriculture, conservation agriculture, and sustainable land management practices, these frameworks contribute to mitigating greenhouse gas emissions, enhancing carbon sequestration, and building resilience to climate-related risks. They also support efforts to protect and restore ecosystems, conserve biodiversity hotspots, and promote sustainable livelihoods for rural communities.

International policy frameworks and regulatory mechanisms in agriculture play a crucial role in promoting sustainable development, ensuring food security, and addressing global challenges. By providing guidance, standards, and incentives for sustainable practices, these frameworks facilitate cooperation, coordination, and collective action among countries to build a more resilient, equitable, and sustainable agricultural sector for the benefit of present and future generations.

14.5 Innovation and technology adoption

Innovation and technology adoption play a pivotal role in transforming agricultural practices, enhancing productivity, and ensuring sustainability. This section delves into how agricultural institutions and policies influence the adoption of innovative technologies and practices in agriculture, thereby driving agricultural development and addressing key challenges.

Research institutions, extension services, and public-private partnerships are instrumental in promoting technology transfer and innovation diffusion within the agricultural sector. Research institutions conduct studies and experiments to develop new technologies and practices that address specific agricultural challenges such as improving crop yields, conserving water resources, or reducing postharvest losses. Extension services play a crucial role in disseminating these innovations to farmers, providing training, technical assistance, and advisory services to help farmers adopt and implement new technologies effectively. Public-private partnerships facilitate collaboration between government agencies, research institutions, and private companies to accelerate innovation and technology adoption in agriculture.

Case studies play a crucial role in providing real-world examples and practical insights into how innovation can be effectively harnessed to address key challenges in agriculture. These case studies often highlight successful initiatives and strategies that have been implemented in various contexts to enhance agricultural productivity, resilience, and sustainability.

One example of a case study could focus on a public-private partnership aimed at promoting the adoption of precision agriculture technologies. Precision agriculture utilizes advanced technologies such as GPS-guided machinery, remote sensing tools, and data analytics to optimize resource use and improve crop management practices. The case study may illustrate how a collaboration between a government agency, agricultural research institution, and private sector companies has facilitated the adoption of precision agriculture practices among farmers. It could showcase how these technologies have enabled farmers to make data-driven decisions regarding seeding, irrigation, and crop protection, resulting in improved yields, reduced input costs, and enhanced environmental sustainability.

Another case study might highlight an extension program that educates smallholder farmers on climate-smart agricultural practices. Climate-smart agriculture encompasses a range of practices that help farmers adapt to and mitigate the impacts of climate change while enhancing productivity and sustainability. The case study could focus on how an extension program, supported by a NGO or a government agency, provides training and technical assistance to farmers on implementing practices such as conservation agriculture, agroforestry, or water-saving irrigation techniques. It could demonstrate how these practices have helped farmers improve soil health, conserve water, sequester carbon, and diversify their income streams, ultimately enhancing their resilience to climate variability and improving food security in their communities.

Overall, case studies serve as valuable tools for showcasing innovative approaches, sharing best practices, and inspiring action in the agricultural sector. By highlighting successful initiatives and strategies, case studies provide actionable insights and practical guidance to policymakers, practitioners, and stakeholders seeking to drive positive change and sustainable development in agriculture.

Innovation and technology adoption are essential drivers of agricultural development, helping farmers overcome challenges, increase productivity, and improve livelihoods. Understanding how agricultural institutions and policies influence the adoption of innovative technologies is crucial for designing effective strategies to promote agricultural innovation and foster sustainable agricultural development.

14.6 Addressing emerging challenges

The agricultural sector faces a myriad of emerging challenges including climate change, resource scarcity, and shifting consumer preferences. This section delves into how agricultural institutions and policies are adapting to address these challenges, exploring innovative approaches and policy interventions aimed at promoting sustainability and resilience in agriculture.

One key focus is on promoting climate-smart agriculture, which involves practices that help mitigate greenhouse gas emissions, adapt to climate change impacts, and sustainably increase agricultural productivity. Policy interventions may include incentives for adopting climate-smart practices such as conservation agriculture, agroforestry, and water-efficient irrigation techniques. These practices not only help farmers adapt to changing climatic conditions but also contribute to reducing the carbon footprint of agriculture.

Another important consideration is sustainable intensification, which seeks to increase agricultural productivity while minimizing environmental impacts and enhancing resource efficiency. This approach involves promoting practices such as integrated pest management, precision agriculture, and organic farming, which optimize inputs, reduce waste, and promote ecosystem health. Policy interventions may include support for R&D, extension services, and financial incentives to encourage the adoption of sustainable intensification practices.

In addition, inclusive rural development is a priority, aiming to ensure that agricultural policies and institutions benefit all stakeholders including smallholder farmers, women, and marginalized communities. This may involve measures to improve access to land, credit, and markets for smallholder farmers as well as investments in rural infrastructure, education, and healthcare. By promoting inclusive rural development, agricultural policies can help address social inequalities, reduce poverty, and foster sustainable development in rural areas.

Case studies serve as detailed examinations of real-life examples or experiences that demonstrate how countries are incorporating sustainability principles into their agricultural policies and institutions. These case studies provide insights into specific strategies and initiatives that countries are implementing to enhance resilience and adapt to changing environmental and socioeconomic conditions in the agricultural sector.

For instance, one case study might focus on how a country has adopted agroecological zoning as a strategy to manage water resources more effectively, particularly in regions facing increasing water scarcity due to factors such as climate change or over-extraction. Agroecological zoning involves delineating different zones within a country based on factors such as climate, soil type, and water availability and then developing tailored agricultural practices and policies for each zone. The case study could illustrate how this approach has helped optimize water use, minimize water wastage, and improve overall water management in agricultural areas, contributing to greater resilience in the face of water scarcity challenges.

Another case study may highlight a government's partnership with local communities to develop climate-resilient agriculture strategies. This could involve engaging farmers, indigenous groups, and other stakeholders in participatory decision-making processes to identify and implement practices that are well-suited to their specific agro-ecological regions and socioeconomic contexts. The case study might showcase examples of community-led initiatives such as the adoption of drought-resistant crop varieties, the promotion of agroforestry systems, or the implementation of soil conservation measures. By empowering local communities to take ownership of resilience-building efforts, governments can foster more adaptive and sustainable agricultural systems that are better equipped to withstand the impacts of climate change and other environmental stressors.

Overall, these case studies provide valuable insights into the practical application of sustainability principles in agriculture, highlighting innovative approaches and successful strategies that can inform policy development, capacity-building efforts, and investment decisions aimed at building resilience and sustainability in agricultural systems.

Addressing emerging challenges requires a multifaceted approach that combines policy interventions, technological innovations, and stakeholder engagement. By promoting climate-smart agriculture, sustainable intensification, and inclusive rural development, agricultural institutions and policies can help build resilience, enhance sustainability, and ensure food security for future generations.

This chapter has provided a comprehensive overview of the role of agricultural institutions and policies in addressing global agricultural challenges. Through examining various aspects such as international organizations, bilateral agreements, policy frameworks, and technology adoption, we have gained valuable insights into how agricultural governance influences the sustainability and resilience of agricultural systems.

It is clear that addressing global agricultural challenges requires concerted efforts and collaboration at both national and international levels. International cooperation, knowledge sharing, and capacity building are essential for promoting sustainable agriculture and ensuring food security worldwide. By working together, countries can exchange best practices, leverage resources, and develop innovative solutions to address shared challenges such as climate change, resource scarcity, and food insecurity.

In moving forward, it is imperative for policymakers, researchers, and practitioners to prioritize the development of more effective and equitable agricultural institutions and policies. This includes investing in research and innovation, strengthening governance structures, and promoting inclusive decision-making processes. By adopting a holistic and collaborative approach, we can build a more sustainable and resilient agricultural sector that meets the needs of both present and future generations.

14.7 Some of the institutions and policies of Pakistan

Here are key aspects of institutions and policies:

Governance and institutional frameworks: Establishing effective governance structures, institutional frameworks, and regulatory bodies support sustainable agriculture and resource management. These institutions ensure compliance with environmental regulations, coordinate stakeholders, and provide policy guidance.

Land-use planning and management: Policies and land-use planning strategies help regulate land use, minimize land degradation, and preserve natural ecosystems. They guide sustainable land management practices, protect valuable agricultural land, and prevent encroachment into ecologically sensitive areas.

Agricultural extension services: Agricultural extension services provide technical support, training, and knowledge transfer to farmers. These services disseminate information on sustainable agricultural practices, climate-smart techniques, and resource management strategies. Extension services bridge the gap between research institutions and farmers, supporting the adoption of best practices.

Financial incentives and support: Governments and financial institutions can provide financial incentives, subsidies, and credit facilities to promote sustainable agriculture and resource management. These incentives encourage the adoption of sustainable practices, facilitate access to resources, and support farmers in transitioning to more sustainable farming systems.

Research and innovation: Policies that prioritize research and innovation in agriculture drive sustainable practices and technological advancements and investments in research institutions.

An overview of agricultural institutions and organizations in Pakistan

Pakistan has several agricultural institutions and organizations that play a crucial role in the development and advancement of the agricultural sector.

Here is an overview of some key institutions:

Ministry of National Food Security and Research: The Ministry of National Food Security and Research is responsible for formulating policies, planning, and implementing programs related to agriculture, food security, and research. It oversees various departments and organizations involved in agricultural development.

Pakistan Agricultural Research Council (PARC): PARC is the apex agricultural research organization in Pakistan. It conducts and coordinates agricultural research, develops technologies, and provides technical support to improve agricultural productivity and sustainability. PARC collaborates with national and international research institutions.

Pakistan Agricultural Research Institute (PARI): PARI is a leading agricultural research and education institute in Pakistan. It focuses on agricultural research, training, and capacity building. PARI conducts research on crop improvement, plant protection, soil management, and agricultural engineering.

Agriculture extension services: Agriculture extension services in Pakistan are provided by the provincial agriculture departments. These services aim to transfer agricultural knowledge and technologies to farmers, promote best practices, and enhance productivity. They play a vital role in disseminating information and providing advisory services to farmers.

National Rural Support Program (NRSP): NRSP is a nonprofit organization working for rural development in Pakistan. It provides support to farmers in various areas including microfinance, agricultural inputs, market access, and capacity building. NRSP focuses on empowering rural communities and improving their livelihoods.

Agricultural policies and their impact on integrated agriculture

Agricultural policies in Pakistan influence the development and implementation of integrated agriculture approaches. The government's policies and programs can support or hinder the adoption of integrated agriculture practices.

Here are some key policy areas and their impact:

Agricultural diversification: Policies that encourage diversification of agriculture beyond traditional crops can promote integrated agriculture. Supporting the cultivation of high-value crops, agroforestry, and livestock-crop integration can enhance resource utilization, income diversification, and sustainable production practices.

R&D: Policies that prioritize R&D in integrated agriculture provide a supportive framework for innovation and technology adoption. Investments in research institutions, collaboration with international partners, and funding for research projects contribute to the advancement of integrated agriculture practices.

Extension services: Policies that strengthen and expand agricultural extension services help disseminate knowledge and best practices related to integrated agriculture. Effective extension services facilitate technology transfer, provide advisory support, and promote awareness and adoption of integrated approaches among farmers.

Market access and value chains: Policies that facilitate market access for integrated agriculture products, promote value addition, and establish efficient value chains can encourage farmers to adopt integrated agriculture practices. Market-oriented policies create incentives for farmers to produce high-quality products and increase profitability.

Environmental sustainability: Policies that promote environmental sustainability and natural resource management align with the principles of integrated agriculture. Regulations on water use, land management, and chemical inputs can encourage farmers to adopt integrated approaches that minimize negative environmental impacts.

Government initiatives and support for agricultural development

The Government of Pakistan has implemented various initiatives and provided support for agricultural development. These initiatives aim to enhance productivity, improve food security, and uplift rural communities.

Here are some key government initiatives:

Prime Minister's Agriculture Emergency Program: Launched in 2019, this program focuses on increasing agricultural productivity, improving water management, promoting mechanization, and providing support to small farmers. It includes initiatives such as the Kamyab Kisan Program, the Olive Tree Plantation Project, and the National Agricultural Transformation Program.

Agriculture Transformation Plan: The Agriculture Transformation Plan aims to modernize the agricultural sector, increase productivity, and enhance value addition. It focuses on improving access to credit, promoting precision agriculture, strengthening supply chains, and supporting agricultural R&D.

Ehsaas Agricultural Emergency Program: This program focuses on providing financial and technical support to small farmers to enhance their productivity and income. It includes initiatives such as interest-free loans, crop insurance, seed and fertilizer subsidies, and capacity-building programs for farmers.

Agriculture Credit Support Scheme: The government provides credit support schemes and subsidies to farmers to facilitate access to finance for agricultural activities. These schemes aim to improve access to capital, encourage investment in agriculture, and promote modern farming practices.

National Food Security and Research Division: The National Food Security and Research Division, under the Ministry of National Food Security and Research, is responsible for implementing various programs and policies to ensure food security and support agricultural development. It focuses on improving seed quality, enhancing agricultural research, and addressing food security challenges.

International collaborations and partnerships in agriculture

Pakistan actively engages in international collaborations and partnerships in the field of agriculture. These collaborations aim to promote knowledge exchange, technology transfer, capacity building, and research collaboration.

Here are some examples:

International Fund for Agricultural Development (IFAD): Pakistan collaborates with IFAD to implement agricultural development projects, improve rural livelihoods, and enhance food security. IFAD provides financial support and technical expertise to initiatives such as the IFAD Rural Development Project and the Pakistan Poverty Alleviation Fund.

Food and Agriculture Organization (FAO): Pakistan works closely with FAO to address food security challenges, promote sustainable agriculture, and enhance agricultural practices. FAO provides technical assistance, expertise, and support for initiatives such as capacity building, policy development, and technology transfer.

International Centre for Agricultural Research in the Dry Areas (ICARDA): Pakistan collaborates with ICARDA on R&D projects related to dryland agriculture, livestock management, and natural resource management. These collaborations focus on improving productivity, resilience, and sustainability in dryland farming systems.

International Crops Research Institute for the Semi-Arid Tropics (ICRISAT): Pakistan partners with ICRISAT on R&D projects related to crop improvement, soil management, and climate-smart agriculture. These collaborations aim to enhance agricultural productivity, promote sustainable farming practices, and address challenges specific to semi-arid regions.

South-South Cooperation: Pakistan participates in South-South cooperation initiatives with other developing countries to exchange knowledge, experiences, and best practices in agriculture. These collaborations focus on areas such as capacity building, technology transfer, and policy development.

List of international agriculture research organizations and their headquarters

- Food and Agriculture Organization (**FAO**)
  Headquarters: Rome, Italy

- International Rice Research Institute (**IRRI**)
  Headquarters: Los Baños, Philippines

- International Maize and Wheat Improvement Center (**CIMMYT**)
  Headquarters: Texcoco, Mexico

- International Livestock Research Institute (**ILRI**)
  Headquarters: Nairobi, Kenya

- International Center for Agricultural Research in the Dry Areas (**ICARDA**)
  Headquarters: Beirut, Lebanon

- International Crops Research Institute for the Semi-Arid Tropics (ICRISAT)
  Headquarters: Patancheru, Telangana, India

- World Agroforestry Centre (**ICRAF**)
  Headquarters: Nairobi, Kenya

- International Water Management Institute (IWMI)
  Headquarters: Colombo, Sri Lanka

- WorldFish
  Headquarters: Penang, Malaysia

- Center for International Forestry Research (**CIFOR**)
  Headquarters: Bogor, Indonesia

- International Food Policy Research Institute (**IFPRI**)
  Headquarters: Washington, D.C., United States

- Bioversity International
  Headquarters: Rome, Italy

- International Fertilizer Development Center (**IFDC**)
  Headquarters: Muscle Shoals, Alabama, United States

- International Center for Tropical Agriculture (**CIAT**)
  Headquarters: Palmira, Colombia

- International Potato Center (**CIP**)
  Headquarters: Lima, Peru

- International Center for Agricultural Research in the Dry Areas (**ICARDA**)
  Headquarters: Beirut, Lebanon

- International Institute of Tropical Agriculture (**IITA**)
  Headquarters: Ibadan, Nigeria

- Africa Rice Center (AfricaRice)
  Headquarters: Abidjan, Côte d'Ivoire

- International Livestock Research Institute (**ILRI**)
  Headquarters: Nairobi, Kenya

- World Vegetable Center (WorldVeg)
  Headquarters: Tainan, Taiwan

- International Fertilizer Association (**IFA**)
  Headquarters: Paris, France

- Consultative Group for International Agricultural Research (**CGIAR**)
  Headquarters: Montpellier, France

- International Federation of Organic Agriculture Movements (**IFOAM**)
  Headquarters: Bonn, Germany

- International Seed Testing Association (**ISTA**)
  Headquarters: Zürich, Switzerland

- International Commission on Irrigation and Drainage (**ICID**)
  Headquarters: New Delhi, India

- International Union of Soil Sciences (**IUSS**)
  Headquarters: Vienna, Austria

- International Society of Horticultural Science (**ISHS**)
  Headquarters: Leuven, Belgium

- International Network for Bamboo and Rattan (**INBAR**)
  Headquarters: Beijing, China

- International Cocoa Organization (**ICCO**)
  Headquarters: Abidjan, Côte d'Ivoire

- International Centre of Insect Physiology and Ecology (**ICIPE**)
  Headquarters: Nairobi, Kenya

List of some prominent agricultural research organizations in Pakistan

- Pakistan Agricultural Research Council (**PARC**)
  Headquarters: Islamabad

- National Agriculture Research Centre (**NARC**)
  Headquarters: Islamabad

- International Maize and Wheat Improvement Center (CIMMYT) – Pakistan Office
  Location: Islamabad

- International Rice Research Institute (IRRI) – Pakistan Office
  Location: Lahore

- International Crops Research Institute for the Semi-Arid Tropics (ICRISAT) – Pakistan Office
  Location: Islamabad

- Pakistan Agricultural Research Council (PARC) – National Tea and High-Value Crops Research Institute (**NTHRI**)
  Location: Shinkiari, Khyber Pakhtunkhwa

- Pakistan Agricultural Research Council (PARC) – National Institute for Genomics and Advanced Biotechnology (NIGAB)
  Location: National Agricultural Research Centre, Islamabad

- Nuclear Institute for Agriculture and Biology (**NIAB**)
  Location: Faisalabad

- Ayub Agricultural Research Institute (**AARI**)
  Location: Faisalabad

- Barani Agricultural Research Institute (**BARI**)
  Location: Chakwal, Punjab

- Rice Research Institute (**RRI**)
  Location: Kala Shah Kaku, Punjab

- Sugarcane Research and Development Board (**SRDB**)
  Location: Lahore

- Livestock and Dairy Development Board (**LDDB**)
  Location: Islamabad

- Punjab Agriculture Research Board (**PARB**)
  Location: Lahore

- Sindh Agriculture Research Council (**SARC**)
  Location: Karachi

- Balochistan Agriculture Research & Development Center (**BARDC**)
  Location: Quetta

Within the pages of Chapter 15 the journey comes full circle, as we encounter a compendium of definitions and related terms. This glossary serves as a compass that navigates the intricate landscape of integrated agriculture, offering concise explanations for the myriad concepts we've explored. As we conclude this voyage through the tapestry of integrated agriculture, the definitions and terms within this chapter stand as signposts, guiding us toward a deeper comprehension of the interconnected world that sustains us all.

And so, with the final chapter, we draw the curtain on our journey through the intricate realms of integrated agriculture – a journey that has uncovered the symphony of interdependence, the dance of resilience, and the art of harmony. As the echoes of these pages resonate, may the wisdom within serve as a guidepost for sustainable futures, nurturing the planet's vitality and echoing the sentiment that the pursuit of integrated agriculture is a testament to our dedication to a more balanced, bountiful, and vibrant world.

References

Abbasi, S. A., & Butt, A. R. (2010). Pakistan's agriculture: A backward and forward linkage analysis. Pakistan Economic and Social Review, 48(1), 37–58.

Ahmed, M., Ashfaq, M., Khan, A. A., & Ahmed, Z. (2011). An overview of agricultural trends and policies in Pakistan. The Lahore Journal of Economics, 16, 43–62.

Amanullah. (2024). Harvesting Success: A Comprehensive Guide to Agronomy Courses. Punjab Books, Lahore, PK.

Amanullah., Ondrasek, G., Al –Tawaha, A. R. M. S., eds. (2023). Integrated nutrients management: An approach for sustainable crop production and food security in changing climates. Lausanne: Frontiers Media SA. doi: 10.3389/978-2-8325-3169-3

Amanullah, & Khalid, S. (2020). Agronomy – Climate Change & Food Security. IntechOpen, London, United Kingdom.

Amanullah, & Fahad, S. (2018). Nitrogen in Agriculture-Updates. InTechOpen, London, UK.

Amanullah, & Fahad. (2018). Corn – Production and Human Health in Changing Climate. InTech, London, UK.

FAO. (2018a). The Future of Food and Agriculture: Alternative Pathways to 2050. Food and Agriculture Organization of the United Nations, Rome, Italy.

FAO. (2018b). Soil Pollution: A Hidden Reality. FAO, Rome-Italy.

FAO. (2019). The International Code of Conduct for the Sustainable Use and Management of Fertilizers (30 pp.). FAO, Rome, Italy.

FAO. (2017). Unlocking the Potential of Soil Organic Carbon. FAO, Rome-Italy.

FAO. (2016). Soil and Pulses: Symbiosis for Life. FAO, Rome-Italy.

FAO and ITPS. (2016). Voluntary Guidelines for Sustainable Soil Management (VGSSM). Food and Agriculture Organization (FAO), Rome, Italy.

Government of Pakistan. (2019). Agriculture Policy 2018. Ministry of National Food Security and Research, Islamabad, Pakistan.

Government of Pakistan. (2019). Pakistan Economic Survey 2018–2019. Ministry of Finance, Islamabad, Pakistan.

Hafeez, M. M., & Imtiaz, M. (2019). Agriculture sector of Pakistan: The past, present, and future. Journal of Agricultural Science and Technology, 21(3), 625–636.

Hasan, M. F., & Shahbaz, M. (2020). Agricultural productivity and rural-urban inequality nexus in Pakistan: New evidence using district-level data. Sustainability, 12(1), 270.

Iftikhar, M., Hassan, S., & Hussain, A. (2012). Role of institutions in development of agriculture in Pakistan: A review. Pakistan Journal of Agricultural Sciences, 49(3), 299–309.

Janjua, P. Z., & Haider, S. J. (2019). Institutional dimensions of agricultural extension system in Punjab, Pakistan. Journal of Agricultural Extension and Rural Development, 11(8), 175–182.

Javed, M., & Bakhsh, K. (2019). Agriculture and the economy of Pakistan: A macroeconomic overview. International Journal of Financial Research, 10(2), 116–122.
Khan, M. A., & Iqbal, J. (2018). Role of institutions in agricultural development: A case study of Pakistan. European Journal of Agriculture and Forestry Research, 6(3), 25–32.
Malik, M. A., Ashraf, S., & Ahmad, Z. (2019). Institutions and policies in Pakistan's agriculture sector: An overview. Journal of Agriculture and Environmental Sciences, 2(1), 38–49.
Qureshi, M. E., & Mahmood, A. (2015). Pakistan's agricultural growth: An analysis of recent trends. International Journal of Economics, Commerce and Management, 3(7), 51–61.
Qureshi, S. E., & Ahmed, M. (2011). The agricultural sector of Pakistan: An overview. Munich Personal RePEc Archive (MPRA) Paper No. 28858.
Rashid, A., Hameed, S., & Abid, M. (2019). Role of agriculture in the economy of Pakistan. South Asian Studies, 34(2), 383–397.
Rehman, T., Dorward, A., & Balcombe, K. (2007). Constraints on development of the Punjab non-farm sector: The case of agricultural machinery services and trading in Pakistan. World Development, 35(10), 1746–1766.
Siddiqui, R., & Kemal, A. R. (2005). Environment, growth and development in Pakistan: A case of missing linkages. The Pakistan Development Review, 44(4), 711–723.
Siddiqui, R., & Kemal, A. R. (2006). Macro and micro policy environment and farm production in Pakistan. The Pakistan Development Review, 45(4), 739–761.
Tariq, S. A. (2013). Agriculture sector in Pakistan: An analysis of growth trends and prospects. Pakistan Journal of Social Sciences (PJSS), 33(1), 119–130.
Zafar, H. M., & Ali, S. (2019). An evaluation of agriculture policies and their impacts on the agricultural sector of Pakistan. Journal of Agriculture and Environment for International Development, 113(1), 185–200.

Chapter 15
Definitions and related terms: a comprehensive glossary

Agriculture: The practice of cultivating crops, raising livestock, and utilizing natural resources for food, fiber, and other products.

Agronomy: The science and practice of crop production and soil management.

Agroecological agronomy: Agroecological agronomy integrates ecological principles and practices into crop production systems. It emphasizes the understanding of ecological interactions, biodiversity conservation, and the use of ecosystem-based approaches to enhance soil health, pest control, nutrient cycling, and overall sustainability of agricultural systems.

Agroecology: Agroecology is the study of ecological principles applied to agricultural systems. It emphasizes the integration of ecological concepts and practices, such as biodiversity conservation, biological pest control, and nutrient cycling, to achieve sustainable and resilient agriculture.

Agroforestry: Agroforestry is the intentional integration of trees or shrubs with crops or livestock within the same farming system. It combines agricultural production with the benefits of trees, such as soil conservation, water management, biodiversity conservation, and additional income streams.

Aquaforestry: Integrating tree planting with aquaculture to provide shade and habitat for fish.

Aquaponics: Aquaponics combines aquaculture (fish farming) with hydroponics (soilless plant cultivation). It creates a symbiotic system where fish waste provides nutrients for plants, and plants filter and purify the water, creating a sustainable and integrated farming system.

Aquifer: An underground layer of rock that stores and transmits groundwater.

Biochar: Biochar is a type of charcoal produced from the pyrolysis of biomass. It is used as a soil amendment to improve soil fertility, water retention, and carbon sequestration. Biochar can enhance soil health and nutrient availability, contributing to sustainable agriculture.

Biocontrol: Biocontrol refers to the use of beneficial organisms, such as predatory insects, parasitic wasps, or microbial agents, to control pests or diseases in agriculture. It offers a sustainable and environmentally friendly approach to pest management, reducing the reliance on chemical pesticides.

https://doi.org/10.1515/9783111448077-015

Biodiesel: A renewable fuel made from vegetable oils or animal fats.

Biodiversity: Biodiversity refers to the variety of living organisms, including plants, animals, and microorganisms, within a given ecosystem. It plays a crucial role in maintaining ecosystem services, pollination, natural pest control, and overall ecosystem resilience in agricultural landscapes.

Bioeconomy: The bioeconomy refers to the sustainable production and utilization of biological resources, including agriculture, forestry, fisheries, and their conversion into value-added products, energy, and services. It aims to promote economic growth, environmental sustainability, and social well-being.

Bioenergy: Bioenergy is energy derived from biomass, which includes organic materials such as crops, forestry residues, or dedicated energy crops. It can be converted into various forms, including heat, electricity, or biofuels, for use in transportation, heating, and power generation.

Biofertilizers: Biofertilizers are substances containing beneficial microorganisms that enhance soil fertility and plant nutrition. They contribute to sustainable agriculture by fixing nitrogen, solubilizing phosphorus, or promoting plant growth through biological processes, reducing the reliance on synthetic fertilizers.

Biofortification: Biofortification involves enhancing the nutritional content of crops through breeding or biotechnology to address nutrient deficiencies in human diets. It aims to improve the nutritional quality of staple crops by increasing essential vitamins, minerals, or other nutrients.

Biogas: Biogas refers to a renewable energy source produced through the anaerobic digestion of organic materials such as agricultural waste, animal manure, or food scraps. It consists mainly of methane and can be used as a fuel for heating, electricity generation, or cooking.

Bioherbicides: Bioherbicides are derived from natural sources, such as microorganisms or plant extracts, and are used to control weeds in agricultural systems. They provide sustainable weed management options by targeting specific weed species while minimizing environmental impacts.

Biomass: Biomass refers to organic matter derived from plants or other biological sources that can be used as a renewable energy resource. It includes agricultural residues, energy crops, forest residues, or dedicated biomass crops used for bioenergy production.

Biopesticides: Biopesticides are naturally occurring substances derived from living organisms, such as bacteria, fungi, or plants, used to control pests or diseases in agriculture. They offer an alternative to synthetic chemical pesticides and are considered more environmentally friendly.

Bioprocessing: Bioprocessing involves the use of biological agents or processes to convert raw materials into value-added products. In agriculture, it includes activities such as fermentation, enzyme production, or bioconversion of biomass for various applications, including biofuels or bioplastics.

Bioremediation: Bioremediation involves the use of biological organisms or processes to clean up or mitigate environmental pollution or contamination. In agriculture, it can be used to remediate soil or water polluted with pesticides, heavy metals, or other contaminants.

Biosecurity: Biosecurity refers to measures taken to prevent the introduction or spread of harmful organisms, such as pests, diseases, or invasive species, in agricultural systems. It involves practices such as quarantine, disease surveillance, sanitation, and control strategies to protect plants, animals, and ecosystems.

Biotechnology: Biotechnology involves the use of biological systems, organisms, or processes to develop or modify products or technologies for various applications. In agriculture, biotechnology plays a role in crop improvement, genetic engineering, tissue culture, and the development of biopesticides or bioherbicides.

Carbon sequestration: Carbon sequestration refers to the process of capturing and storing atmospheric carbon dioxide (CO_2) in vegetation or soils. Conservation agriculture practices, such as maintaining soil cover and increasing organic matter content, contribute to carbon sequestration, helping mitigate climate change.

Catchment area: The land area that contributes water to a specific point, such as a river or reservoir.

Climate action: Climate action encompasses efforts and initiatives taken to address climate change and its impacts. It involves mitigation actions to reduce greenhouse gas emissions, adaptation measures to cope with changing climatic conditions, and advocacy for sustainable practices and policies.

Climate change: Climate change refers to long-term shifts and alterations in global or regional weather patterns, including changes in temperature, precipitation, wind patterns, and more. It is primarily driven by human activities, such as the burning of fossil fuels and deforestation, leading to increased greenhouse gas concentrations in the atmosphere.

Climate crisis: Climate crisis refers to the urgent and critical state of the Earth's climate system due to the rapid and severe impacts of climate change. It highlights the need for immediate action to mitigate greenhouse gas emissions, protect vulnerable communities, and promote sustainable development.

Climate resilience: Climate resilience is the ability of systems, communities, and ecosystems to withstand and recover from climate change impacts. It involves adapting to changing climatic conditions, reducing vulnerabilities, and building capacity to cope with climate-related challenges.

Climate-smart agriculture: Climate-smart agriculture focuses on adapting to and mitigating climate change impacts in agricultural systems. It involves practices that enhance resilience, such as water management, conservation agriculture, agroforestry, and the use of climate information for decision-making.

Climate-smart agronomy: Climate-smart agronomy focuses on adapting agriculture to climate change and mitigating its impacts. It involves the integration of climate information, resilient crop varieties, appropriate agronomic practices, and innovative technologies to enhance the resilience of agricultural systems, reduce greenhouse gas emissions, and promote sustainable crop production in changing climates.

Compost: Decomposed organic matter used as a soil conditioner and fertilizer.

Conservation agriculture: Conservation agriculture is an approach to farming that emphasizes minimizing soil disturbance, maintaining soil cover, and practicing crop rotation. It aims to enhance soil health, reduce erosion, improve water management, and promote overall sustainability in agricultural systems.

Conservation agriculture: Conservation agriculture is an approach to farming that aims to promote sustainable and environmentally friendly practices by minimizing soil disturbance, maintaining soil cover, and practicing crop rotation. It helps enhance soil health, reduce erosion, improve water management, and promote overall sustainability in agriculture.

Conservation agronomy: Conservation agronomy focuses on the application of sustainable agricultural practices that promote soil conservation, water management, and biodiversity preservation. It aims to optimize crop production while minimizing negative environmental impacts and promoting long-term soil health and fertility.

Conservation biology: Conservation biology is a scientific discipline that focuses on the study and preservation of biodiversity and the protection of species, ecosystems, and natural resources. It involves research, conservation planning, and management strategies to mitigate threats and promote the sustainable use of natural resources.

Conservation buffer strips: Conservation buffer strips are areas of land planted with vegetation, such as grasses or native plants, along the edges of fields or water bodies. They act as buffers to filter sediment, absorb excess nutrients, and reduce pesticide runoff, protecting water quality and promoting biodiversity.

Conservation corridor: A conservation corridor is a linear area of protected land that connects fragmented habitats and allows for the movement of species. It facilitates genetic exchange, maintains biodiversity, and promotes ecosystem resilience in landscapes affected by human activities.

Conservation easement: A conservation easement is a legal agreement between a landowner and a conservation organization or government entity to protect natural resources, restrict certain types of development, and maintain the conservation values of the land in perpetuity.

Conservation farming: Conservation farming refers to agricultural practices that integrate soil and water conservation techniques, such as contour plowing, terracing, and cover cropping. It aims to minimize soil erosion, enhance soil health, and promote sustainable crop production while conserving natural resources.

Conservation genetics: Conservation genetics is a field that applies genetic principles and techniques to the study and conservation of endangered species and populations. It involves assessing genetic diversity, understanding population dynamics, and developing strategies to preserve genetic integrity and ensure long-term survival.

Conservation management: Conservation management refers to the planning and implementation of practices and strategies to protect and preserve natural resources, habitats, and ecosystems. It involves monitoring, restoration, and sustainable use of resources while considering ecological, social, and economic factors.

Conservation planning: Conservation planning is the process of identifying, prioritizing, and implementing strategies to conserve and manage natural resources effectively. It involves assessing biodiversity, habitat quality, threats, and stakeholder engagement to develop conservation goals and action plans.

Conservation Reserve Program (CRP): The Conservation Reserve Program is a US federal program that provides financial incentives to landowners for converting highly erodible or environmentally sensitive agricultural land into conservation areas. It aims to reduce soil erosion, improve water quality, and enhance wildlife habitat.

Conservation science: Conservation science is an interdisciplinary field that combines ecological, biological, and social sciences to study and address conservation issues. It involves research, monitoring, and analysis to understand ecological processes, assess threats, and develop evidence-based conservation strategies.

Conservation stewardship: Conservation stewardship involves responsible and sustainable management of natural resources by individuals, organizations, or communities. It encompasses actions and practices that protect, restore, and sustainably use ecosystems and promote long-term environmental health.

Conservation tillage: Conservation tillage refers to reduced or no-till agricultural practices that minimize soil disturbance during planting and seedbed preparation. It helps retain crop residues, maintain soil structure, and reduce erosion, contributing to improved soil health and conservation of natural resources.

Conservation trust: A conservation trust is an organization or entity dedicated to acquiring and managing land or conservation easements for the protection of natural resources, wildlife habitat, and open space. It works to safeguard ecologically important areas and promote conservation values.

Conservation volunteer: A conservation volunteer is an individual who participates in activities aimed at conserving and preserving natural resources, habitats, or wildlife. They contribute their time, skills, and efforts to initiatives such as habitat restoration, wildlife monitoring, and environmental education.

Conservation zone: A conservation zone is an area designated for the protection and conservation of natural resources, wildlife, or ecosystems. It may have restrictions or regulations in place to limit certain activities and promote conservation goals.

Cover crops: Cover crops are non-commercial crops grown primarily to cover the soil between cash crop seasons. They help protect the soil from erosion, improve soil fertility through nitrogen fixation and nutrient cycling, enhance water infiltration, suppress weeds, and promote biodiversity.

Crop modeling and agronomy: Crop modeling and agronomy involve the use of computer models and simulations to predict and optimize crop growth and yield under different environmental and management scenarios. It helps farmers and researchers make informed decisions regarding crop selection, planting dates, fertilizer application, and other management practices for improved productivity and resource efficiency.

Crop physiology and agronomy: Crop physiology and agronomy is an interdisciplinary field that combines the study of plant physiology and agronomy principles. It focuses on understanding the physiological processes in crops, such as photosynthesis, nutrient uptake, and growth, and applying this knowledge to optimize crop management practices for enhanced productivity and sustainability.

Crop rotation: Crop rotation is the practice of growing different crops successively in a specific sequence on the same piece of land. It helps break pest and disease cycles, improve soil health, enhance nutrient cycling, and promote overall crop productivity and sustainability.

Decarbonizing agriculture: Decarbonizing agriculture refers to the process of reducing or eliminating carbon emissions associated with agricultural activities. This involves adopting practices and technologies that minimize the release of greenhouse gases, such as carbon dioxide (CO_2), methane (CH_4), and nitrous oxide (N_2O), while also enhancing carbon sequestration in soils and vegetation. Decarbonizing agriculture is essential for mitigating climate change, as the agricultural sector is a significant contributor to greenhouse gas emissions globally. This transition involves implementing sustainable farming practices, improving resource efficiency, transitioning to renewable energy sources, and restoring ecosystems to enhance their carbon storage capacity.

Desertification: The process of fertile land becoming desert due to various factors, including human activities.

Digital agriculture: Digital agriculture leverages digital technologies, such as sensors, drones, and artificial intelligence, to enhance efficiency and decision-making in agriculture. It enables data-driven insights for optimizing irrigation, monitoring crop health, and improving resource management.

Diversification: The strategy of varying agricultural activities to reduce risks and enhance sustainability.

Drought resistance: Ability of plants to survive and thrive in conditions of low water availability.

Dryland farming: Agricultural activities in regions with low and erratic rainfall.

Eco-friendly: Practices that are environmentally responsible and sustainable.

Ecosystem services: Ecosystem services refer to the benefits that ecosystems provide to human well-being. Conservation agriculture enhances ecosystem services such as soil erosion control, water filtration, carbon sequestration, biodiversity conservation, and climate regulation.

Environmental conservation practices: Environmental conservation practices are actions and strategies that aim to protect and conserve natural resources, biodiversity, and ecosystems. These practices may include habitat restoration, protected area management, sustainable land use, and waste reduction.

Environmental conservation: Environmental conservation refers to the protection and preservation of natural resources, habitats, and ecosystems. It involves efforts to prevent environmental degradation, promote biodiversity, and ensure the sustainable use of resources.

Environmental education: Environmental education is the process of imparting knowledge and awareness about environmental issues, conservation, and sustainable practices. It aims to promote environmental literacy, responsible behavior, and engagement in environmental stewardship.

Environmental ethics: Environmental ethics refers to the moral principles and values that guide human interaction with the environment. It involves considering the rights of nature, the responsibility to protect ecosystems, and the ethical implications of human actions on the environment.

Environmental impact assessment (EIA): Environmental impact assessment is a process that assesses the potential environmental consequences of proposed development projects. It helps identify and mitigate potential adverse effects on the environment and promotes sustainable development.

Environmental impact mitigation: Environmental impact mitigation involves actions taken to minimize or offset the negative environmental effects of human activities. It includes measures such as pollution control, habitat restoration, and the implementation of sustainable practices.

Environmental justice: Environmental justice is the fair and equitable distribution of environmental benefits and burdens among different social groups. It advocates for equal access to a clean and healthy environment, regardless of race, income, or other social factors.

Environmental monitoring: Environmental monitoring involves the systematic collection and analysis of data to assess environmental conditions, identify changes, and detect potential threats or pollution. It helps in evaluating the effectiveness of conservation measures and informing environmental management decisions.

Environmental policy: Environmental policy encompasses laws, regulations, and guidelines that govern human activities and their impacts on the environment. It sets standards for pollution control, resource management, and conservation efforts.

Environmental pollution: Environmental pollution refers to the introduction of harmful substances or contaminants into the environment, leading to negative impacts on ecosystems, human health, and natural resources. It includes pollution of air, water, soil, and noise pollution.

Environmental resilience: Environmental resilience refers to the ability of ecosystems and communities to withstand and recover from disturbances or changes. It involves maintaining biodiversity, promoting ecosystem health, and building adaptive capacity to cope with environmental challenges.

Environmental restoration: Environmental restoration refers to activities aimed at returning degraded or damaged ecosystems to their original or improved ecological condition. It involves reestablishing native vegetation, restoring wetlands, and rehabilitating habitats to enhance ecosystem functioning.

Environmental stewardship: Environmental stewardship involves taking responsibility for the care and management of natural resources. It includes sustainable practices, conservation efforts, and promoting environmental awareness and action.

Environmental sustainability indicators: Environmental sustainability indicators are measures used to assess and track progress toward sustainable development goals. They provide quantitative or qualitative data on various environmental factors, such as air quality, water availability, or biodiversity.

Environmental sustainability: Environmental sustainability is the practice of using resources in a way that meets present needs without compromising the ability of future generations to meet their own needs. It involves responsible management of resources, reduction of pollution, and promotion of ecosystem health.

Erosion: The process of soil being washed away or blown by wind, leading to the loss of fertile topsoil.

Erosion control: Techniques to prevent or minimize soil erosion, including terracing and vegetation planting.

Eutrophication: Excessive nutrients, often from agricultural runoff, causing overgrowth of aquatic plants and degradation of water bodies.

Fallow: Period of resting or leaving land uncultivated to allow soil fertility to recover.

Farmer training and education: Farmer training and education programs play a crucial role in promoting and implementing conservation agriculture practices. They provide farmers with knowledge, skills, and tools to understand the principles and benefits of conservation agriculture and adopt appropriate techniques on their farms.

Fertilizer: Substances applied to soil to provide essential nutrients to plants for healthy growth.

Fishery: The industry involved in catching, processing, and selling fish.

Forage agronomy: Forage agronomy deals with the science and management of forage crops used for livestock feed. It includes the selection of appropriate forage species, cultivation techniques, and strategies to optimize yield, quality, and nutritional value of forage crops for efficient livestock production.

Fossil fuel: Non-renewable energy sources like coal, oil, and natural gas.

Global warming: The long-term increase in Earth's average surface temperature due to greenhouse gas emissions.

Greenhouse gas: Gases, such as carbon dioxide and methane, that contribute to the greenhouse effect and global warming.

Green manure: Plants grown and incorporated into the soil to improve fertility and structure.

Gully erosion: Deep and wide channels formed by concentrated water flow, leading to severe soil loss.

Hedgerow: Rows of trees or shrubs planted along field edges to reduce soil erosion and provide habitat for beneficial organisms.

Hybrid seeds: Seeds produced by crossbreeding two different varieties or species to achieve desired traits.

Hydrological cycle: The continuous movement of water on, above, and below the surface of the Earth.

Hydroponics: Hydroponics is a method of growing plants in a soilless medium with a nutrient-rich water solution. It allows for efficient water use, controlled nutrient delivery, and year-round crop production, making it suitable for urban and resource-limited environments.

Insecticide: A substance used to control or eliminate insects that damage crops.

Integrated agriculture: Integrated agriculture involves the combination of various agricultural activities, such as crop production, livestock rearing, and aquaculture, within a single farming system. It aims to optimize resource utilization, minimize waste, and create synergistic interactions among different components.

Integrated aquatic resource management: Integrated aquatic resource management involves the sustainable management of freshwater and marine resources, including fisheries, aquaculture, and aquatic ecosystems. It considers ecological, social, and economic aspects to promote the long-term health and productivity of aquatic environments.

Integrated aquifer management: Integrated aquifer management focuses on the sustainable management of groundwater resources. It involves monitoring groundwater levels, controlling abstraction rates, promoting recharge mechanisms, and addressing pollution and overexploitation issues.

Integrated carbon management: Integrated carbon management focuses on the systematic reduction, capture, and storage of carbon emissions to mitigate climate change impacts. It includes strategies such as energy efficiency, renewable energy, afforestation, and carbon sequestration technologies.

Integrated climate change adaptation: Integrated climate change adaptation refers to the development and implementation of strategies and actions that address the impacts of climate change across various sectors and at different scales. It involves considering climate risks and vulnerabilities, integrating adaptation measures into planning processes, and promoting resilience.

Integrated coastal management: Integrated coastal management focuses on the sustainable management of coastal areas, considering various aspects such as coastal ecosystems, tourism, fisheries, pollution control, and shoreline protection. It aims to balance economic development with environmental conservation and community well-being.

Integrated coastal protection: Integrated coastal protection involves the implementation of measures to protect coastal areas from erosion, storm surges, and sea-level rise. It may include the use of engineered structures, beach nourishment, dune restoration, and ecosystem-based approaches to ensure coastal resilience.

Integrated coastal zone management: Integrated coastal zone management encompasses the integrated planning and management of coastal areas. It considers environmental, social, and economic aspects to balance conservation, sustainable development, and resilience in coastal ecosystems.

Integrated coastal zone planning: Integrated coastal zone planning involves the systematic and coordinated management of coastal areas. It integrates ecological, social, and economic considerations to balance development activities, protect coastal ecosystems, and ensure the sustainable use of resources.

Integrated community development: Integrated community development focuses on the holistic development of communities, encompassing social, economic, and environmental aspects. It involves the integration of livelihood opportunities, infrastructure development, social services, and community participation to enhance well-being and reduce inequalities.

Integrated crop-livestock systems: Integrated crop-livestock systems combine crop cultivation with livestock production within the same farming system. The integration of crops and livestock allows for nutrient recycling, efficient resource utilization, and diversified income sources.

Integrated data analytics: Integrated data analytics refers to the utilization of advanced data analytics techniques to analyze and integrate data from multiple sources and domains. It aims to extract insights, identify patterns, and support evidence-based decision-making.

Integrated digital solutions: Integrated digital solutions involve the integration of digital technologies, such as artificial intelligence, internet of things, and data analytics, to provide comprehensive and interconnected solutions for various sectors and domains.

Integrated disaster management: Integrated disaster management involves the coordination and integration of various measures and actions to prevent, prepare for, respond to, and recover from natural or man-made disasters. It encompasses activities such as risk assessment, early warning systems, emergency response planning, and post-disaster recovery.

Integrated disaster risk reduction: Integrated disaster risk reduction focuses on the systematic and coordinated efforts to reduce the risks and impacts of natural and human-induced disasters. It involves measures such as early warning systems, preparedness planning, community engagement, and infrastructure resilience.

Integrated ecosystem management: Integrated ecosystem management involves the holistic and inclusive management of ecosystems, considering ecological processes, biodiversity conservation, sustainable resource use, and the needs of local communities. It promotes the integration of ecological, economic, and social dimensions for the long-term health and sustainability of ecosystems.

Integrated energy planning: Integrated energy planning involves the coordinated and optimized planning of energy resources, including fossil fuels, renewable energy sources, and energy efficiency measures. It aims to ensure a reliable, affordable, and sustainable energy supply while minimizing environmental impacts.

Integrated environmental assessment: Integrated environmental assessment combines various components of environmental assessment, including environmental impact assessment, strategic environmental assessment, and cumulative impact assessment, into a comprehensive framework. It aims to evaluate the environmental implications of development projects, policies, or plans in a holistic manner.

Integrated environmental monitoring: Integrated environmental monitoring encompasses the systematic collection, analysis, and reporting of data on various environmental parameters. It involves the integration of monitoring efforts across air, water, soil, biodiversity, and other environmental components to assess environmental quality and inform decision-making.

Integrated farming: Integrated farming is a holistic approach that combines various agricultural practices such as crop cultivation, livestock rearing, aquaculture, and agroforestry within a single farming system. It aims to maximize resource utilization, enhance productivity, and promote sustainability.

Integrated food safety systems: Integrated food safety systems encompass the coordination and integration of various measures and regulations to ensure the safety and quality of food throughout the food supply chain. It involves monitoring, inspection, traceability, and risk management to protect consumer health.

Integrated food systems: Integrated food systems focus on the entire food value chain, from production to consumption. They aim to promote sustainable agricultural practices, ensure food safety, reduce food waste, and enhance the nutritional quality of food.

Integrated forest management: Integrated forest management refers to the comprehensive planning and sustainable management of forest resources. It involves balancing ecological, economic, and social considerations to conserve biodiversity, promote timber production, protect forest ecosystems, and support local communities.

Integrated health systems: Integrated health systems involve the coordination and integration of primary healthcare services, specialized care, and public health initiatives. They aim to ensure comprehensive and seamless healthcare delivery, promote preventive care, and address health inequities.

Integrated healthcare delivery: Integrated healthcare delivery involves the coordination and integration of various healthcare services, including primary care, specialized care, preventive care, and support services. It aims to provide comprehensive and seamless healthcare to individuals and communities.

Integrated information systems: Integrated information systems refer to the integration of different data sources, databases, and information technologies to enable seamless sharing, analysis, and utilization of information within an organization or across multiple entities.

Integrated knowledge systems: Integrated knowledge systems involve the combination of diverse knowledge sources, including scientific, indigenous, and local knowledge, to enhance understanding and decision-making related to natural resource management, conservation, and sustainable development.

Integrated land-use planning: Integrated land-use planning involves the systematic consideration and coordination of different land uses, such as agriculture, residential, industrial, and conservation, to optimize land utilization, minimize conflicts, and promote sustainable development.

Integrated nutrient management: Integrated nutrient management combines organic and inorganic fertilizers, crop residues, and other nutrient sources to provide balanced and adequate nutrition to crops while minimizing nutrient losses and environmental impacts.

Integrated pest and vector management: Integrated pest and vector management focuses on the holistic management of pests, including insects, diseases, and vectors, using a combination of prevention, monitoring, biological control, and targeted use of pesticides. It aims to minimize pesticide use, protect human health, and maintain ecological balance.

Integrated pest and weed management: Integrated pest and weed management combines multiple strategies, including biological control, cultural practices, and judicious use of pesticides, to effectively manage pest and weed populations while minimizing adverse environmental and health impacts.

Integrated pest control programs: Integrated pest control programs employ a combination of pest monitoring, biological control methods, cultural practices, and judicious use of pesticides to manage pest populations effectively while minimizing environmental and human health risks.

Integrated pest management (IPM): IPM is an approach that combines various pest control strategies, including biological control, cultural practices, and judicious use of pesticides, to manage pests effectively while minimizing environmental and health risks.

Integrated remote sensing and geographic information systems **(GIS)**: Integrated remote sensing and GIS involve the integration of remote sensing data and spatial analysis techniques to study and monitor various aspects of the environment, such as land cover, vegetation, water resources, and climate change impacts.

Integrated renewable energy systems: Integrated renewable energy systems involve the integration of various renewable energy sources, such as solar, wind, biomass, and hydropower, to ensure a reliable and sustainable energy supply. They may include energy storage, smart grid technologies, and demand-side management to optimize energy production and consumption.

Integrated risk assessment: Integrated risk assessment combines various types of risks, including environmental, health, social, and economic risks, into a comprehensive assessment framework. It aims to understand the interdependencies and impacts of different risks to support informed decision-making and risk management strategies.

Integrated river basin management: Integrated river basin management involves the holistic and participatory management of water resources within a river basin. It considers the interconnectedness of upstream and downstream areas, balancing water allocation, ecosystem protection, and stakeholder involvement for sustainable water management.

Integrated rural development: Integrated rural development encompasses a multidimensional approach that addresses the economic, social, and environmental needs of rural areas. It integrates various sectors such as agriculture, education, healthcare, infrastructure, and livelihoods to promote holistic and sustainable development.

Integrated smart grid systems: Integrated smart grid systems involve the integration of advanced communication, monitoring, and control technologies into electricity grids to enable efficient energy management, demand response, and integration of renewable energy sources.

Integrated socioecological systems: Integrated socioecological systems refer to the interconnected and interdependent relationships between social systems and ecological systems. It recognizes the reciprocal influences and feedbacks between human activities and the environment, highlighting the importance of integrated approaches for sustainability.

Integrated spatial planning: Integrated spatial planning involves the coordinated and strategic planning of land use, infrastructure development, and environmental considerations. It aims to optimize resource utilization, minimize conflicts, and promote sustainable development within a defined geographic area.

Integrated supply chain management: Integrated supply chain management involves the coordination and optimization of all activities involved in the production, transportation, and distribution of goods or services. It aims to streamline operations, minimize costs, and improve overall supply chain efficiency.

Integrated tourism development: Integrated tourism development focuses on the sustainable and inclusive development of tourist destinations. It involves the integration of infrastructure development, environmental conservation, community engagement, and tourism planning to maximize socioeconomic benefits and minimize negative impacts.

Integrated transportation systems: Integrated transportation systems focus on the efficient and coordinated management of different modes of transportation, including road, rail, air, and waterways, to enhance mobility, reduce congestion, and improve connectivity within a region or country.

Integrated urban agriculture: Integrated urban agriculture involves the cultivation of crops, rearing of livestock, and other agricultural practices within urban areas. It aims to promote local food production, reduce food miles, and enhance urban sustainability and resilience.

Integrated urban development: Integrated urban development focuses on the comprehensive planning and management of urban areas. It involves the integration of physical, social, economic, and environmental aspects to create sustainable, inclusive, and livable cities.

Integrated urban resilience: Integrated urban resilience focuses on enhancing the ability of cities to withstand and recover from shocks and stresses, including natural disasters, climate change impacts, and socioeconomic disruptions.

Integrated urban transport systems: Integrated urban transport systems focus on developing efficient, sustainable, and multimodal transportation networks within urban areas. They involve the integration of public transport, cycling infrastructure, pedestrian-friendly designs, and traffic management to improve mobility and reduce congestion.

Integrated urban water management: Integrated urban water management involves the comprehensive management of water resources within urban areas, including water supply, wastewater treatment, stormwater management, and water conservation measures. It aims to ensure sustainable and resilient water services in urban environments.

Integrated urban water supply and sanitation: Integrated urban water supply and sanitation systems aim to provide reliable and sustainable water supply and sanitation services to urban areas. They involve the integration of water treatment, distribution networks, wastewater management, and sanitation facilities to ensure safe water access and proper sanitation practices.

Integrated waste management systems: Integrated waste management systems aim to optimize waste reduction, recycling, and disposal practices. They involve the integration of waste segregation, recycling facilities, waste-to-energy conversion, and proper landfill management to minimize waste generation and environmental impacts.

Integrated waste recycling: Integrated waste recycling refers to a comprehensive approach that involves the collection, segregation, processing, and recycling of different types of waste materials to minimize waste generation and promote resource conservation.

Integrated waste-to-energy systems: Integrated waste-to-energy systems involve the conversion of waste materials, such as organic waste or biomass, into energy through processes such as anaerobic digestion, composting, or thermal conversion. It enables the recovery of energy from waste while reducing environmental impacts.

Integrated water governance: Integrated water governance involves the coordination and collaboration among multiple stakeholders, including government agencies, communities, and civil society, to manage water resources effectively. It aims to ensure equitable water allocation, participatory decision-making, and sustainable water management.

Integrated water management: Integrated water management involves the coordinated planning, allocation, and conservation of water resources to meet the needs of various sectors, including agriculture, industry, and domestic use. It focuses on sustainable water use, efficient irrigation practices, and the protection of water quality and ecosystems.

Integrated water pollution control: Integrated water pollution control involves the management and control of different sources of water pollution, including industrial discharges, agricultural runoff, and domestic sewage. It employs a range of measures to prevent, reduce, and treat water pollution to protect water quality and ecosystems.

Integrated water-energy nexus: The water-energy nexus approach recognizes the interdependencies and trade-offs between water and energy resources. It aims to optimize resource use, promote energy efficiency in water systems, and enhance water efficiency in energy production.

Integrated water-energy-food Nexus: The water-energy-food nexus approach considers the interdependencies and synergies among water resources, energy production, and food security. It aims to optimize resource management and address trade-offs to ensure sustainable development and resilience in the face of water, energy, and food security challenges.

Integrated water-energy-food-land nexus: The integrated water-energy-food-land nexus approach considers the interdependencies and trade-offs among water, energy, food, and land resources. It aims to develop policies and strategies that optimize resource use, enhance resilience, and achieve sustainable development goals.

Integrated watershed development: Integrated watershed development aims to promote sustainable land and water management practices within a watershed. It involves the integration of soil conservation, water harvesting, afforestation, and community participation to improve water availability, reduce erosion, and enhance ecosystem services.

Invasive species: Non-native species that outcompete and disrupt native ecosystems.

***Jatropha*:** A plant with oil-rich seeds used for producing biodiesel.

Jhum cultivation: Traditional shifting cultivation practiced in parts of Asia, involving clearing and burning of land.

Job diversification: Engaging in various income-generating activities to enhance rural livelihoods.

Kharif: Monsoon season in South Asia, during which crops like rice, cotton, and maize are typically cultivated.

Kiln: An oven or furnace used to bake or dry materials, such as clay for pottery.

Kilowatt-hour (kWh): A unit of energy equivalent to one kilowatt of power used for one hour.

Kitchen garden: A small area near a house where vegetables and herbs are cultivated for household consumption.

Lagoon: A shallow body of water separated from the sea by sandbars, often used for aquaculture.

Legume: Plants that belong to the pea family and have nitrogen-fixing properties.

Livelihood: Means of securing the necessities of life, often through employment or economic activities.

Livestock extension: Educational services provided to farmers on livestock management practices.

Megaclimate: Megaclimate refers to the long-term climate conditions over an extensive geographic region or the entire planet. It encompasses the broader patterns and trends in temperature, precipitation, and atmospheric circulation that influence the Earth's climate system.

Mesoclimate: Mesoclimate refers to the climate conditions that prevail over a larger geographical area, typically smaller than a region but larger than a local microclimate. It takes into account factors like topography, land use, and air masses, which can create unique climate patterns.

Microclimate: Microclimate refers to the localized climate conditions within a specific area, such as a garden, forest, or urban environment. It is influenced by factors like topography, vegetation cover, and human activities, and can differ from the larger regional climate.

Monocropping: Planting a single crop on a large area of land.

Monoculture: Cultivating a single crop on a piece of land, which can lead to decreased biodiversity and increased vulnerability.

Monsoon: A seasonal wind pattern that brings heavy rainfall, particularly to South Asia.

Mulching: Mulching involves the application of organic or inorganic materials on the soil surface to conserve moisture, suppress weed growth, and moderate soil temperature. Mulches can be made from materials such as straw, wood chips, plastic films, or crop residues.

Natural resources: Elements of the environment used by humans for various purposes, including air, water, and minerals.

Niche farming: Specialized agricultural activities focused on specific crops or products.

Nitrogen fixation: Conversion of atmospheric nitrogen into forms that can be used by plants, often facilitated by symbiotic bacteria.

Nutrient agronomy: Nutrient agronomy focuses on the management of soil nutrients and their efficient use in crop production. It involves assessing soil nutrient levels, understanding nutrient requirements of crops, and implementing balanced fertilizer application strategies to maximize nutrient uptake, minimize losses, and promote sustainable nutrient management.

Nutrient cycling: Nutrient cycling involves the efficient use and recycling of nutrients within agricultural systems. It includes practices such as nutrient management planning, balanced fertilization, cover cropping, and the use of organic amendments to optimize nutrient availability for crops while minimizing nutrient losses.

Organic agriculture: Organic agriculture involves the production of crops and livestock using natural methods without synthetic pesticides, fertilizers, or genetically modified organisms. It focuses on maintaining soil health, biodiversity, and ecological balance while meeting organic certification standards.

Organic certification: Verification that agricultural products meet specific organic standards.

Organic farming: Agricultural method that avoids synthetic pesticides and fertilizers, focusing on sustainable practices.

Organic matter: Decomposed plant and animal materials in the soil that contribute to its fertility and structure.

Overgrazing: Grazing livestock excessively, leading to degradation of vegetation and soil.

Paleoclimate: Paleoclimate refers to the study of past climates and climatic conditions on Earth before the availability of instrumental records. It involves analyzing geological, biological, and chemical evidence to understand historical climate changes and variations.

Pasture: An area of land used for grazing livestock.

Permaculture: Permaculture is an ecological design system that aims to create sustainable and self-sufficient agricultural systems. It emphasizes principles such as observation, diversity, and natural patterns to create productive and resilient ecosystems.

Perennial crops: Plants that live for multiple growing seasons, as opposed to annuals.

Pesticide: Chemical substances used to control pests and diseases in agriculture.

Precision agriculture: Precision agriculture utilizes advanced technologies, such as GPS, remote sensing, and data analytics, to optimize the use of inputs and improve crop production. It involves site-specific management practices, allowing farmers to make informed decisions about seeding, fertilizing, and irrigation.

Precision agronomy: Precision agronomy involves the use of advanced technologies, such as remote sensing, global positioning systems (GPS), and data analytics, to precisely manage and monitor crop production. It aims to optimize resource use, enhance crop yields, and minimize environmental impacts by applying inputs in a site-specific and tailored manner.

Quarantine: A period during which plants, animals, or other goods are isolated to prevent the spread of disease.

Quiescent phase: Dormant period in the life cycle of plants or insects.

Quinoa: Nutrient-rich grain crop that originated in the Andes and is known for its high protein content.

Quality seed: High-quality seeds that are genetically pure, disease-free, and have good germination rates.

Rainfed agriculture: Crop cultivation solely dependent on rainfall, without irrigation.

Rainwater harvesting: Collecting and storing rainwater for future use.

Regenerative agriculture: Regenerative agriculture is an approach that aims to restore and improve the health of agricultural systems. It emphasizes practices such as cover cropping, crop rotation, composting, and holistic management to enhance soil fertility, increase biodiversity, and promote carbon sequestration.

Riparian: Pertaining to the areas along the banks of rivers and streams.

Rural development: Process of improving living standards and well-being in rural areas through economic, social, and environmental interventions.

Silvopasture: Integrating trees with pasture or grazing land to benefit both livestock and ecosystem.

Soil amendments: Soil amendments are materials added to the soil to improve its physical, chemical, or biological properties. Common soil amendments include organic matter, compost, lime, gypsum, and mineral fertilizers.

Soil and water agronomy: Soil and water agronomy involves the study and management of soil-water interactions in agricultural systems. It addresses topics such as soil water availability, irrigation techniques, drainage systems, soil moisture monitoring, and water-use efficiency to optimize water management and maintain soil health for sustainable crop production.

Soil classification: Soil classification is the systematic categorization of soils based on their physical, chemical, and morphological properties. It helps in understanding soil characteristics, identifying soil types, and making informed decisions regarding soil management practices.

Soil conservation practices: Soil conservation practices are measures implemented to protect and preserve soil resources. These include contour plowing, terracing, mulching, crop rotation, cover cropping, no-till farming, and the establishment of buffer strips.

Soil conservation: Soil conservation refers to the practices and techniques used to prevent soil erosion, improve soil fertility, and protect the quality of soil resources. It includes measures such as contour plowing, terracing, cover cropping, and the use of conservation tillage methods.

Soil degradation: Soil degradation refers to the decline in soil quality and health due to factors such as erosion, nutrient depletion, compaction, salinization, or pollution. It can have negative impacts on agricultural productivity, water quality, and ecosystem functioning.

Soil ecosystem: The soil ecosystem refers to the complex interactions between soil organisms, plants, and the physical and chemical properties of the soil. It is a dynamic system that supports various ecological processes and plays a vital role in sustaining life on Earth.

Soil erosion control: Soil erosion control measures focus on preventing or reducing the loss of topsoil caused by wind or water erosion. Practices such as contour plowing, terracing, and grassed waterways help minimize soil erosion, preserve soil fertility, and protect downstream water bodies.

Soil erosion: Soil erosion is the process of the detachment, transportation, and deposition of soil particles by natural forces such as wind, water, or gravity. It can lead to the loss of topsoil, reduced soil fertility, and environmental degradation if not properly managed.

Soil fertility: Soil fertility refers to the ability of a soil to provide essential nutrients in adequate quantities and appropriate forms for plant growth. It is influenced by factors such as nutrient content, pH, organic matter, and microbial activity in the soil.

Soil health management: Soil health management focuses on maintaining or improving the physical, chemical, and biological properties of the soil. It involves practices such as organic matter addition, soil nutrient management, erosion control, and reduced soil compaction to promote optimal soil conditions for plant growth.

Soil health: Soil health refers to the capacity of a soil to function effectively as a living ecosystem that sustains plants, animals, and microorganisms. It encompasses factors such as soil structure, nutrient content, organic matter levels, and biological activity.

Soil microorganisms: Soil microorganisms are microscopic organisms, including bacteria, fungi, protozoa, and nematodes, that inhabit the soil. They play important roles in nutrient cycling, organic matter decomposition, disease suppression, and overall soil health.

Soil moisture: Soil moisture refers to the amount of water present in the soil. It plays a crucial role in plant growth, as it affects nutrient availability, root development, and overall crop productivity.

Soil nutrients: Soil nutrients are essential elements required by plants for their growth and development. They include macronutrients (such as nitrogen, phosphorus, and potassium) and micronutrients (such as iron, zinc, and copper).

Soil organic matter: Soil organic matter refers to the decomposed and partially decomposed organic materials present in the soil. It contributes to soil fertility, nutrient holding capacity, water retention, and the formation of soil aggregates.

Soil pH: Soil pH is a measure of the acidity or alkalinity of the soil. It influences nutrient availability to plants and affects soil microbial activity. Different plants have different pH requirements for optimal growth.

Soil testing: Soil testing involves the analysis of soil samples to assess their nutrient content, pH level, and other characteristics. It helps determine the soil's fertility status and guides the application of appropriate fertilizers and soil amendments.

Sustainable agriculture: Sustainable agriculture refers to farming practices that promote long-term environmental, economic, and social sustainability. It involves methods that optimize resource use, reduce negative environmental impacts, and prioritize the well-being of farmers and local communities.

Sustainable construction: Sustainable construction involves designing and constructing buildings and infrastructure that minimize resource consumption, reduce waste, and promote energy efficiency. It focuses on using environmentally friendly materials, employing energy-saving technologies, and incorporating green building principles.

Sustainable consumption: Sustainable consumption refers to making informed choices and decisions that reduce the environmental and social impacts of consumption patterns. It involves minimizing waste, choosing eco-friendly products, and adopting responsible consumer behaviors.

Sustainable development: Sustainable development is a concept that emphasizes meeting the needs of the present generation without compromising the ability of future generations to meet their own needs. It involves balancing economic growth, social well-being, and environmental protection.

Sustainable Development Goals: The Sustainable Development Goals or Global Goals are a collection of 17 interlinked objectives designed to serve as a "shared blueprint for peace and prosperity for people and the planet, now and into the future."

Sustainable energy: Sustainable energy refers to energy sources and systems that have minimal negative environmental impacts and can be replenished naturally. It includes renewable energy sources such as solar, wind, hydro, geothermal, and biomass.

Sustainable fisheries: Sustainable fisheries aim to manage fish populations and aquatic ecosystems in a way that ensures their long-term health and productivity. It involves responsible fishing practices, ecosystem-based management, and conservation measures to prevent overfishing and protect marine biodiversity.

Sustainable forestry: Sustainable forestry involves managing forests in a way that balances timber production, biodiversity conservation, and ecosystem services. It focuses on maintaining the health and resilience of forest ecosystems while meeting the demands for wood products.

Sustainable intensification: Sustainable intensification focuses on increasing agricultural productivity while minimizing environmental impacts. It involves the adoption of practices and technologies that enhance efficiency, optimize resource use, and reduce the need for expansion into new areas.

Sustainable land management: Sustainable land management involves the conservation and sustainable use of land resources while maintaining ecosystem functions and services. It includes practices such as soil conservation, reforestation, erosion control, and sustainable land-use planning.

Sustainable packaging: Sustainable packaging refers to the use of environmentally friendly materials and design practices that minimize waste, reduce carbon footprint, and promote recycling or composting. It aims to reduce the environmental impact of packaging throughout its lifecycle.

Sustainable supply chain: A sustainable supply chain involves managing the flow of goods and services in a way that minimizes environmental impacts, promotes social responsibility, and ensures economic viability. It focuses on sustainable sourcing, responsible manufacturing, and ethical business practices.

Sustainable tourism: Sustainable tourism focuses on minimizing the negative environmental, social, and cultural impacts of tourism while maximizing the benefits for local communities and preserving natural and cultural heritage. It promotes responsible travel practices and sustainable destination management.

Sustainable transport: Sustainable transport refers to transportation systems and modes that minimize environmental impacts, reduce congestion, and promote efficient mobility. It includes options such as public transit, cycling, walking, and the use of clean and fuel-efficient vehicles.

Sustainable urban planning: Sustainable urban planning focuses on creating cities and urban areas that are environmentally friendly, socially inclusive, and economically vibrant. It involves strategies such as compact development, green infrastructure, sustainable transportation, and mixed land use to promote livability and sustainability.

Sustainable waste management: Sustainable waste management involves reducing, reusing, recycling, and properly disposing of waste in an environmentally responsible manner. It aims to minimize waste generation, promote resource recovery, and reduce the environmental impacts of waste disposal.

Sustainable water management: Sustainable water management involves the equitable and efficient use of water resources while protecting water quality and ecosystems. It includes practices such as water conservation, watershed management, and integrated water resource management.

Terracotta: Fired clay material used for making pots, tiles, and other decorative items.

Terracing: Creating stepped platforms on hilly terrain to prevent soil erosion and water runoff.

Tillage: The preparation of soil for planting, including plowing and harrowing.

Topography: The physical features of a landscape, including elevation and slope.

Undernutrition: A condition resulting from inadequate consumption of essential nutrients, leading to poor health.

Urban agriculture: Urban agriculture refers to the practice of growing food and raising livestock within urban areas. It can include rooftop gardens, community gardens, and vertical farming, providing fresh food, improving food security, and promoting local sustainability.

Urban sprawl: Unplanned and rapid expansion of urban areas.

Urbanization: The process of increasing population and development in urban areas.

Vegetative propagation: Reproduction of plants using parts like stems, leaves, or roots.

Value chain: Series of activities involved in producing, processing, and distributing agricultural products to consumers.

Vermicomposting: Process of using earthworms to decompose organic matter and produce nutrient-rich compost.

Vermifiltration: A method of wastewater treatment that involves passing it through a bed of earthworms.

Vertical farming: Vertical farming involves growing crops in vertically stacked layers or towers using artificial lighting and controlled environmental conditions. It maximizes land use, reduces water consumption, and enables year-round production in urban settings.

Water conservation practices: Water conservation practices aim to optimize water use and reduce water loss in agricultural systems. Techniques such as efficient irrigation methods (e.g., drip irrigation), water harvesting, and soil moisture monitoring help minimize water waste and improve water-use efficiency.

Water footprint: The total amount of water used to produce goods and services.

Water harvesting: Collecting and storing rainwater to be used for irrigation and other purposes.

Waterlogging: Saturation of soil with water, often leading to reduced plant growth.

Weed agronomy: Weed agronomy deals with the study and management of weeds in agricultural systems. It involves understanding weed biology, ecology, and the development of effective weed control strategies, including integrated weed management, herbicide application techniques, and cultural practices to minimize weed competition and preserve crop yields.

Weed management: Strategies to control and manage the growth of unwanted plants that compete with crops.

Xenobiotics: Chemical substances that are foreign to an organism's natural biochemical processes.

Xeriscaping: Landscaping method that conserves water through the use of drought-resistant plants and efficient irrigation.

Xerophyte: Plants adapted to arid environments with low water availability.

Xylotomy: The study of wood structure and anatomy.

Yield: The quantity of agricultural products produced per unit of land or other resource.

Yield gap: The difference between potential and actual crop yields due to various constraints.

Yield potential: The maximum amount of crop a plant can produce under ideal conditions.

Yield stability: Consistency in crop production, regardless of environmental variations.

Zero-budget natural farming: A sustainable farming approach that eliminates external inputs and relies on natural processes.

Zero emission: A state where no harmful pollutants are released into the environment.

Zero grazing: Keeping livestock confined in specific areas to prevent overgrazing and environmental degradation.

Zero hunger: The goal of ending all forms of hunger and malnutrition.

Index

https://doi.org/10.1515/9783111448077-016

www.ingramcontent.com/pod-product-compliance
Lightning Source LLC
Chambersburg PA
CBHW061155240926
51846CB00006B/1021
* 9 7 8 3 1 1 1 4 4 8 0 1 5 *